Energy
and
The Atmosphere

A Physical–Chemical Approach

Energy
and
The Atmosphere

A Physical–Chemical Approach

Ian M. Campbell

School of Chemistry
University of Leeds

JOHN WILEY & SONS LTD

London · New York · Sydney · Toronto

Library of Congress Cataloging in Publication Data:
Campbell, Ian M
    Energy and the atmosphere.
    Bibliography: p.
    Includes index.
    1. Atmosphere. 2. Atmospheric chemistry. 3. Fuel.
I. Title.
QC866.C27   551.5′1   76−57689
ISBN 0 471 99482 0 (cloth)
ISBN 0 471 99481 2 (paper)

Typeset in IBM Press Roman by Preface Ltd, Salisbury, Wilts
Printed in Great Britain by Unwin Brothers Ltd.,
The Gresham Press, Old Woking, Surrey.

Contents

Preface

The current surges of interest in Energy and the Atmospheric Environment have generated a demand by students of sciences and engineering for inclusion of these topics in their courses. This book is a response intended for the primary and intermediate levels, aimed at providing a broad-based and unified account. Accordingly, aspects of planetary, engineering, and biological science, and of fuel technology, are incorporated into central cores of physical and chemical knowledge, with the object of drawing out essential interrelationships. The book is intended to present a simplified overview; the resultant selection of material is restricted, therefore, and it is hoped that not too many readers will disagree with the choice, emphasis, and balance achieved. It is recognized that there are many books which cover the component parts of the content in more depth and more rigorously: but it is felt that there exists a gap in present booklists for a text giving a sense of perspective and relating one topic to another at a level suitable for a reader making fairly early contact with the theme. For example, the conflicting objectives in solving the 'Energy Crisis' and producing the 'Cleaner Atmospheric Environment' are discussed with the intention of creating in the reader's mind a framework upon which materials from more specialized texts can be hung at a later stage, if desired. To this end, some of the more specialized texts are listed in the further reading suggestions at the end of each chapter.

The book presumes a knowledge of elementary thermodynamics, reaction kinetics, and the quantum theory of matter. However some fundamental aspects of outstanding importance in the present context received more detailed attention in the relevant chapters, including the concept of redox processes, basic photochemical theory, and the basis of term symbols identifying different electronic states of atoms and molecules. The book makes little attempt to advance beyond what is well established, particularly in the areas of atmospheric chemical conversions and the quantitative modelling thereof, but seeks to convey only in simplified terms a sense of what are likely to be critical considerations. For example, in assessments of rates or extents of energy or chemical conversions, development will often be made only for a presumed homogeneous layer of the atmosphere under overhead sun conditions in order to point up key processes. My belief in the value of approximate 'spot' calculations in this connection will become clear.

The S.I. system of units is used throughout, except insofar as, for example, the pressure unit of a 'standard atmosphere' is evidently more convenient to use at points in a discussion of atmospheric phenomena. Also, in conjunction with conventional units of $mol\,dm^{-3}$ for concentration, it will sometimes be convenient to use length units expressed in dm rather than m or its tridecadic fractions and multiples.

The impetus to write this book has been provided by my teaching of aspects of energy production and conversion, and of atmospheric chemistry, to undergraduate chemists and master's degree candidates at Leeds University. The fragmented nature of reasonably comprehensive booklists has been a severe handicap in this.

Referencing has been kept to a minimum to avoid a 'review' appearance. Specific references are made within a chapter, while more general references appear in the lists at the end.

The structure of the book can be likened to an investigative visit to the Earth from space; the incoming flight focuses attention upon energy in the main, the outgoing flight upon chemical conversions.

In Chapter 1 the Earth is compared with the other planets in our solar system and the origin of the various atmospheres is discussed. Chapter 2 considers the dissipation of solar radiant energy on Earth and the balancing reradiation characteristics. Closely related is the development of gross physical structure in the atmosphere; the basis of this is the topic of Chapter 3. The next three chapters deal principally with energy conversion at the Earth's surface. The photosynthetic processes responsible for prehistoric and current fixation of solar energy are considered from the physical-chemical point of view in Chapter 4. In Chapter 5, the reader will find a survey of fuels and their combustion characteristics, the overwhelming source of our energy now and in the near future. Both thermodynamic and kinetic aspects are discussed against the typical time scales involved, with reference to the generation and emission of pollutant species. Chapter 6 examines the operating modes of common power-generating systems, with emphasis upon the efficiencies of fuel—energy conversion and the related pollutant emission characteristics. This chapter advances from current energy technology to the discussion of novel technology which may be on the verge of entry into service.

Chapters 7 and 8 commence discussion of chemical conversions, dealing mainly with interactions between the Earth's surface and the lower atmosphere. The major element cycles and potential anthropogenic perturbation on an overall global basis are dealt with in Chapter 7. Chapter 8 is devoted to the localized phenomena of photochemical conversion in urban environments, instanced by 'photochemical smog'. In Chapter 9 the chemical conversion cycles in the upper atmosphere are assessed, identifying the major elementary reactions and, in particular, seeking an insight into the potential for depletion of the stratospheric ozone layer and the possible biological consequences. The concluding short Chapter 10 rounds matters off with a brief account of ionospheric chemistry.

Although the book is written by a physical chemist and will be most relevant to those in this discipline, students of other sciences and applied sciences should find considerable parts of relevance and interest.

I.M.C.

Chapter 1

The Earth as a Planet

1.1 The Vital Factors

The Earth is unique as a planet; that goes without saying since life exists here alone as far as we know within our solar system. What is life? To reduce the answer to fundamental scientific definition we may say that it is but one of a set of instances of departure from a full thermodynamic and chemical equilibrium. Were thermodynamic criteria to hold sway without perturbation, then life should not exist itself nor should our atmosphere contain the oxygen to sustain life. What is the perturbing influence? That can be specified exactly as inflowing energy in the form of radiation within an exceedingly narrow portion of the electromagnetic spectrum between 300 nm and 800 nm, that is the light from the Sun. A set of fragile coincidences combine to make the perturbation effective but not overly so. The Sun just happens to have a surface brightness temperature of approximately 6000 K, which makes the maximum intensity in terms of photon output within the continuous spectrum lie at around wavelength 600 nm; alternatively the maximum output on an energy basis occurs around 450 nm. Both of these wavelengths lie within the perturbing region of the electromagnetic spectrum. The coincidence is the more staggering when we realize that temperatures well above one million degrees Kelvin must exist within the Sun below the visible surface!

But why should radiation with wavelength between 300 and 800 nm have such significance? The answer stems from the quantum nature of the light. On the basis of Planck's Law

$$\epsilon = h\nu \tag{1.1}$$

where ϵ is the energy content of one quantum of light of frequency ν and h is Planck's constant equal to 6.625×10^{-34} J s. The frequency of light is related to its wavelength (λ) through the velocity of light $c = 3 \times 10^{8}$ m s^{-1} (in air) as

$$\nu = c/\lambda \tag{1.2}$$

Thus equation (1.1) may be recast as

$$\epsilon = hc/\lambda \tag{1.3}$$

Insertion of $\lambda = 300$ nm $= 3 \times 10^{-7}$ m, and $\lambda = 8 \times 10^{-7}$ m into this equation gives $\epsilon = 6.63 \times 10^{-19}$ J and $\epsilon = 2.48 \times 10^{-19}$ J respectively. These are vanishingly small amounts of energy when it is considered that the combustion of a single gram of carbon in air releases close to 30,000 J. But it is here that a second coincidence operates: It just happens that the energy required to excite an electron bound within a molecular orbital to one of the empty orbitals at higher energy is precisely of the order of 10^{-19} to 10^{-18} J. The important consequence is that the solar radiation impinging upon a terrestrial molecule can be absorbed in a primary photochemical act and the electromagnetic energy becomes converted into potential chemical energy available to induce secondary reactions. Moreover the typical energy required to break chemical bonds in molecules lies in much the same range of microscopic energy. Hence visible light, as it may be termed generally, has exactly the right scale of energy per light quantum or photon to give rise to the possibility of photochemistry, that is chemical reactions driven by the energy of sunlight. The quantum theory does not permit the accumulation and aggregation of photon energies to produce electron excitation, as is demonstrated by the sharp threshold for the photoejection of electrons from a metal surface irradiated with monochromatic (i.e. single wavelength) light in the Photoelectric Effect. What chance then for life to have evolved had the bulk of the Sun's energy output been of quanta in the energy ranges 10^{-20} to 10^{-21} J, or 10^{-16} to 10^{-17} J, or alternatively had the energies required to excite electrons in molecules or break chemical bonds been in the same ranges?

A third coincidence is that the Earth orbits the Sun at an average distance of approximately 93 million miles $(1.49 \times 10^{11}$ m). The radius of the Earth is 6.4×10^6 m. The disc for interception of solar radiation therefore has an area of 1.3×10^{14} m^2. The surface area of a sphere centred on the Sun containing the Earth is $4\pi(1.49 \times 10^{11})^2 = 2.8 \times 10^{23}$ m^2. Hence the Earth intercepts only about one-billionth part (1 billion $= 10^9$) of the total solar radiant flux. As it turns out, this energy influx to the planet is just sufficient to produce the average ambient temperature of 288 K. The delicacy of this energy balance is evident when it is considered that below 273 K water would be permanently frozen, while with an average closer to 350 K a scalding steamy atmosphere would be produced. Either of these alternatives would surely have stultified the development of life. Hence we can be aware of the highly privileged position of Earth in the solar system.

We could go on to develop further vital coincidences, such as the ultraviolet light filtering action of the stratospheric ozone layer which removes harmful radiation and preserves the created life, but such topics will be dealt with in detail in later chapters. For the present we shall digress to consider the other planets of our solar system and the factors thereon which forbid life as we know it.

1.2 The Other Planets

Table 1.1 summarizes relevant physical data for the planets of our solar system. In this table we pay particular attention to relative parameters, referred to the Earth as standard. Listed are relative mass (M_r), relative density (D_r), relative planetary radius (r_r), relative distance from the Sun (R_r) and the relative radiation

Table 1.1. Physical parameters of the planets

	M_r	D_r	r_r	R_r	L_r/A	T_v/K
Mercury	0.05	1	0.4	0.387	6.678	450
Venus	0.8	0.95	1	0.723	1.914	235
Earth	1	1	1	1	1	288
Mars	0.11	0.727	0.5	1.523	0.431	220
Jupiter	318	0.242	11	5.2	0.037	140
Saturn	95	0.129	9	10	0.011	120
Uranus	15	0.273	4	20	0.0027	85
Neptune	17	0.364	4	30	0.0011	65
Pluto	0.1(?)	?	0.5(?)	40	0.0006	40

Data are generally available and are collected from various sources.

interception ratio per unit disc area (L_r/A). This last parameter is simply the value of R_r^{-2}. Also listed is the temperature (T_v) of the visible surface.

We see that the planets divide into two general categories on the basis of density. The inner planets, Mercury, Venus, Earth, and Mars, have relatively high densities, indicating that most of their bulk is made up by rock-like materials. The outer planets, Jupiter, Saturn, Uranus, and Neptune, have rather low densities, consistent with the accepted theory that most of their bulk is made up by hydrogen and helium.

We may concentrate upon Jupiter as representative of the outer giant planets. Spectroscopic observations of the light back-reflected from the cloud layer have revealed absorption lines of molecular hydrogen, methane and ammonia. By symmetry the hydrogen molecule has no normal (dipole) absorption in the infrared region of the electromagnetic spectrum, but it is so abundant in the optical path into, and back out of, Jupiter's atmosphere that quadrupole and some pressure induced dipole bands are developed. In fact it is the overtone bands which have been observed, for example, the (3,0) band in the vicinity of 815 nm wavelength and the (4,0) band near 637 nm. The strengths of the absorption bands have provided estimates of the relative abundances of these gases in the transparent upper atmosphere. The resultant $H_2 : CH_4$ ratio is very close to that expected if Jupiter had an overall H-to-C ratio of about 3000, which is just that in the Sun itself. As all plausible models of Jupiter predict that the $H_2:CH_4$ abundance ratio above the cloud layer should closely reflect the H:C ratio for the planet as a whole, this is prime evidence that Jupiter has the same elemental abundance ratios as the Sun. Inevitably this leads to he postulation that this giant planet was formed originally with the solar composition and has retained all its matter since. Although corresponding measurements on the $NH_3:H_2$ relative abundance ratio yields an H:N ratio not markedly different from the solar abundance ratio, this may be given less weight than the H:C ratio since NH_3 is considered to be partially condensed in the cloud tops. Methane on the other hand is not expected to form a condensed phase, nor are any other organic molecules predicted to exist in any significant

Table 1.2. Solar abundances
of the major elements

Element	Atomic abundance (relative to silicon)
Hydrogen	28236
Helium	1779
Oxygen	16.6
Carbon	10.0
Nitrogen	2.4
Neon	2.1
Silicon	1.00
Iron	0.09

Source: Data compiled from measurements in the following papers: D. L. Lambert, *Observatory*, **87**, 199, 288 (1967); D. L. Lambert and B. Warner, *Monthly Notices of the Royal Astronomical Society*, **138**, 213 (1968); B. Warner, *Monthly Notices of the Royal Astronomical Society*, **138**, 229 (1968).

abundance. Table 1.2 shows the solar abundance ratios for elements of major significance and such a composition may hence be applied to Jupiter.

The postulation that Jupiter has a solar composition implies that very little hydrogen can have escaped to space over the planet's lifetime. A fundamental justification of this view comes from the kinetic theory of gases. A gas molecule in lower regions of the atmosphere will have no possibility of escaping from the gravitational field since, even if it does possess momentarily a high enough velocity, collisions with other molecules will rob it of its high kinetic energy and deflect it from the necessary outward trajectory. On the other hand, molecules in higher regions will suffer a much lower frequency of collisions because of the reduced total pressure and may escape if they have a sufficiently large vertical velocity component. Since a few of the molecules in the equilibrium velocity distribution will have much higher than average velocities, it is not necessary for the mean molecular velocity of a gas at the temperature of the upper atmosphere to approach the minimum escape velocity for that gas in order to be lost to space at a significant rate. Calculation suggests that, if the mean molecular velocity is around one third of the escape velocity, then the loss rate will correspond to the escape of one half of the atmospheric content of that gas within a period of weeks rather than years. At the same time the distribution of velocities is such that if the mean molecular velocity is about one fifth of the escape velocity, then the gas will be lost to space only over a time scale of a few hundred million years. In fact a realistic limit for the mean molecular velocity of one-sixth or less of the escape velocity may be imposed as the criterion for the effective retention for all time of a gas by a planet since its creation.

We may now approach this situation more quantitatively. The equilibrium distribution of scalar velocity, c, is given by the Maxwell–Boltzmann function expressing the fraction of molecules having velocities between c and $c + dc$, where dc is an infinitesimal increment:

$$\frac{dN}{N} = 4\pi \left(\frac{m}{2\pi kT}\right)^{3/2} \exp(-mc^2/2kT)\, c^2\, dc \tag{1.4}$$

Here m is the mass of the molecule, k is the Boltzmann constant (1.38×10^{-23} J molecule^{-1} K^{-1}), T is the temperature, and N is the total number of molecules in the assembly.:

The average or mean velocity is then given by:

$$\bar{c} = \frac{\int_0^\infty c\, dN}{\int_0^\infty dN} = \frac{1}{N} \int_0^\infty c\, dN = 4\pi \left(\frac{m}{2\pi kT}\right)^{3/2} \int_0^\infty e^{-mc^2/2kT}\, c^3\, dc \tag{1.5}$$

Integrating and reducing yields:

$$\bar{c} = (8kT/\pi m)^{1/2} \tag{1.6}$$

The fraction of molecules, $F(c)$, with velocities above multiples of the average velocity, $\bar{c}$, may then be calculated by integrating equation (1.4) between limits of the defined velocity c, and infinity. The results of such calculations are summarized in Figure 1, where $F(c)$ is plotted against $c/\bar{c}$. From this figure we may read off a value of $F(c) = 10^{-18.7}$ for $c/\bar{c} = 6$. If c is equated to the escape velocity in the upper atmosphere of a planet, we see that only 2×10^{-19} of the molecules will even momentarily have the potential to escape. On the other hand if the average velocity is one third of the escape velocity we can read off that about one molecule in only ten thousand will now have the potential for escape; on the face of it an increase of nearly 15 orders of magnitude in the probability over the $c/\bar{c} = 6$ case. Thus we see the force of our above criterion for the complete retention of a planetary atmosphere.

The escape velocity from a planet of mass M at a distance r from the centre is $(2GM/r)^{1/2}$, where G is the gravitational constant (6.67×10^{-11} N m^2 kg^{-2}).

Therefore the gas retention criterion imposed above demands:

$$6(8kT/\pi m)^{1/2} \leqslant (2GM/r)^{1/2} \tag{1.7}$$

for the planet to possess its original abundance of a gas. Evidently the appearance of m in the denominator on the left side means that hydrogen is the gas most likely to have escaped. Moreover, considering that hydrogen may well be atomized by solar short-wavelength radiation above the cloud layer of Jupiter, the best test is to use m equal to the mass of one hydrogen atom, i.e. $m = 0.001/6 \times 10^{23}$ kg. The

6

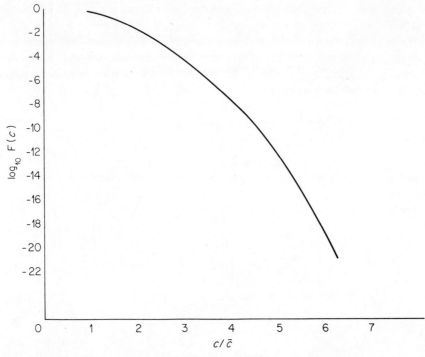

Figure 1. Plot of $\log_{10} F(c)$ versus $c/\bar{c}$ for an ideal gas, where $F(c)$ is the fraction of molecules possessing a scalar velocity c greater than the average scalar velocity $\bar{c}$

other parameters have values of $T = 140$ K, $M = 1.9 \times 10^{27}$ kg, and $r = 69,600$ km. Thus equation (1.7) produces the inequality:

$$1.03 \times 10^4 \leqslant 6.04 \times 10^4 \ (\text{both in m s}^{-1}).$$

This justifies the postulation that no significant amounts of material will have escaped from Jupiter, or for that matter from the outer planets, Saturn, Uranus, and Neptune, from similar considerations. This accounts for the low densities on the basis of near complete retention of the original hydrogen and helium contents. However the lower masses and higher effective temperatures of the inner planets, Mercury, Venus, Earth, and Mars, have allowed the escape of atmospheric gases to a much greater and more significant extent. Mercury, with the smallest mass and highest effective temperature, allows rapid escape even of carbon dioxide; it is therefore no surprise that Mercury is virtually devoid of an atmosphere. Similar considerations apply to Mars, although it does possess a thin atmosphere. In the cases of Earth and Venus, the primeval atmosphere has been lost but the planetary masses and temperatures are sufficient to have permitted substantial retention of the secondary atmospheres resulting from interior outgassings. Hence we can appreciate the main reason for the different atmospheres possessed by the inner and outer planets.

Returning to Jupiter for the moment, we ask whether a model can be advanced which will allow us an insight below the screening clouds. Is this truly a cold planet as suggested perhaps by the visible surface temperature of 140 K? That statement needs some qualification. The temperature of 140 K is deduced on the basis of the H_2 and CH_4 absorption bands and the rotational temperatures contained therein. However microwave and radiowave observations reveal disc brightness temperatures (from the black body continuum emission) which extend up to 300 K. This phenomenon is ascribed to the penetration through the frozen ammonia haze by these longer wavelength radiations originating from deeper layers. This suggests much higher temperatures within Jupiter's cloud-enshrouded atmosphere and has led to the postulation of a vertical temperature gradient determined by the conditions of adiabatic equilibrium.

Basically, adiabatic equilibrium produces such an atmospheric temperature gradient as would be obtained if each layer of gases had started at high pressure and high temperature at the solid surface, and had undergone an adiabatic expansion to the pressure appropriate to its actual altitude above the surface. Convected parcels of gases will in fact undergo such a process; it has been established that the atmosphere of Jupiter shows evidence of deep convective motion, which is likely to be one source of its banded appearance.

Pursuit of the adiabatic model indicates a true planet surface temperature of around 2000 K and a pressure of about 2×10^5-times atmospheric at ground level on Earth. This is hardly the cold planet suggested by the visual observations! The chemical structure of the Jovian atmosphere can then be developed by super-imposition of solar abundance ratios of the elements. In view of the high altitude clouds, it is not surprising that the gross chemical structure of the atmosphere is determined by thermodynamic criteria alone. Figure 2 shows the schematic representation of the principal cloud layers of Jupiter predicted on the basis of the volatility and stability of chemical compounds.

One further point needs an answer. In Table 1.1 we see that the solar flux per unit disc area for Jupiter is roughly 4% of that for Earth. In view of the extensive perturbation of our atmosphere by the shorter wavelength solar radiation, why is there not a similar, if reduced effect on Jupiter even if it is restricted to the transparent upper atmosphere? It would be expected, for example, that ammonia and methane would be subject to photodissociation. Laboratory studies have shown the effectiveness of the processes:

$$NH_3 + h\nu(\lambda \leqslant 220 \text{ nm}) \longrightarrow NH_2 + H$$

$$CH_4 + h\nu(\lambda \leqslant 160 \text{ nm}) \longrightarrow CH_3 + H$$

The association of the radical fragments is likely at the low temperatures of the upper atmosphere. Hence hydrazine (N_2H_4) and ethane (C_2H_6) might be expected to form and to freeze out. Calculations show that on a time scale of hundreds of millions of years, most of the ammonia and methane should have been so converted to heavier molecules and frozen out into the cloud layers. But let us consider the fate of the frozen heavier molecules. Gravitational attraction of their aggregates will

8

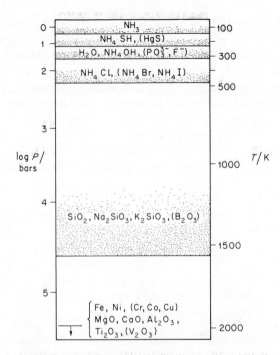

Figure 2. Schematic representation of the principal cloud layers predicted for Jupiter. (The altitude scale is strongly non-linear: the 1000 K level is about 500 km below the NH_3 clouds while the 2000 K level is about 3800 km deeper.) Parentheses enclose minor constituents. From J. S. Lewis, 'Observability of spectroscopically active compounds in the atmosphere of Jupiter', *Icarus*, **10**, 406 (1969). Reproduced by permission of Academic Press, Inc., New York

draw them down into the lower regions of the atmosphere where convective downdraughts will bring them in contact with temperatures high enough to effect pyrolysis, thus producing amino and methyl radicals in the presence of excess high temperature hydrogen gas in the main. The resultant reconversion into ammonia and methane then closes a cycle which produces no permanent chemical effect. Upward convective motions will take the reformed molecules back to maintain the H:C:N ratios of the upper atmosphere. The cycle for ammonia is illustrated schematically in Figure 3.

Hence the picture which emerges for Jupiter is of a near perfect chemical and physical equilibrium planet, virtually unperturbed by the solar flux. The other outer planets may be viewed similarly. Jupiter radiates more energy than it receives from the Sun by a factor of about 3; the source may be gravitational compaction of material within the planet. Calculations show that Jupiter has nearly the maximum size for a planet; had it been somewhat larger then it would have become a sun in its own right.

Before we leave the outer planets, we may consider the anomaly of Titan, the largest of the moons of Saturn. Titan is only 87% more massive than our own moon, with a diameter 40% larger. The visible surface temperature of Titan is around 80 K but even with this low temperature the moon would not have been expected to be capable of retaining an atmosphere. But there is no doubt that Titan, alone amongst the satellites of Saturn, does possess a considerable atmosphere, perhaps with a surface pressure as much as one tenth that at the Earth's surface. Methane bands are clearly developed in the spectrum of the light reflected from Titan and the atmosphere is also certain to contain a major hydrogen component with some ammonia. At the temperature of 80 K, estimated for Titan's exosphere, even hydrogen atoms have a mean velocity of only 1.3×10^3 m s^{-1}. However, because Titan's orbital speed around Saturn is 5.6×10^3 m s^{-1}, a hydrogen atom of average velocity, easily able to escape Titan's small gravitational field, will not possess the Saturnian escape velocity of about 8×10^3 m s^{-1} at Titan's orbit. Hence current thinking considers that the gases undoubtedly lost by Titan are forced to orbit Saturn, forming a gaseous torus at Titan's orbit – yet another, if invisible, ring of Saturn! Titan therefore is thought to retain its atmosphere by continuous interchange of gas with the gaseous torus in which it moves.

Another interesting aspect of Titan is that the solid surface can be seen and is distinctly coloured dark red. This is considered to be the result of irradiation by the short wavelength solar flux, inducing photodissociation and polymerization of

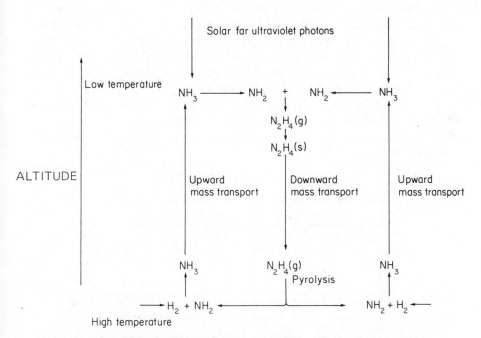

Figure 3. Simplified representation of the atmospheric ammonia cycle on Jupiter

methane and ammonia by the mechanisms mentioned for Jupiter. Presumably the products fall to the surface and accumulate thereon. This shows clearly that the perturbation of an atmosphere by solar radiation can be effective, even at that vast range, particularly since laboratory irradiations of simulated Titan atmospheres have produced a reddish polymeric material with similar optical properties to Titan's surface. Analysis of the laboratory products has indicated the formation of straight chain alkanes with NH_2 and probably OH and C=O groups, the oxygen being derived from water ice incorporated into the original mixtures, but by far the most exciting products detected were amino acids. Moreover chemical similarities have been found between the polymeric material and material recovered from Pre-Cambrian sediments on Earth. It then becomes not unrealistic to link what is happening chemically on Titan today with the pre-history of Earth itself. Therefore it is hardly surprising that there is great interest in achieving a Titan fly-by ancillary to the current Pioneer spaceflight missions on course past Jupiter and Saturn.

Now we return to consider the inner planets. Perhaps we recognize the abnormality of Earth and Venus amongst the planets in possessing atmospheres which contain major components based upon oxygen, free O_2 in the case of Earth and CO_2 in the case of Venus. We can now understand that, even if the four inner planets started with solar compositions (conjecture), escape of the gases to space will have substantially modified the elemental abundances. With Earth the theory of the loss of atmospheric components cannot revolve around escape velocity arguments alone. The highest regions of the atmosphere of our planet have very high temperatures indeed, e.g. 1500 K at 400 km altitude, due to ionospheric phenomena. At such altitudes, therefore, any gas molecule should be able to escape from the gravitational pull of the Earth with ease, but the pressure in these highest regions is extremely low and there is effectively a temperature inversion, i.e. increasing temperature with altitude from about 85 km. Therefore mass transport from the lower zones must be the more important determining factor for loss of gases by the Earth.

However, the position overall is much more complex as can be deduced from Table 1.3. This represents the Earth's elemental composition.

It is outstanding that hydrogen does not appear in this list of the 15 most abundant elements.

Table 1.3. The overall elemental composition of the Earth (% by weight)

Fe	35.39	S	2.74	Co	0.20	Ti	0.04
O	27.79	Ni	2.70	Na	0.14	P	0.03
Mg	17.00	Ca	0.61	Mn	0.09	Cr	0.01
Si	12.64	Al	0.44	K	0.07		

Source: B. H. Mason, *Principles of Geochemistry*, 1958, p. 50. Reproduced with permission of John Wiley & Sons Inc., New York.

Table 1.4. Masses of the shells of the Earth's structure

	Mass x 10^{-24}/kg	% of total mass
Atmosphere	5×10^{-6}	0.00009
Hydrosphere	1.41×10^{-3}	0.024
Crust	0.024	0.4
Mantle	4.075	68.1
Core	1.876	31.5

Source: B. H. Mason, *Principles of Geochemistry*, 1958, p. 41. Reproduced with permission of John Wiley & Sons Inc., New York.

Table 1.4 indicates the underlying reason why this is so, notwithstanding the fact that approximately two-thirds of the Earth's surface is covered with water.

The hydrosphere which contains virtually all of the hydrogen is hence only 0.024% of the total planetary mass. It is quite clear that the composition of the mantle predominantly determines the gross composition of Earth — hence the high abundance of the heavier elements in Table 1.4.

Any relationship between the solar composition data of Table 1.2 and the terrestrial composition data of Table 1.3 is obscure to say the least. Even elements like iron and silicon, which might be expected to be relatively involatile and hence retained almost completely by the Earth, are in the ratio of 2.8 as opposed to 0.09 in the solar composition. As a consequence we are forced to concern ourselves only with what is implied for the evolution of our atmosphere by the geological record preserved within the rocks of the crust. Beyond that, Earth's evolution is the province of astrophysical theories which are outside the scope of this book.

1.3 Evolution of the Atmosphere of the Earth

The age of the Earth is estimated to be at least 4.6×10^9 years. The oldest rocks of sedimentary origin date back to 3.5×10^9 years ago. This is important since sedimentary rocks are necessarily laid down from extensive sheets of water. The dating above therefore represents a lower limit to the existence of widespread liquid water on the surface. Again, in view of the world-wide distribution of early Pre-Cambrian fossils, and since the existence of life implies the prevalence of liquid water, we conclude that seas of sorts must have existed for at least half of the Earth's present lifetime. We can go further from the fossil record. The higher plants and animals are known to be relatively intolerant of excessive fluctuations of the atmospheric oxygen and carbon dioxide levels. Consequently we may deduce that over the last fifth of Earth's lifetime the atmosphere must have remained fairly close to the present composition.

The existence of liquid water is significant in the sense that it sets upper and lower limits to the Earth's surface temperature those aeons ago (1 aeon = 10^9

years). Unless the atmosphere was formerly much denser than now, and therefore had a higher surface pressure, the lower limit is the more interesting. It implies a surface temperature, over a large part of the planet at least, above 273 K. The backward extrapolation of estimated atmospheric loss rates through time supports this contention. The surface temperature of the Earth is determined primarily by the solar flux incident upon it, but solar theory suggests that the luminosity of the Sun has increased over geological time to its present level; the value for the primeval Earth is estimated to have been some $40 \pm 10\%$ lower. It is then somewhat surprising that the surface temperature has not changed by more than 20 to 30 K, if that.

Let us first consider the simple radiant equilibrium temperature of a planet and how it may be calculated in principle.

The effective temperature of a planet is the temperature of a black body, or perfect radiator, of the same dimensions as the planet, which receives the same amount of radiation per unit time from the Sun as is absorbed by the planet. The planet must reradiate on the average at exactly the rate at which it absorbs to attain equilibrium.

The Planck radiation law specifies the spectral distribution of black body radiation. If T is the effective temperature, λ the wavelength, and k is Boltzmann's constant, then the radiance, $E(\lambda, T)$ in watts per unit area emitted into unit solid angle (1 steradian) is given by:

$$E(\lambda, T) = \frac{c_1}{\lambda^5} \{\exp(c_2/\lambda T) - 1\}^{-1} \, d\lambda \qquad (1.8)$$

The constants c_1 and c_2 have values 3.739×10^5 W m^{-2} sr^{-1} nm^4 and 1.438×10^7 nm K respectively. Figure 4 shows the resultant black body radiances as functions of wavelength for the approximate solar visible temperature (6000 K) and for the approximate terrestrial surface temperature of 300 K. It is evident that the 300 K spectrum contains a much larger fraction of near infrared radiation than does the 6000 K spectrum.

We may also refer to Wien's law, which specifies the wavelength, λ_{max}, at which a perfect black body emits its maximum radiant energy:

$$\lambda_{max} = C/T \qquad (1.9)$$

If λ_{max} is expressed in nm and T in Kelvin, $C = 2.897 \times 10^6$. Thus for solar radiation with $T = 6000$ K:

$$\lambda_{max} = 2.897 \times 10^6 / 6000 = 483 \text{ nm}$$

while for the terrestrial reradiation:

$$\lambda_{max} = 9667 \text{ nm}.$$

These are evidently in agreement with the curves of Figure 4.

When the radiance spectrum is integrated over all wavelengths, the total radiant energy flux emitted, $E(T)$, is obtained. This is related to T through the

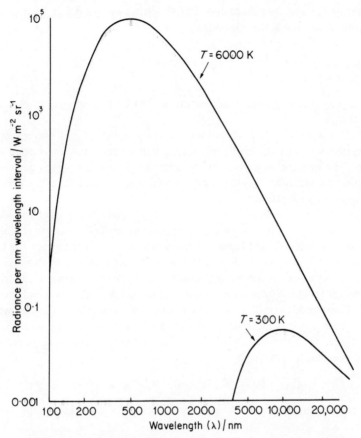

Figure 4. Plots of black-body radiances as functions of wavelength for temperatures of 6000 K (approximately the solar visible surface temperature) and 300 K (approximately the average terrestrial surface temperature)

Stefan–Boltzmann law:

$$E(T) = \sigma T^4 \qquad (1.10)$$

where σ is known as the Stefan–Boltzmann constant and has value 5.672×10^{-8} for T in Kelvin and $E(T)$ in W m^{-2}.

Now of course it is unrealistic to consider the Earth's surface as a perfect black body emitter. In fact the typical emissivity of a material like sand is only around 0.6 in the infrared region as opposed to unity for a true black body. If the terrestrial surface is considered as a so-called 'grey' body, i.e. an imperfect black body with emissivity less than unity but constant as a function of wavelength, then the actual radiance will be that of a true black body but at a lower temperature than the actual temperature. This lower temperature is referred to as the brightness temperature (T_B) of the grey body. T_B is always quoted for the standard wavelength of 650 nm by convention. For an emissivity, $\epsilon = 0.6$ and a true surface temperature of say 288 K the relationship:

$$\ln \epsilon = \frac{c_2}{650} \left\{ \frac{1}{T} - \frac{1}{T_B} \right\}$$

leads to a predicted brightness temperature of 284 K. This appears to be a rather small difference.

Kirchhoff's law states that for all media or surfaces the ratio of the emissivity to the absorptivity is unity. In effect this means that a near transparent body will be incapable of radiating effectively. On the other hand a surface which can absorb a high fraction of radiation incident upon it will approximate to a black body in reradiating with a high efficiency.

We may consider the Earth's atmosphere as perfectly transparent to incoming solar radiation and also in a first approximation as perfectly transparent towards the reemission in the infrared from the terrestrial surface. Just outside the Earth's atmosphere 1.353×10^3 J s^{-1} are incident on a surface of area 1 m^2 normal to the solar rays. This number is termed the Solar Constant. With our present model all of this penetrates to the surface where some of the energy is absorbed and the rest is reflected. The albedo, a, is defined as the fraction of the incident radiation which is reflected; hence $1 - a$ is the fraction absorbed. The Earth intercepts solar radiation in a disc of area πr^2, with $r = 6.4 \times 10^6$ m^2. Hence the total energy absorbed by the Earth per second (R_E) is given by

$$R_E = (1 - a)\pi (6.4 \times 10^6)^2 (1.353 \times 10^3) = 1.763 \times 10^{17}(1 - a) \text{ J s}^{-1}$$

The average value of a is approximately 0.35, so that R_E becomes 1.15×10^{17} J s^{-1}.

For equilibrium exactly this amount of energy must be reemitted per second, according to the form of the Stefan—Boltzmann equation (1.10). Thus the rate of energy reradiation is given by $4\pi r^2 \sigma T^4$, where $4\pi r^2$ is the surface area of the approximately spherical Earth. On equation to R_E we obtain:

$$T^4 = (1 - a) \times 1.353 \times 10^3 / 4\sigma$$

from which $T = 250$ K only. Obviously such a temperature would not allow the existence of liquid water now, far less throughout the course of geological time.

It should be noted that the total spherical surface area, $4\pi r^2$, can only be used in the reradiant energy equation when light and dark hemispheres have nearly the same temperature, usually because of a relatively dense atmosphere and rapid planetary rotation. If, on the other hand, the dark side becomes very cold, then only the hemispherical surface area, $2\pi r^2$, is used, as would be the case for Mercury and Mars.

The deficit between the mean temperature of 250 K for a transparent-atmosphere Earth and the actual mean temperature of 288 K cannot be explained by the heat flow from the interior of the Earth as this only amounts to 2×10^{-5} of the solar constant. Rather we must consider that the Earth's atmosphere does have

Table 1.5. Major components
of dry air at sea level (volume-%)

Nitrogen	78.084
Oxygen	20.948
Argon	0.934
Carbon dioxide	0.033[a]
Neon	0.00182
Helium	0.00052
Methane	0.0002

[a]Estimated for 1977.
Source: Generally available, e.g. in
U.S. Standard Atmosphere Tables,
1962.

components, for example, carbon dioxide, which have significant absorption bands within the reradiation spectrum.

Table 1.5 shows the present major composition of our atmosphere. These components with the exception of methane have constant mixing ratios up to about 100 km altitude.

Figure 5 shows the reradiation spectrum of the Sahara Desert, as detected from a satellite, with the principal absorbing species indicated. It is evident that the

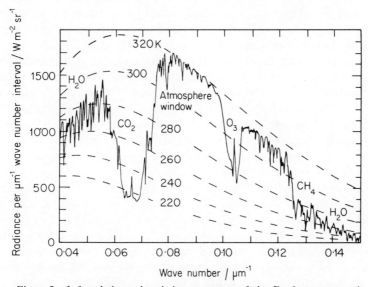

Figure 5. Infrared thermal emission spectrum of the Earth as measured from a satellite over the Sahara Desert. Radiances of black-bodies at several temperatures are shown as dashed curves. The principal atmospheric absorption bands are indicated by the formula of the species concerned. From R. A. Hanel, B. Schlachman, D. Rogers and D. Vanous, 'Nimbus 4 Michelson Interferometer', *Applied Optics*, **10**, 1376 (1971). Reproduced by permission of the Optical Society of America

atmosphere is by no means transparent for the Earth's reradiant spectrum, although it is substantially so for the incoming visible radiation.

Let us reason out the fate of an infrared quantum emitted from the surface of the Earth when the atmosphere has a potential for absorbing that quantum. Let us suppose that at some altitude this quantum is absorbed by an infrared active molecule such as CO_2. The resultant vibrationally-excited molecule has a finite lifetime before it reradiates the quantum in fluorescence. This fluorescence is in random directions as opposed to the presumed vertical trajectory of the original quantum. Accordingly there is a substantial probability that it will be directed back towards the surface. Of course, if it so happens that it is still directed upwards on reemission, then a second, third and so on absorption–fluorescence event at higher altitudes makes it increasingly likely that the quantum will be turned back. The net result of such action by an assembly of atmospheric molecules will be an effective trapping of energy within the surface–atmosphere system; hence the equilibrium temperature will be raised above that for equilibrium with a fully transparent atmosphere. The magnitude of the rise will depend upon the degree of atmospheric opacity, but it is considered that this trapping phenomenon is responsible for the bulk of the 38 K excess of the mean temperature of the Earth over the transparent atmosphere value calculated above.

The energy-trapping phenomenon is termed the Greenhouse Effect, the analogy being drawn with the heat-containing action of the glass walls of a greenhouse. However this analogy should not be pursued too closely; the greenhouse glass does not act only by absorption of infrared reradiation but also by preventing the convective dissipation of heat and providing insulation through poor heat conductivity to the outside air.

We have now come to a qualitative rationalization of the present-day mean temperature of Earth, but how can we explain a roughly similar mean temperature over the geological time scale, when the solar constant was up to 50% lower than it is now? The only conceivable explanation is that a much larger greenhouse effect must have been operating then, raising the mean surface temperature well above a transparent-atmosphere value of as little as 230 K. There are two outstanding points to be made which can help us to believe that this was likely, and point to carbon dioxide and water vapour as the principal active species.

First we must accept the hypothesis that the inner planets lost their primordial atmospheres at a very early stage. Consideration of the arguments made for the outer planets in conjunction with the inevitable high temperature of the juvenile Earth and the relatively low escape velocities for the low-mass planets makes this credible. The atmospheres possessed now by Earth and Venus must, as a consequence, have been created by outgassing from the interiors. In the case of Venus the high surface temperature (750 K) coupled with the almost complete absence of water vapour, and therefore liquid, can be shown on the basis of thermodynamic data to preclude much incorporation of CO_2 into the crustal material. Hence, as CO_2 has been outgassed from the interior of Venus into the atmosphere, it has largely accumulated to produce the present surface pressure of around 75 atmospheres with 95% of the atmosphere being CO_2. Amongst others, silicate equilibria such as

$$CaMgSi_2O_6 + CO_2 \rightleftharpoons MgSiO_3 + CaCO_3 + SiO_2$$

will be ineffective in buffering the CO_2 accumulation at 750 K. So there is good reason to believe that the present mass of CO_2 in the atmosphere of Venus represents close to the integrated flux of outgassing. The recent spectroscopic and spaceflight investigations of the Venusian atmosphere have allowed deduction that the total CO_2 content is roughly equal in magnitude to the total potential amount of CO_2 on Earth, here largely withdrawn as carbonates in limestone and chalk. In itself this is a coincidence which suggests strongly that the solid phases of juvenile Earth and Venus held roughly the same concentrations of dissolved CO_2. Were the present carbon content of the Earth's crust reconverted to CO_2, then our atmosphere would become rather similar to that of Venus, ignoring for the moment the vaporization of the seas.

Now we look back into Earth's fossil record to find that carbonate or shell-forming life was not present until Cambrian times (0.65×10^9 years ago or one-eighth of the geological time scale). Since interior outgassing would be expected to be a reasonably continuous phenomenon, then in the absence of carbonate-forming organisms, one is forced to conclude that Earth's atmosphere must have in earlier times contained much more CO_2 than at present. Consequently a larger greenhouse effect would have operated, which could have produced the required 50 to 60 K enhancement of surface temperature. The first appearance of photosynthesizing plant life known at present occurred some 2 aeons ago in the form of the blue-green algae, although bacteria of sorts probably appeared before. Hence it might be conjectured that these algae were responsible, through their photosynthesizing action, for the generation of an oxygen content in the atmosphere which permitted the evolution of higher life forms and led to the fixing of much of the CO_2.

We must be aware that there is another viewpoint advocated by scientists who, acknowledging the necessity for a greenhouse effect on the Pre-Cambrian Earth, propose that ammonia in the atmosphere (also infrared active), rather than carbon dioxide, was the effective component. There is some evidence apparently supporting the existence of such a necessarily reducing atmosphere but the overall picture is less satisfying. Readers are referred to the papers by Meadows and by Sagan and Mullen in the list at the end of this chapter for further development of the counterarguments.

The other major constituents of our atmosphere, with the exception of the rare gases, are all in a state of detailed dynamic balance. For example, the physical water cycle turns over an amount equal to the volume of the oceans in 50,000 years or so. Nitrogen is removed from the atmosphere mainly by biological fixation and restored by the action of denitrifying bacteria. The relative rates of these processes are such as to keep most of the nitrogen as N_2 in the atmosphere. Were the Earth a purely thermodynamic chemical equilibrium system, then most of the nitrogen would be present in the aqueous phase, mainly as the nitrite ion. The annual photosynthesis of an estimated 8×10^{13} kg of organic carbon means the annual liberation of some 2×10^{14} kg of O_2 and the fixation of about 3×10^{14} kg of CO_2. This amount of O_2 is some 0.02% of the Earth's free oxygen so that it takes around 5000 years for a complete turnover of the atmospheric O_2 (Section 7.3).

Generally the chemical balance of the atmosphere is predominantly established by the actions of the hydrosphere and the biosphere. Clearly the amount of any constituent will depend upon the relative rates of each part of the cycle. Man's activities generate additional terms in such cycles; if these are large enough they can perturb the detailed natural balance.

1.4 Projections for Future Evolution of the Earth's Atmosphere

Our Sun continues to evolve and the solar constant will continue its slow increase. Hence the effective temperature for a transparent-atmosphere Earth will increase and the actual surface temperature likewise. The actual temperature will rise the faster because of the increased water-vapour content of the lower atmosphere which will enhance the greenhouse effect. Eventually a so-called Runaway Greenhouse will bring this planet to the state already reached by Venus. There the huge CO_2 content of the atmosphere traps reradiation so effectively as to produce the greenhouse enhancement of some 300 to 400 K, the runaway situation. Under these conditions the Earth could achieve a surface pressure of 300 atmospheres of steam.

However this cataclysm is nowhere near at hand for Earth. According to various calculations it will require a solar constant some 50% higher than at present to cause a runaway greenhouse to develop on our planet; this is at least 3 aeons into the future. However, this is on the assumption of natural changes in the atmosphere; Man must be cautious in the perturbations which he brings about lest hostile conditions are produced by lack of forethought at a much earlier date.

BIBLIOGRAPHY

1. 'Composition and Structures of Planetary Atmospheres,' by D. M. Hunten, *Space Science Reviews*, **12**, 539–599 (1971).
2. 'Observability of Spectroscopically Active Compounds in the Atmosphere of Jupiter,' by J. S. Lewis, *Icarus* **10**, 393–409 (1969).
3. 'Red Clouds in Reducing Atmospheres,' by B. N. Khare and C. Sagan, *Icarus*, **20**, 311–321 (1973).
4. 'Earth and Mars: Evolution of Atmospheres and Surface Temperatures,' by C. Sagan and G. Mullen, *Science*, **177**, 52–56 (1972).
5. 'The Origin and Evolution of the Atmospheres of the Terrestrial Planets,' by A. J. Meadows, *Planetary Space Science*, **21**, 1467–1474 (1973).

Chapter 2

The Natural Energy Balance of the Earth

We have already made some progress towards an understanding of the natural energy balance of our planet in the course of Chapter 1. In Section 1.3 we dealt with the basic concepts of radiative energy balance and saw the entry of the greenhouse effect as an additional consideration.

In this chapter we shall look at the present Earth and consider in more detail the dissipation and conversion of solar radiance as it passes down through the atmosphere. We can be sure that the density of light flux at the surface will be less than would correspond to the solar constant: scattering by molecules and aerosols, absorption by molecules, diffusion and reflection by clouds will be attenuating factors. At the same time reradiation will be modified by the convection and turbulence of the atmosphere, the evaporation of water and subsequent precipitation, exchange of radiation between ground and air and vice versa and the radiative properties of the atmosphere itself.

2.1 Factors Modifying the Solar Irradiance

(i) Molecular Scattering

The fundamental theory of light scattering by molecules was expounded by Lord Rayleigh in 1871, with reference to gases where the molecular dimensions are small compared with the wavelengths of visible radiation. Regarding the light as an electromagnetic wave, electrons within the molecule are perturbed by its passage and oscillate about their equilibrium positions with the same frequency as the exciting beam. Consequently transient dipoles are induced in the molecules, which then act as secondary scattering centres by reemitting the transiently absorbed energy in all directions. This phenomenon is to be distinguished from absorption—fluorescence situations since no true absorption of photons actually occurs.

Rayleigh showed that for gases the extent of scattering of light of wavelength λ was inversely proportional to the inverse fourth power of λ. Hence it is to be expected that the 350-nm component of incoming solar radiation will be scattered some 28-times more effectively than the 800-nm component. This is the phenomenon which gives the sky its blue colour: non-impinging rays of the solar light have their blue violet components scattered preferentially towards the observer.

It is predicted that the effect of scattering by molecules of the atmosphere will show its greatest development towards shorter wavelengths: at 350 nm for example, some 50% of the incoming solar radiation is expected to be scattered out of the direct beam. However, we must note that sky brightness makes a significant contribution to the light flux density received at the surface: when an isolated dark cloud covers the Sun, it does not go dark! In fact in the spectral ultraviolet region, the sky brightness contributes about half of the flux density to ground. Correspondingly the net attenuation due to Rayleigh scattering in this spectral region is only of the order of 25% on the average.

(ii) Particulate Scattering

In this form of scattering, the particle dimensions are large compared with the wavelengths of visible light. The resultant phenomenon is known as Mie scattering. Both solid and liquid aerosol particles are effective; hence light clouds will attenuate the incoming solar radiation basically by scattering.

As for Rayleigh scattering, the largest effects are found at the shortest wavelengths; this may be illustrated by two very different situations. Following the colossal volcanic explosion of Krakatoa in 1883, brilliant red sunsets were seen all over the world due the the injection of fine dust particles into the higher atmosphere. The interpretation is that the dust density in the long atmospheric path of the rays of the setting Sun scattered out the blue end of the solar spectrum, leaving the red effect. In the second instance, the production of colloidal sulphur by acidification of a sodium thiosulphate solution gives rise to a bluish opalescence off the line of a projected beam of white light leaving a red colour in the transmitted beam.

Particles of larger size, say more than 30-times the wavelength of green light in diameter, scatter light uniformly as white light, i.e. with virtually no dependence on wavelength.

The scattering of solar radiation by atmospheric aerosols is rather variable, even excluding cloud effects. The primary factors determining the net effect are the actual aerosol column density (highly variable) and its distribution as a function of altitude, the size distribution of the particles, and their absorbing and refractive characteristics. The principal layers of aerosol particles occur in the stratosphere (above 11 km altitude) but under typical conditions, the particulate scattering effect is very much smaller than Rayleigh scattering. Because of this and the inherent variability, the aerosol attenuation is usually ignored in terrestrial solar irradiance calculations.

Similarly, cloud is difficult to take into account because of its variability. Not only does it diffuse the incoming light but it also introduces reflection characteristics. Thin high cloud such as cirrostratus and altocumulus can produce an effective 10 to 20% reduction in the ground flux density: lower and thicker clouds such as stratocumulus may result in a 60% or more attenuation effect. More will be said of this in Section 2.2 to follow.

(iii) Molecular Absorption

At the outset of this subsection we must define and relate some of the parameters in common usage for expressing the extent of absorption as a function of wavelength of the radiation.

The Absorption Cross-section (given the symbol σ and as such not to be confused with the Stefan–Boltzmann constant (equation (1.10), having the same symbol) of a molecule for absorption of radiation is defined by the Beer–Lambert equation in the form:

$$\frac{I_L(\lambda)}{I_0(\lambda)} = e^{-\sigma NL} = 10^{-\sigma NL/2.303} \tag{2.1}$$

where $I_L(\lambda)$ is the intensity of light of wavelength λ transmitted through pathlength L of the medium containing concentration N of the absorbing species expressed in molecules per unit volume. $I_0(\lambda)$ is the incident intensity.

The Decadic Absorption Coefficient, given the symbol ϵ, is defined by the Beer–Lambert equation in the form:

$$\frac{I_L(\lambda)}{I_0(\lambda)} = 10^{-\epsilon cL} \tag{2.2}$$

where c is the concentration of the absorbing species expressed in moles per unit volume. The units of ϵ, expressing c in $mol\, dm^{-3}$ and L in dm, will be $dm^3\, mol^{-1}\, dm^{-1}$, which we can contract to $dm^2\, mol^{-1}$.

The Optical Thickness, given the symbol τ, of the overall medium is defined by the Beer–Lambert equation now written as:

$$\frac{I_L(\lambda)}{I_0(\lambda)} = e^{-\tau} \tag{2.3}$$

This parameter is useful for expressing the attenuation characteristics of a fairly invariant medium, such as the path through the whole atmosphere to the ground. In this instance, it can be broadened to include a so-called Rayleigh Optical Thickness expressing the extent of molecular scattering, to be combined with the various absorption optical thicknesses.

Thus σ (in m^2) correlates to ϵ (in $dm^2\, mol^{-1}$) through the equation:

$$\sigma = 3.60 \times 10^{-26}\, \epsilon \tag{2.4}$$

It will turn out that, in dealing with the passage of solar radiation through the Earth's atmosphere and its attenuation, we need to consider only two absorbing species, which are O_2 and O_3. Figure 6 shows the absorption cross-sections of these molecules as a function of wavelength. These absorptions are such as to remove all of the solar radiation of wavelengths less than 300 nm before it can reach the Earth's surface. Oxygen is mainly responsible for removal of wavelengths below 190 nm while ozone provides the complement. The absorption coefficients of ozone across the visible region are too small to produce noticeable absorption.

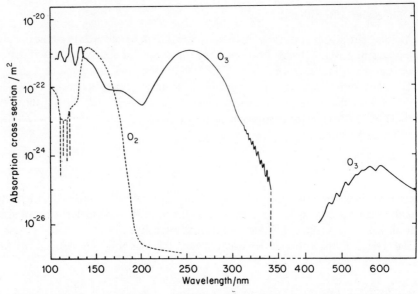

Figure 6. The variations with wavelength of the absorption cross-sections of molecular oxygen and ozone. The detailed variation for molecular oxygen in the region 100 to 130 nm is ignored but the lines indicating cross-sections of 10^{-24} m² or less are shown. From L. Thomas and M. R. Bowman, 'Atmospheric penetration of ultraviolet and visible solar radiations during twilight periods', *Journal of Atmospheric and Terrestrial Physics*, **31**, 1311 (1969). Reproduced by permission of Pergamon Press Ltd

A commonly used tabulation form, particularly for ozone absorption, invokes yet another variant of the Beer–Lambert law. The strength of absorption at a particular wavelength is expressed in terms of a coefficient, δ, with the dimensions of reciprocal length:

$$\frac{I_l(\lambda)}{I_0(\lambda)} = 10^{-\delta l} \tag{2.5}$$

The length parameter, l, is defined as the column height (or equivalent thickness) of ozone (pure) at 298 K and 1 atmosphere (standard) pressure which would produce the same extent of absorption in a vertical path. For the ozone in the upper atmosphere, this situation can be visualized as if all the ozone was collected and brought down to the Earth's surface to be spread out in a uniform layer. It turns out that this layer would be only approximately 3 mm thick – on the face of it a not very reassuring defence of life on Earth from the lethal radiation below 300 nm wavelength! But its effectiveness can be gauged from Figure 7, which shows the reduction of ultraviolet radiation intensities between 190 nm and 320 nm wavelengths by ozone as a function of the parameter l above. At 290 nm and below, the 3 mm column height is quite adequate to reduce the penetrating intensities by more than four orders of magnitude for normal incidence, apart from the partial window between 190 and 210 nm.

The absorption coefficients ϵ and δ across the ultraviolet region for ozone are listed in Table 2.1, with other data for a vertical path through the ozone layer based upon the ozone concentration versus altitude profile in Figure 23. The calculations of Table 2.1 ignore for the moment the scattering effects which will enhance the attenuations of the direct ultraviolet radiation.

One interesting point to emerge from Table 2.1 is that most of the absorption of ultraviolet radiance by ozone occurs well above an altitude of 30 km, even though the maximum concentration of ozone is located near by 25 km. In fact the maximum volume rate for absorption of solar energy by ozone is around an altitude of 50 km; the conversion of radiation into heat energy thus achieved creates a maximum in the temperature profile at this altitude, identified by the term Stratopause (Chapter 3).

Absorption of radiation by molecular oxygen and the resultant degradation into heat for wavelengths less than 190 nm occurs predominantly at higher altitudes still. However the total heat energy so produced is much smaller than that coming from ozone absorption and makes little mark on the temperature profile as a function of altitude.

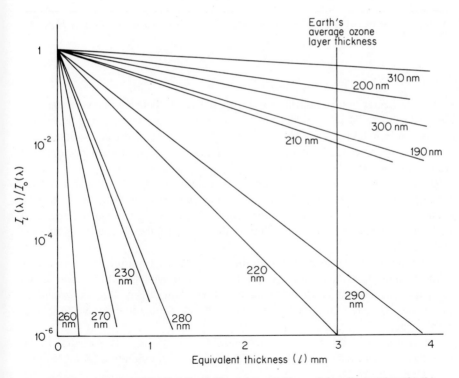

Figure 7. The penetration ($I_l(\lambda)/I_0(\lambda)$) of ultraviolet radiations of decadic wavelengths (λ) through an ozone column of equivalent thickness l mm, according to the absorption coefficients given in Table 2.1

Table 2.1. Radiation absorption data for ozone and the ozone layer

Wavelength/nm	$\epsilon/dm^2 \, mol^{-1}$	δ/mm^{-1}	Altitude for 10^6 attenuation/km	Fraction of solar irradiance transmitted by ozone layer
190	1350	0.60	–	0.0158
200	628	0.28	–	0.145
210	1460	0.65	–	0.0112
220	4480	2.00	–	1×10^{-6}
230	11,950	5.33	29.3	
240	21,210	9.46	33.6	
250	28,820	12.9	35.3	
260	28,280	12.6	35.2	$<10^{-6}$
270	20,380	9.09	33.4	
280	10,430	4.65	28.0	
290	3409	1.52	–	2.7×10^{-5}
300	919	0.41	–	0.0589
310	269	0.12	–	0.436
320	74.0	0.033	–	0.796
330	18.8	8.4×10^{-3}	–	0.944
340	4.5	2×10^{-3}	–	0.986
350	0.76	3×10^{-4}	–	0.998

Absorption coefficient data read off graphs given by M. Griggs, *Journal of Chemical Physics*, **49**, 857 (1968).

(iv) Solar Angle

At a particular point upon the surface of the Earth there are two superimposed angular variations of solar incidence. The first is the daily passage of the Sun across the sky: the second is the annual movement of the Sun from vertical incidence over the Tropic of Cancer (23.5°N) through the Equator to vertical incidence over the Tropic of Capricorn (23.5°S) 6 months later and then back again.

The solar angle is defined in a plane containing the point on the Earth's surface, the centre of the Earth and the centre of the Sun, as Z in Figure 8. Z may be equated to the angle subtended at the centre of the Earth, denoted θ, because of the huge difference between the radius of the Earth and the distance from the Sun to the Earth.

The solar zenith angle is therefore the deviation from normal incidence at the point of impingement on the Earth's surface of the solar rays at local noon. Dawn and sunset are defined as when the solar rays pass tangentially across the point on the surface. The effective length of the day will then appear to depend on the solar zenith angle.

Let us now consider the actual irradiance at a part of the surface when the solar angle has a certain value of Z. Figure 9 then represents the situation for a presumed horizontal flat surface.

Let l be the length in the angular plane of the surface and l' be the hypothetical length of a surface at normal incidence receiving the same flux. It is then evident

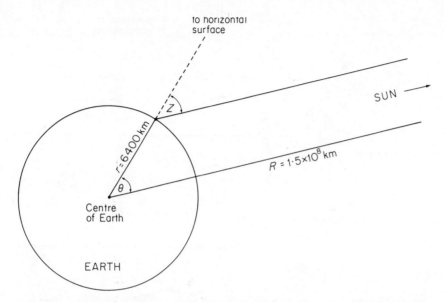

Figure 8. Geometric definition of the solar angle (Z), equated to the angle subtended by the Sun at the centre of the Earth (θ)

that the irradiance at the actual surface is reduced by a factor $l'/l = \cos Z$ compared with the hypothetical surface at normal incidence. Thus for a horizontal surface in Britain at latitude 53.5° N (around Sheffield, Manchester, Liverpool) at noon on Midsummer's Day, the irradiance appropriate to $Z = 53.5 - 23.5 = 30°$ is reduced by $\cos 30 = 0.866$ compared to Los Angeles where the solar zenith angle is then close to zero.

At the same time the pathlength through the atmosphere, and in particular through the stratospheric ozone layer, will have become extended by a factor $\sec Z = (\cos Z)^{-1}$ which may be illustrated by a similar geometrical construction to that in Figure 9. For the British locations above, the atmospheric pathlengths will be longer than that for Los Angeles by the factor $\sec 30 = 1.155$.

The integrated daily flux density at the surface will be of interest from the point of view of the development of photochemical phenomena [such as photosynthetic energy storage (Chapter 4) and photochemical smog formation (Chapter 8)]. In order to assess this we first need to develop equations for the solar angle at a given time of day, at a specified time of the year, and for a defined latitude. A related requirement is the length of the day. Let F represent the latitude in degrees of our location of interest and D in degrees, the Earth's inclination from the vertical (23.5° for the summer solstice). Let H represent the hour angle, the difference between the meridian (longitude) at the defined time of day and that at noon; this amounts to 15° per hour from noon.

The well known relationship to obtain the solar angle, Z, is then:

$$\cos Z = \sin F \cdot \sin D + \cos F \cdot \cos D \cdot \cos H \qquad (2.6)$$

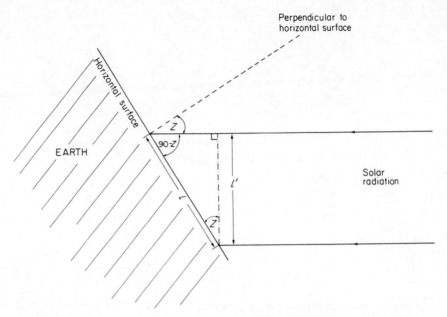

Figure 9. Geometry of the impingement of solar radiance at the Earth's surface for solar angle Z

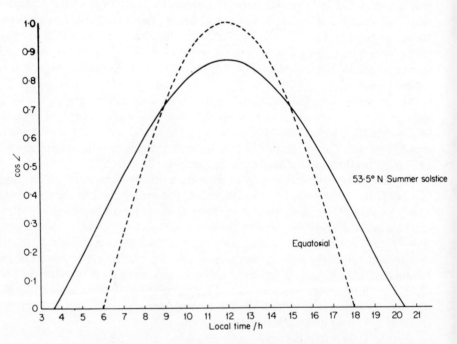

Figure 10. Extraterrestrial diurnal variation of the flux density of solar radiation on a horizontal surface for the Equator and at latitude 53.5° N at the summer solstice

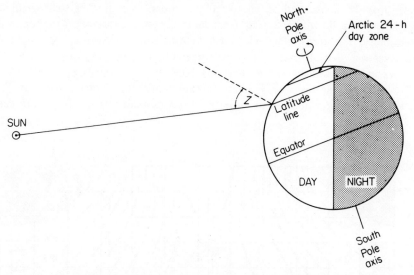

Figure 11. Diagram illustrating day (blank) and night (shaded) durations during the northern hemisphere summer solstice. The lines are latitudes

At dawn or sunset $\cos Z = 0$; if we let H_s be the hour angle under these conditions, then it may be calculated from the relationship:

$$\cos H_s = -\tan F \cdot \tan D \qquad (2.7)$$

Application of this equation (2.7) to the British Midsummer's Day situation, with $F = 53.5°$ and $D = 23.5°$ gives sunrise and sunset removed 8.37 on either side of noon, that is a 16.74-hour period of daylight. Values of $\cos Z$ as a function of time of day for this British location and also for Los Angeles are plotted in Figure 10.

Table 2.2. Calculated time from sunrise to sunset as a function of latitude (Refraction of the rays by the atmosphere will extend daylight above the times given below.)

| Latitude/degrees | Winter solstice | | Summer solstice | |
	h	min	h	min
90	0		~6 months	
80	0		~4 months	
70	0		~2 months	
60	5	33	18	27
50	7	42	16	18
40	9	08	14	52
30	10	04	13	56
20	10	48	13	12
10	11	25	12	38

Source: Calculated from equation (2.7).

These may be considered as relative plots of daily variation of solar irradiance incident upon a horizontal surface (parallel to the ground surface) at the top of the atmosphere. The area under each plot is then proportional to the integrated solar irradiance *per diem*. Perhaps it is some surprise to find that the surface at 53.5°N receives an integrated irradiance about 10% larger than the surface at 23.5°N.

Now despite the necessarily greater attenuation factor deriving from increased scattering and absorption in the longer atmospheric path to reach Britain, it seems

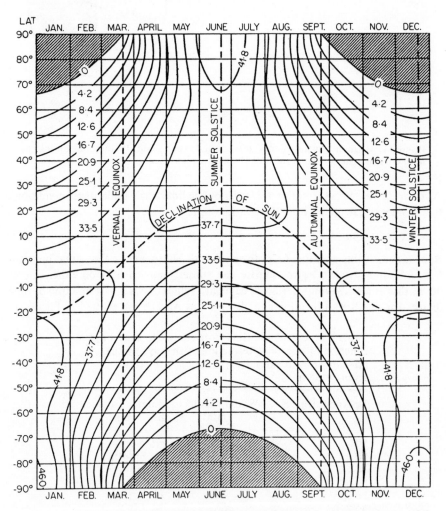

Figure 12. Contour map of the variations of the total daily solar radiance at the top of the atmosphere as functions of latitude and time of year. The shaded areas represent zones of continuous darkness, the contour values are expressed in MJ m⁻² on the basis of the solar constant of 1.353 kW m⁻². From *Smithsonian Meteorological Tables*, 6th edition, Table 134, p. 419. Reproduced by permission of the Smithsonian Institution, Washington, D.C.

reasonable to conclude that the total solar energy received during the summer solstice is approximately the same as that received at the Equator all year round, both on a daily basis. In this we ignore cloud cover. Figure 11 emphasizes that this is largely the result of the longer day at the British latitude.

It is therefore to be emphasized that phenomena which rely upon the integrated solar irradiance for their development are as feasible in Britain in summer on this basis as at lower latitudes. This situation becomes even more surprising for higher latitudes as is shown up by Table 2.2. It is clear that in the lands of the midnight Sun the integrated daily irradiances can be considerably larger than for the Equator. For example, at $73.5°$ N the noon $\cos Z$ factor is 0.643, but this is more than compensated for by the 24-h daylight for more than 2 months.

However ambient surface temperatures are more dependent upon the instantaneous irradiance and therefore do decrease with increasing latitude.

The total daily average solar irradiance incident on a horizontal surface at the top of the atmosphere, E_t, is obtained by integrating the instantaneous irradiance $S \cos Z$, where S is the solar constant, from sunrise to sunset. As a result we obtain:

$$E_t = 430.72(H_s \sin F \cdot \sin D + \cos F \cdot \cos D \cdot \sin H_s)/\text{W m}^{-2} \qquad (2.8)$$

where H in the first term is in radians. For $53.5°$ N at the summer solstice $H_s = 2.216$ rad and the calculation yields $E_t = 447$ W m^{-2}. Figure 12 shows the daily integrated solar irradiance at the top of the atmosphere as a function of latitude and time of the year. The peak E_t of 574 W m^{-2} or 49.6 MJ m^{-2} occurs at the South Pole of all places, on December 22 by virtue of the 24-h period of daylight. The slight assymetry from southern to northern hemispheres arises from the variation of the Earth's distance from the Sun, ranging from 1.0344 on July 5 to 0.9674 on January 3 relative to the average value.

2.2 Dispersal of Solar Irradiance in the Earth–Atmosphere System

(i) Direct Irradiance

The solar radiation intercepted by the Earth is not entirely degraded to other forms of energy. Nor is it necessarily restricted to that part of the surface on which it is incident; the horizontal transfer of heat results in the phenomenon of weather and produces ocean currents.

Hence we are now in a position to resolve E_t above into a sum of energy dissemination terms:

$$E_t = R_a + R_c + R_s + S_a + S_c + A_a + A_c + A_s \qquad (2.9)$$

In this equation R terms denote reflection back to space, S terms denote scattering into space and A terms denote absorption of the radiant energy. The subscripts indicate the medium: a corresponds to the atmosphere (molecules and particulates), c corresponds to clouds, and s to the solid or liquid surface. Evidently only the A terms in equation (2.9) are effective in absorbing the energy.

Table 2.3. Median values for visible light albedos as percentages

	Albedo		Albedo
Surfaces		*Clouds*	
Horizontal water surface	5	Cumiliform	70–90+
(low solar angle)		Stratus	60–85
Fresh snow	85	Altostratus	40–60
Sand desert	30	Cirrostratus	40–50
Green meadow	15	*Planets*	
Deciduous forest	15	Earth	34–42
Coniferous forest	10	Moon	6–7
Crops	20	Mars	16
Dark soil	10	Venus	76
Dry earth	20	Jupiter	73

Data collected from a variety of sources and averaged.

The atmosphere does not absorb much of the solar radiation in total. The primary absorption is by ozone, but this is mainly restricted to the ultraviolet component which is relatively weak compared to the visible. However portions of near infrared radiation are absorbed by carbon dioxide, water vapour and ozone vibrational bands, but again these are displaced from the peak irradiance as a

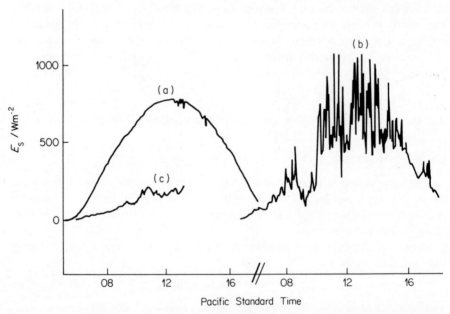

Figure 13. Daily variation of solar radiation flux density (E_s) for stations off the Oregon coast, USA at 45° N during July and August. (a) Cloudless day; (b) day with well-broken cloud; (c) day with full cloud cover. From R. K. Reed and D. Halpern, 'Insolation and net long-wave radiation off the Oregon coast', *Journal of Geophysical Research,* **80,** 841 (1975), copyrighted by the American Geophysical Union and reproduced with permission

function of wavelength. As a result, A_a only amounts to 15 to 20% of E_t. Individual cloud systems absorb (as opposed to reflect or scatter) only about 10% of the incident radiation: the statistical average of cloud cover over the globe then makes A_c amount to only 3% at most of E_t. It is then clear that the atmospheric gases produce considerably greater absorption than do clouds.

A_s is governed by the particular surface albedo, the fraction of the incident irradiance which is directly reflected back. Table 2.3 sets out some typical albedo data for various types of surface.

Averaged over the Earth, cloud cover produces an R_c term which is about 25% of E_t. However the distribution is not uniform: R_c is greatest in middle and higher latitudes and least in the subtropics.

As has been implied in Section 2.1, a substantial fraction of the scattered incoming radiation reaches the Earth's surface as diffuse sky radiation. As a result, S_a is a comparatively small term (around 6% of E_t) and R_s and A_s contain this additional component to direct sunlight. Similar considerations apply to the light which is diffused and to some extent reflected by clouds. Figure 13 shows daily profiles of net solar radiances measured off the Oregon coast (45° N) in July and August for a cloudless day, a day with well broken cloud and for a day with complete cloud cover.

A surprising point from Figure 13 is that the peak instantaneous irradiances

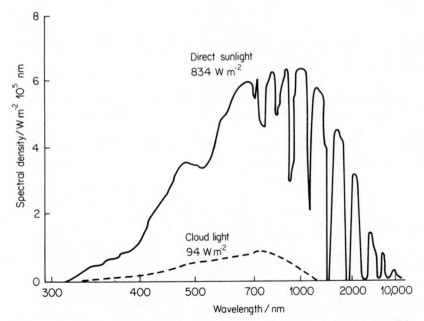

Figure 14. Spectral density distribution as a function of wavelength for direct sunlight and for cloud-diffused light. From J. P. Kerr, G. W. Thurtell and C. B. Tanner, 'An integrating pyranometer for climatological observer stations and mesoscale networks', *Journal of Applied Meteorology*, **6**, 688 (1967). Reproduced by permission of the American Meteorological Society

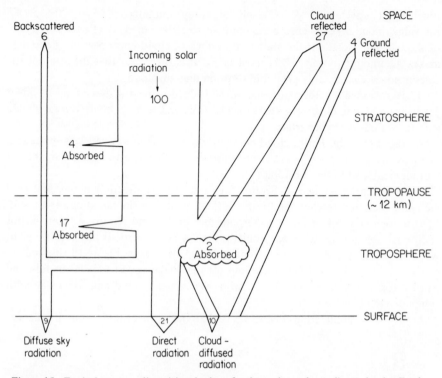

Figure 15. Typical average disposition budget for incoming solar radiance in the Earth–atmosphere system over Britain. (Numbers are arbitrary energy units relative to an extraterrestrial irradiance of 100)

occurred with broken cloud rather than with a cloudless sky, presumably because of light reflected from clouds adding to the direct irradiance when no cloud was in the direct path.

Figure 14 shows the spectral density distribution as a function of wavelength for direct sunlight and for cloud-diffused light. It can be seen that there is no dramatic redistribution between the two profiles.

We may now attempt to produce a diagram summarizing the average disposition of incoming solar irradiance, with particular reference to the cloudier middle to upper latitudes. This is shown as Figure 15.

It can be seen from Figure 15 that some 40% of the incoming solar radiation is eventually absorbed by the surface. About the same amount is reflected or scattered back into space, unused in the Earth's heat balance.

Of interest, particularly for the induction of photochemical smog, is the cloudless sky situation. An average energy budget for this might be:

Incoming radiant energy	=	100
Stratospheric absorption	=	4
Tropospheric absorption	=	17
Back-scattering	=	6
Radiant energy incident at surface	=	73

Surface reflection = 11
Energy absorbed at the surface = 62

Thus the irradiance at the surface on a cloudless day is over 70% of E_t.

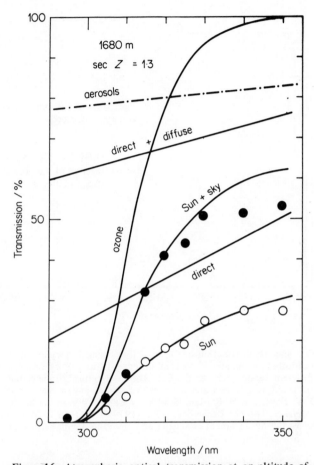

Figure 16. Atmospheric optical transmission at an altitude of 1680 m for a solar zenith angle of 40° and a total vertical ozone equivalent thickness of 2.5 mm. The ozone contribution is represented by the line marked 'ozone'. The dot and dash line marked 'aerosols' represents their contribution for an average amount. The transmission computed from the total air column and scattering cross-section is represented by the line marked 'direct', while an apparent transmission evaluated by assuming that 50% of the scattered light reaches the ground is represented by the line marked 'direct + diffuse'. The curve 'sun' indicates the resultant transmission for the direct solar radiation and can be compared with the measured values (○). The curve 'sun + sky' shows the resulting transmission if the scattering component 'direct + diffuse' is used and can be compared with measured values (●). Reproduced by permission of the National Research Council of Canada from M. Ackerman, *Canadian Journal of Chemistry*, **52**, 1505–1509 (1974)

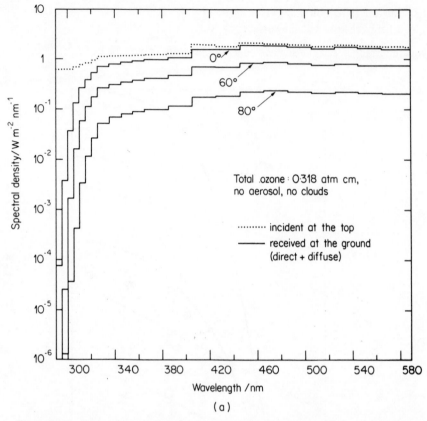

Figure 17(a). Spectral distribution of the solar flux normally incident at the top of the atmosphere (dotted curve), and of the total (direct + diffuse) flux received at the ground for three different zenith angles of the Sun. The total vertical ozone column equivalent thickness is taken as 3.18 mm and aerosols and clouds are assumed absent. From P. Halpern, J. V. Dave, and N. Braslau, 'Sea-level solar radiation in the biologically active spectrum', *Science,* **186** (27 December), 1205 (1974). Copyright 1974 by the American Association for the Advancement of Science and reproduced with permission

On the photochemical aspects of solar radiation, most interest attaches to the transmission of the portion of the solar spectrum between 300 and 400 nm wavelength. Figure 16 shows a representative situation for this radiation for a solar zenith angle of $40°$ with a cloudless sky.

Figure 17(a) shows the irradiance as a function of wavelength across the entire visible region, emphasizing the effect of local solar angle upon the spectral distribution of the combined direct + diffuse light received at the surface for a cloudless sky and ignoring the effects of aerosol scattering.

Figure 17(b) illustrates the calculated variation in spectral distribution of the surface irradiance resulting from the interposition of an average atmospheric aerosol distribution for local solar angles of $0°$ and $80°$. For $Z = 0°$ the direct component of the irradiance is one order of magnitude larger than the diffuse component in the

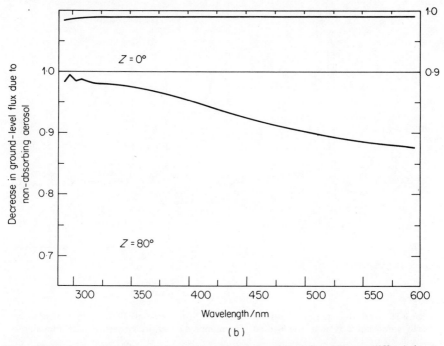

Figure 17(b) Spectral variations of the decrease in ground-level flux (direct + diffuse) due to the presence of non-absorbing aerosol for solar zenith angles of 0° and 80° with average aerosol concentrations and a vertical ozone column equivalent thickness of 3.18 mm. From P. Halpern, J. V. Dave, and N. Braslau, 'Sea-level solar radiation in the biologically active spectrum', *Science*, **186**, (27 December), 1206 (1974). Copyright 1974 by the American Association for the Advancement of Science and reproduced with permission

visible region but approaches the same order of magnitude in the ultraviolet region. The apparent enhancement of the shorter wavelength components for $Z = 80°$ reflects the decreased proportion of the direct component of irradiance; it exceeds the diffuse irradiance only for wavelengths above 470 nm.

(ii) Surface–Atmosphere Energy Transfer

In Figure 4 we see that the Earth's reradiation spectrum is mainly in the infrared portion of the electromagnetic spectrum. In Section 1.3 we mentioned the greenhouse effect resulting from the infrared absorbing characteristics of atmospheric molecules like CO_2 and H_2O. It is also apparent that clouds will enhance this trapping effect by back reflection of infrared radiation.

In general, the atmosphere emits more infrared radiation towards the surface than it does to space. This is because the net downward flux originates in the lower and warmer layers of the atmosphere than does the net upward flux; almost 60% of the counter radiation comes from the lowest 0.1 km and about 90% from the lowest 1.5 km under clear sky conditions. Most of the infrared emission from the

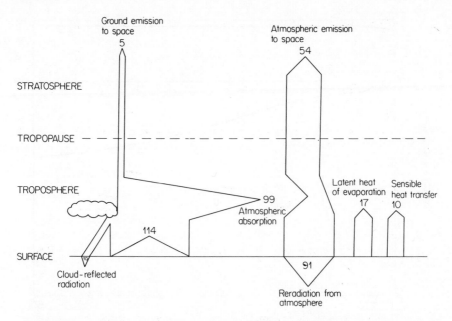

Figure 18 Typical average disposition budget for Earth's reradiation and heat transfer in the Earth–atmosphere system over Britain. The numbers are arbitrary energy units, the same as used in Figure 15

surface which escapes directly to space does so through the so-called 'atmospheric window' between 8500 and 11,000 nm where none of the atmospheric gases absorb strongly.

Figure 18 represents a summary of the heat budget complementary to that of Figure 15. Added to this is the heat expanded in evaporation of water and the heat transferred to the atmosphere from the ground by contact and convection. The energy units, though arbitrary, are the same as for Figure 15. The net loss from the surface is 40 energy units, which exactly balances the gain from solar irradiance shown in Figure 15. Another viewpoint is that the atmosphere loses just as much radiative energy as the solid surface gains in a given period of time, perhaps averaged over a year. Thus the whole Earth system is kept with a radiation balance.

The overall view must be taken as such, since the detailed radiation balance will generally not be held either seasonally or annually at a particular location. For example, at latitudes higher than $40°$ the radiative deficit of the atmosphere exceeds the surplus at the surface. Conversely at latitudes below $40°$ the radiative surplus of the surface exceeds the deficit of the atmosphere. However, we know that an energy balance is maintained in total; this demands a horizontal transfer of heat towards higher latitudes and is achieved by the general circulation of the atmosphere and the release of latent heat in precipitation. Ocean currents also contribute, an outstanding case being the mild climate maintained along the northwestern shores of Europe by the Gulf Stream.

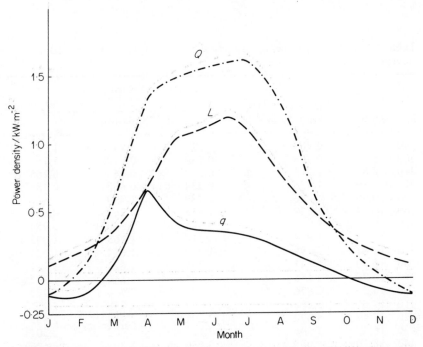

Figure 19. Typical seasonal variation of the terms Q, q, and L for a Western European location

For land, the important energy balance terms are Q, the difference between radiative heat gained by the surface and the net radiative heat lost by the atmosphere in the locality; q, the sensible heat transferred to the air by contact and convection, and L, the latent heat taken up by evaporation of water at the surface less that released by condensation of atmospheric moisture as precipitation. Thus for a daily energy balance;

$$Q = q + L \qquad (2.10)$$

Figure 19 shows the typical variation on an annual basis of these terms expected for a Western European location.

2.3 Quantitative Evaluation of the Surface Solar Irradiance

For the needs of subsequent chapters, we should derive the solar irradiance as a function of wavelength for ground level, with some indication of the effect of solar angle. We have seen in the preceding sections of this chapter that the extraterrestrial solar irradiance as a function of wavelength will be modified before the radiation impinges upon the surface of the Earth.

Table 2.4 lists values of the solar irradiances averaged over 10-nm wavelength intervals on an energy flux density basis (parameter E) and on a photon flux

Table 2.4. Solar irradiances as a function of wavelength (λ/nm) per 10-nm interval centred on λ

λ	E_t^q/W m^{-2}	$(I_t \times 10^{-19})$/ photons m^{-2} s^{-1}	Solar angle		
			0°	20°	40°
			$(I_s \times 10^{-19})$/photons m^{-2} s^{-1}.		
290	4.82	0.703	1.22×10^{-5}	2.94×10^{-6}	1.54×10^{-7}
300	5.14	0.776	0.030	0.0192	7.15×10^{-3}
310	6.89	1.075	0.302	0.264	0.165
320	8.30	1.336	0.734	0.668	0.486
330	10.59	1.758	1.239	1.099	0.837
340	10.74	1.837	1.358	1.262	0.961
350	10.93	1.925	1.468	1.380	1.054
360	10.68	1.934	1.514	1.405	1.093
370	11.81	2.199	1.759	1.634	1.221
380	11.20	2.141	1.747	1.624	1.271
390	10.98	2.155	1.786	1.660	1.302
400	14.29	2.876	2.422	2.253	1.773
410	17.51	3.612	3.092	2.878	2.271
420	17.47	3.692	3.201	2.984	2.359
430	16.39	3.546	3.113	2.902	2.301
440	18.10	4.007	3.550	3.309	2.627
450	20.06	4.542	4.079	3.807	3.030
460	20.66	4.782	4.332	4.044	3.223
470	20.33	4.808	4.375	4.084	3.259
480	20.74	5.009	4.598	4.297	3.438
490	19.50	4.808	4.433	4.148	3.318
500	19.42	4.886	4.549	4.359	3.410
510	18.82	4.829	4.515	4.225	3.392
520	18.33	4.796	4.503	4.218	3.387
530	18.42	4.912	4.637	4.339	3.492
540	17.83	4.844	4.592	4.292	3.462
550	17.25	4.774	4.526	4.244	3.434
560	16.95	4.776	4.552	4.264	3.457
570	17.12	4.910	4.699	4.406	3.565
580	17.15	4.976	4.777	4.475	3.625
590	17.00	5.047	4.855	4.552	3.677
600	16.66	5.029	4.853	4.546	3.675
650	15.11	4.942			
700	13.69	4.822			
750	12.35	4.660			
800	11.09	4.464			

[a]Extraterrestrial solar irradiance (E_t) data derived from M. P. Thekaekara, 'Extraterrestrial solar spectrum, 3000–6100 Å, at 1-Å intervals', *Applied Optics*, **13**, 518–522 (1974). Reproduced by permission of the Optical Society of America.

Table 2.5. Rayleigh optical thicknesses for the atmosphere

λ/nm	τ	λ/nm	τ	λ/nm	τ	λ/nm	τ
290	1.36	370	0.51	450	0.23	530	0.12
300	1.19	380	0.46	460	0.21	540	0.11
310	1.04	390	0.42	470	0.20	550	0.11
320	0.92	400	0.38	480	0.18	560	0.10
330	0.81	410	0.34	490	0.17	570	0.09
340	0.72	420	0.31	500	0.15	580	0.085
350	0.64	430	0.28	510	0.14	590	0.079
360	0.57	440	0.26	520	0.13	600	0.074

Source: Calculated on the basis of an inverse fourth power dependence on the wavelength from values quoted for wavelengths of 290 and 600 nm by P. Halpern, J. V. Dave, and N. Braslau, 'Sea-level solar radiation in the biologically active spectrum', *Science*, **186**, 1204–1207 (1974).

density basis (parameter I). The wavelengths are in the middle of the 10-nm interval. The first columns give the extraterrestrial values, E_t and I_t, for normal incidence; the following columns give estimated values for the photon flux density at the surface, I_s, for solar angles $Z = 0°$, $20°$, and $40°$, assuming average ozone concentrations and distributions as a function of altitude in the optical path. The diffuse component (sky radiance) is taken into account by making the assumption that half of the radiation scattered by molecules in the atmosphere eventually reaches the surface; Figure 16 and the paper by Ackermann cited at the end of this chapter provide good justification for this approximation. The calculated Rayleigh optical thicknesses as a function of wavelength for the atmospheric path normal to the surface are listed in Table 2.5. Aerosol concentrations and distributions in the atmosphere are highly variable; consequently particulate diffusion has been ignored in Table 2.4. Calculations suggest that aerosol scattering is relatively unimportant for solar angles of $40°$ or less ($\leqslant 20\%$ at 300 nm and $\leqslant 10\%$ at 600 nm are reasonable limits for this attenuation). However Figure 17(b) draws attention to the important redistribution in the spectrum of direct + diffuse solar radiation resulting from an average aerosol presence for high solar angles. As already mentioned, the major part of the irradiance is diffuse at the shorter wavelengths at high solar angles so that the photochemical consequences of particulate scattering could be significant under these conditions.

2.4 The Scale of Solar Energy

The Earth's disc intercepts only 1 part in 2 billion (2×10^9) of the total radiant energy released by the Sun. In order to appreciate the enormous order of magnitude of this seemingly negligible fraction of the solar irradiance, Table 2.6 may be examined. This is a table of energies, all expressed relative to the solar energy intercepted daily by the Earth's disc. The daily receipt of solar energy by

Table 2.6. Relative energies involved in various natural and anthropogenic phenomena

	Relative energy
Energy of the Earth's rotation	10^7
Energy released in the 1976 Chinese earthquake	33.6
Combustive energy stored in the Earth's coal reserves	13.1
Combustive energy stored in the Earth's oil reserves (including tar sand and shale oil)	1.2
Solar energy intercepted daily by Earth	1
Combustive energy stored in the Earth's natural gas reserves	0.9
Latent heat absorbed by Spring snow/ice melting	0.1
Presently known North Sea oil reserves	0.02
Annual latent heat released by precipitation over U.K.	0.02
Annual USA consumption of energy (1970)	0.005
Annual British consumption of energy (1972)	0.0006
Annual dissipation of wave energy on British coasts	0.0002
Total energy content of annual grain crop in Britain	0.00004
Energy released in Krakatoa explosion of 1883	10^{-5}
Einstein energy equivalent of 1 gram of matter	6×10^{-9}
Energy of average lightning stroke	10^{-13}

Earth is 1.49×10^{22} J. As we have seen only about 40% of this reaches the surface to be absorbed there.

We see, with some initial astonishment perhaps, that if all the solar energy received by the Earth could be collected, then it would take only some 13 days to accumulate the energy equivalent of all the known coal reserves, by far the most abundant fuel. Accepting the proviso that only irradiance incident at the surface should be considered in this connection and taking general account of day and night, cloud cover, etc., it would take, even then, only a matter of months to receive the equivalent energy of the coal reserves.

However, the general problem with solar energy is its dispersity and hence difficulty of collection. Nature has devised photosynthesis to this end. As will be discussed later (Chapter 4), the photosynthetic mechanism of the plants under normal conditions can only convert around 1% of the radiant energy into stored chemical energy. In Section 1.3, the annual photosyntheiss of organic carbon was given as about 8×10^{13} kg. Hence it turns out that the energy equivalent of the total coal reserves is produced by worldwide photosynthesis in about 100 years or so. This may be compared with the geological time span of many hundreds of millions of years in order to appreciate the minuteness of the Earth's energy 'savings' account as compared the the 'current' account.

Man has invented the solar cell, through which solar radiant energy may be converted to electrical energy with a working efficiency of around 11%. The annual integrated solar irradiance for Britain is some 3×10^9 J m.$^{-2}$. Hence, in theory, about 40% of the area of Scotland, if covered with solar cells, could supply the

British energy need. However the impracticability of such a proposal can be realized immediately in the enormous cost of covering the area of 3×10^{10} m^2 with solar cells at a current cost of £3500 per m^2.

Accordingly any scheme to harvest the energy of the Sun is driven back to using nature's way. The feasibility of this approach will receive fuller attention in Chapter 4.

BIBLIOGRAPHY

1. *'Physical Climatology'*, by W. D. Sellers, University of Chicago Press, 1965, Chapters 1—8.
2. 'The Photochemistry of Air Pollution', by P. A. Leighton, Academic Press, New York, 1961, Chapter II.
3. 'Solar Ultraviolet Flux below 50 km', by M. Ackermann, *Canadian Journal of Chemistry*, **52**, 1505—1509 (1974).
4. 'Sea-level Solar Radiation in the Biologically Active Spectrum', by P. Halpern *et al.*, *Science*, **186**, 1204—1208 (1974).
5. 'The Middle Ultraviolet reaching the Ground', by A. E. S. Green *et al.*, *Photochemistry and Photobiology*, **19**, 251—259 (1974).

Chapter 3

The Gross Structure of the Earth's Atmosphere

The pressure of the atmosphere at sea level is never far from its standard value (760 Torr, 1013 mbar, 1.013×10^5 N m^{-2}). At an altitude of 115 km, where coincidentally the ambient temperature is approximately the same as the mean surface temperature, the average density (and hence the total pressure) is some eight orders of magnitude lower. Molecular nitrogen is still the main constituent at 115 km, as evidenced by the mean molecular weight of 28.32; the significant appearance of atomic oxygen (10% of the concentration of N_2) accounts in the main for the decrease from the ground level value of 28.96. Above about 200 km, atomic oxygen becomes the major constituent; the pressure has by this altitude fallen by ten orders of magnitude compared with the ground level pressure and the temperature is around 850 K. Helium eventually supersedes atomic oxygen as the main constituent and above 1000 km or so altitude, the Earth has an outer hydrogen atmosphere. This separation of constituents is a diffusive phenomenon which marks the dominance of molecular diffusion over eddy diffusion. The altitude at which the two modes have roughly equal effectiveness is around 105 km and is termed the Turbopause. It is readily observed by rocket trail released experiments: below 105 km the released trail of gases develops an indicative globular eddy structure within a few seconds and the eddies grow rapidly with time; above 105 km the released trail shows a smooth, non-turbulent, growth. Gravitational separation of the constituent atmospheric gases takes place, therefore, above the turbopause, leading to the increasing preponderance of lighter gases and a decrease of the mean molecular weight with altitude.

Figure 20 shows the U.S. Standard Atmosphere profiles of temperature and mean molecular weight, together with the ratio of the pressure to the standard pressure (P/P_0), as functions of altitude up to the approximate altitude of the turbopause. We shall adopt the U.S. Standard Atmosphere as the typical one.

The decreasing density of gases with increasing altitude in general means that almost one half of the total mass of the atmosphere lies in the altitude range 0 to 5 km. The density of the atmosphere at surface level is 1.3×10^9 μg m^{-3} typically, and falls to 0.3, 10^{-3}, and 10^{-5} μg m^{-3} at altitudes of 200, 600, and 1000 km respectively. Even at 20 km (the proposed cruise altitude of Concorde) the density is only just under 7% of that at the Earth's surface.

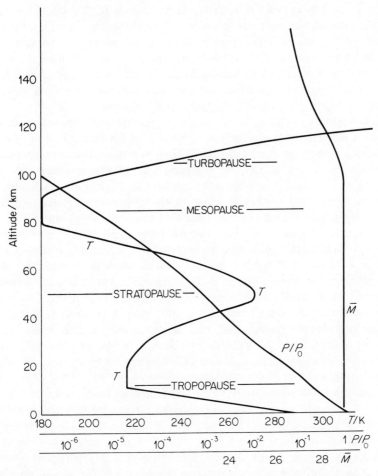

Figure 20. Gross physical structure of the Earth's atmosphere. T is temperature, p/p_0 is the relative pressure to that at sea-level and $\bar{M}$ is the average molecular weight

In this chapter, therefore, we must look at the physical structure of the atmosphere and examine in some detail the physical effects which generate this structure. For future reference we should wish to subdivide the atmosphere into distinct layers so that we can more easily refer to specific ranges of altitude.

3.1 Atmospheric Layer Resolution

Reversals in the temperature gradient with altitude are used as the primary basis for dividing layers: the boundaries are referred to as *Pauses* with an identifying prefix. The temperature at a defined altitude below the turbopause is reasonably constant and only fluctuates to a relatively small extent due to solar heat by day and the 11-year solar cycle of activity. At an altitude of 200 km the variations about the mean are generally still less than 50%, but at higher altitudes variations of a factor of ten may occur. However, it should be noted that because of the diminutive heat capacities of huge volumes at these very high altitudes, the concept of temperature really only has meaning in the context of the thermal velocities of sparse ions, atoms and molecules. Hence temperatures such as 1500 K at 400 km altitude commonly quoted must be taken with this in mind.

It is worthwhile digressing for a moment to consider one of the methods which has been used to measure the temperature profile of the atmosphere as a function of altitude, namely infrared radiometry from satellites. The infrared radiation emitted by molecules of CO_2 in the band near 15000 nm wavelength is not interfered with by emissions from other species; CO_2 has the added advantage of a constant mixing ratio until very close to the turbopause. On account of the pressure-dependent broadening of spectral lines, radiations emitted by CO_2 molecules close to ground level have line widths of about $10 \, m^{-1}$ in wavenumber (reciprocal wavelength) units, whereas molecules at higher altitudes may radiate with line widths of only $0.1 m^{-1}$ because they are much less pressure broadened. Kirchhoff's law (Section 1.3) states that the absorptivity at a particular wavelength is equal to the emissivity at that wavelength. This means that the CO_2 band in absorption shows the same linewidths as the band in emission at any altitude, assuming local thermodynamic equilibrium as is invariably the case. Hence there is a decreasing ability for CO_2 molecules to absorb in the spectral wings of the 15,000-nm band with increasing altitude; consequently, vertically directed radiation from CO_2 molecules at lower altitudes escapes to space with a higher probability the greater the displacement from the centre of the band fundamental structure.

The measurement of the irradiance received across a set of wavelengths in the wings of the structure of the 15000-nm band of CO_2, using a radiometer mounted on a satellite, provides in principle sufficient information to specify the temperature as a function of altitude. The emission at any particular altitude is governed by the Planck radiation function (equation 1.8) for a given wavelength, dependent on the temperature, multiplied by the emissivity for that wavelength, which is simply identical with the absorption coefficient from Kirchhoff's law. The falling pressure with increasing altitude governs the spread of the emission into the wings.

The TIROS 7 (launched in 1963) and Nimbus 4 (launched in 1970) satellites carried radiometers. Computer analysis of the data received using appropriate weighting functions, derived from a model of band spectral characteristics, produced the temperature profile as a function of altitude.

Returning to examine Figure 20 more closely, we see that temperature falls with altitude from ground level up to about 11 km, where the first reversal of temperature gradient occurs. This is termed the *tropopause* and the layer below the *troposphere*.

Above the tropopause the temperature rise is slow up to about 25 km (only about 20 K increase in 25 km) and then more rapid between 25 and 40 km (around 40 K), finally reaching a maximum and another reversal point close to 50 km. This is termed the *stratopause* and the region between this and the tropopause the *stratosphere*.

Above 50 km the temperature decreases quite sharply with increasing altitude until the lowest temperature of any part of the atmosphere is reached near 85 km, typically 180 K, but 135 K has been recorded, the lowest temperature ever seen in the Earth – atmosphere system. The minimum temperature at 85 km defines the *mesopause*, with the region between 50 and 85 km, approximately, being known as the *mesosphere*.

No further reversals of temperature gradient occur at higher altitudes and the region above 85 km is termed the *thermosphere*.

We shall now proceed to look at the atmospheric layers in more detail and at the physical and chemical origins of the temperature profile.

3.2 The Troposphere

The troposphere does not have a uniform thickness. In the tropics the tropopause is elevated to about 15 to 18 km, in mid-latitudes it lies at an average height of 11 km, while in the polar regions it is depressed to as low as 8 or 9 km. To put these altitudes on a more meaningful framework, it may be remarked that the height of Mount Everest is just over 9 km while the height of Ben Nevis is only 1.3 km. Further, the tropopause varies on a daily and seasonal basis – for example, during a 24-hour period its altitude may vary by several kilometers at any one location. However for the purposes of clarity we shall consider an altitude of 11 km as average and invariant.

The tropopause represents more than just a point of temperature gradient reversal. In Figure 21 are shown mixing ratios for carbon monoxide and ozone as a function of altitude: these data were obtained by sampling from an aircraft over the North Sea. It can be seen that in the neighbourhood of the tropopause sharp changes occur in opposite directions for the two trace species.

The troposphere is primarily the zone of weather and has constant atmospheric turbulence. Temperature falls progressively with altitude, averaging -6.5 K km^{-1}, a phenomenon known as the Lapse Rate. It can be understood in simple terms.

When some disturbance causes a volume of air to rise in the atmosphere, the air expands as a consequence of the decreasing pressure and does so adiabatically, that is, it exchanges heat with the surrounding air much more slowly than it expends energy in the work of expansion. Pressure falls with altitude in the troposphere according to the familiar barometric formula:

$$\Delta P/\Delta h = -g \cdot \rho \tag{3.1}$$

where P is pressure, h altitude, g the gravitational constant (9.806 m s^{-2} at sea level and $45°$ latitude), and ρ is the density of the atmosphere within the element Δh of altitude.

For an adiabatic expansion the heat change is zero, so that:

Internal Energy Decrease = Work done in Expansion

46

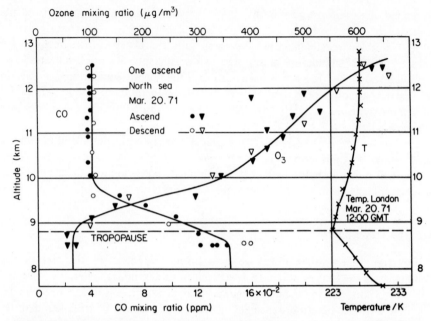

Figure 21. The mixing ratios of carbon monoxide and ozone as functions of altitude, as measured over the North Sea, March 1971. On the right is the temperature profile with altitude over London. From W. Seiler and P. Warneck, *Journal of Geophysical Research*, 77, 3209 (1972). Copyrighted by the American Geophysical Union and reproduced with permission

i.e.

$$C_v \cdot \Delta T = -P \cdot \Delta V \tag{3.2}$$

where C_v is the molar heat capacity at constant volume and ΔV is the change in volume for the 1 mol of air considered.

The ideal gas equation may be applied to air for all practical purposes. For 1 mol of air:

$$P \cdot V = R \cdot T$$

Differentiating we obtain:

$$P \cdot dV + V \cdot dP = R \cdot dT$$

which in the rearranged incremental form becomes:

$$-P \cdot \Delta V + R \cdot \Delta T = V \cdot \Delta P \tag{3.3}$$

Combining equations (3.2) and (3.3) and taking $C_p - C_v = R$, where C_p is the molar heat capacity at constant pressure, we obtain:

$$C_p \cdot \Delta T = (C_v + R) \cdot \Delta T = -P \cdot \Delta V + R \cdot \Delta T = V \cdot \Delta P = (M/\rho) \cdot \Delta P \tag{3.4}$$

where M is the molecular weight of air and ρ is the density.

Hence, substituting for ΔP in equation (3.1), we derive:

$$\Delta T/\Delta h = -M \cdot g/C_p \tag{3.5}$$

Between 200 and 300 K we may assume that the vibrational contributions to the heat capacities of N_2 and O_2 are negligible and that from the law of Equipartition of Energy C_p has the value of $7R/2$. Therefore, evaluating equation (3.5):

$$\Delta T/\Delta h = -\frac{9.806 \times 0.02909}{3.5 \times 8.314} = -9.802 \times 10^{-3} \text{ K m}^{-1} \quad \text{or} \quad -9.8 \text{ K km}^{-1}$$

This is the lapse rate for dry air and is greater than the actual lapse rate, averaging -6.5 K km^{-1}. The reason for the disparity is of course that natural air contains water vapour: rising air will cool at very nearly the dry adiabatic lapse rate of -9.8 K km^{-1}, provided that in the process it remains unsaturated with respect to water vapour. However, as the air rises and expands, the temperature will eventually fall to the dew point when the water vapour will begin to condense out as liquid droplets, converting latent heat to sensible heat in the process. It is this release of heat which effectively reduces the lapse rate as compared to dry air.

The quantitative evaluation of the moist air adiabatic lapse rate is a complex calculation and we shall not go into it, but the net result is that when an altitude of 11 km is reached in the U.S. Standard Atmosphere, the temperature has fallen from 288.15 K at sea level to 216.65 K, at the apparent rate of -6.5 K km^{-1}, taken as constant with altitude in this range, as represented in Figure 20.

3.3 The Stratosphere

The origin of the rising temperature from 11 to 50 km in the standard atmosphere is the degradation of a portion of the solar irradiance to thermal energy through the agency of primary absorption by ozone. As we have seen, in Table 2.1, most of the radiation with wavelengths below 300 nm has been absorbed above 30 km altitude.

Figure 22 shows profiles, as a function of wavelength, of the radiant energy incident on a horizontal plane of area 1 m^2 at various altitudes for overhead sun ($Z = 0$) on the basis that absorption by ozone is the only source of attenuation; this is very closely true for the wavelength range above 200 nm being considered. The points shown are the central wavelengths of the 10-nm interval over which the averaging is performed.

It is further assumed that there is no attenuation of any significance in this connection above an altitude of 60 km. Figure 23 shows the typical ozone concentration versus altitude profile adopted in computing Figure 22, which provides justification for the above assumption. The equivalent thickness of the ozone layer is taken as 3 mm.

In Figure 22, the integrated areas between the lines corresponding to the altitudes indicated are measures of the radiant energy absorbed in that altitude range. It is apparent that the most effective absorption between 240 and 270 nm results in near complete elimination of that element of the ultraviolet radiation by

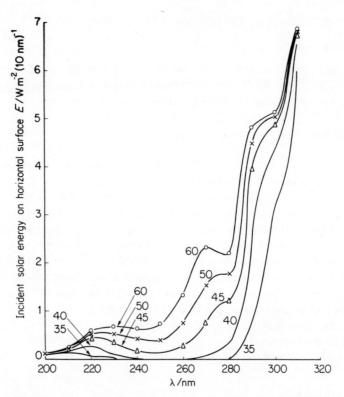

Figure 22. Plots of average middle-ultraviolet irradiance per 10 nm wavelength interval on a 1 m² horizontal surface for overhead Sun as a function of middle wavelength for indicated altitudes, on the assumption that ozone is the only source of attenuation

40 km altitude. This has the effect of removing the central portion of the integrated energy absorption below 40 km in Figure 22 and creating the maximum energy absorption (integrated area) between 45 and 50 km. It is obvious that the integrated area for the altitude layer 10 km above 50 km is only approximately equal to the integrated area for the layer 5 km below 50 km at most. On this basis we would expect less concentrated energy absorption between 50 and 60 km altitude and hence the maximum in the temperature profile to appear somewhere in the vicinity of 50 km (the stratopause).

At this stage, it is worthwhile to examine how calculations of the heating rates of the absorbing atmospheric layers are made and then to go on to consider the reradiative cooling rates which produce the energy balance and hence the stable temperature profile.

The points for each wavelength in Figure 22 give the efficiency of transmission of radiant energy down through the ozone layer to a particular altitude. For example, at 280 nm we read off radiant energy [$E(280)$] values of 2.22 W m⁻² in the range 280 ± 5 nm for the altitude of 60 km (assumed identical with the extraterrestrial value), $1.80\,\mathrm{W\,m^{-2}}$ for 50 km altitude, and $0.355\,\mathrm{W\,m^{-2}}$ for 40 km. This means that the transmitted energy flux densities, $E_{50}(280)$ and $E_{40}(280)$ for altitudes of 50 and 40 km respectively, are 81.1% and 16.0% of the extraterrestrial value, $E_t(280)$, respectively.

We shall adopt the ozone concentration profile shown in Figure 23 as standard and take the vertical ozone column equivalent thickness for S.T.P. conditions to be exactly 3 mm. We define f as the fraction of the total ozone column above an altitude to be specified; the equivalent thickness of the overlying ozone column is then $3f$ mm. Thus from equation (2.5) and the data of Table 2.1 we may calculate the photon flux density incident upon a horizontal surface at altitude h for wavelength λ, $I_h(\lambda)$, on the basis of:

$$I_h(\lambda) = I_t(\lambda) \cdot 10^{-3\delta \cdot f} \tag{3.6}$$

where $I_t(\lambda)$ is the extraterrestrial photon flux density. Table 3.1 shows the values of f calculated on the basis of the ozone profile in Figure 23.

The exactly analogous equation to (3.6) in terms of radiant energy is simply:

$$E_h(\lambda) = E_t(\lambda) \cdot 10^{-3\delta \cdot f} \tag{3.7}$$

Values of $I_t(\lambda)$ and $E_t(\lambda)$ may be derived from Table 2.4.

In the first stage of a calculation, we would calculate the values of $E_h(\lambda)$ from equation (3.7) over the range of wavelengths to give the radiant energy flux

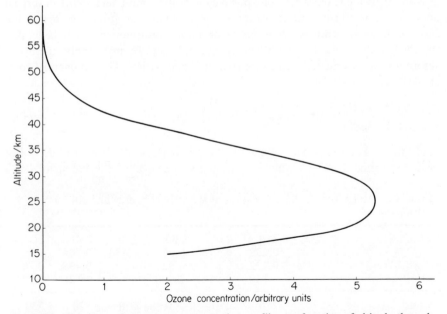

Figure 23. The form of the ozone concentration profile as a function of altitude through the stratosphere. Based upon the profile given by M. Nicolet, 'Aeronomic reactions of hydrogen and ozone', *Aeronomica Acta*, A – No. 79, p. 9 (1970). Reproduced by permission of Professor M. Nicolet

Table 3.1. Standard fractions (f) of the total stratospheric ozone column above altitudes (h) for a vertical path on the basis of Figure 23

f	h/km
0.0068	50
0.018	45
0.057	40
0.166	35
0.253	32.5
0.364	30
0.465	27.5
0.579	25
0.694	22.5
0.803	20
0.957	15

spectrom incident upon a layer at a particular altitude. Let us choose $h = 50$ km for a specimen calculation.

From equation (3.7) and with $f = 6.81 \times 10^{-3}$, using the values of coefficient δ as listed in Table 2.1, the results shown in Table 3.2 are derived.

Now let us focus our attention upon a cube of the atmosphere with 1 m vertical sides located at 50 km altitude. We can be sure that any radiation, even that in the 240 to 270 nm range where ozone has its maximum absorption, will only be weakly absorbed within this 1-m path length. Accordingly we may apply the weak absorption approximation form of the Beer–Lambert law. The exponential form of equation (2.2) will be:

$$\frac{I_L(\lambda)}{I_h(\lambda)} = \frac{E_L(\lambda)}{E_h(\lambda)} = 10^{-\epsilon c L} = e^{-2.303 \epsilon c L} \tag{3.8}$$

where the subscripts L indicate transmission through length L at altitude h. For weak absorption $2.303\epsilon c L$ will be small and we can expand the exponential as a

Table 3.2. Parameters in the calculation of the radiant energy absorbed at altitude 50 km

λ/nm	200	210	220	230	240	250	260
$E_t(\lambda)$/W m^{-2}	0.107	0.229	0.575	0.667	0.630	0.704	1.30
$E_{50}(\lambda)$/W m^{-2}	0.106	0.222	0.525	0.525	0.522	0.409	0.792
$10^5 \times \Delta E_{50}(\lambda)$/W	–	0.130	0.944	2.500	3.477	4.526	8.295

λ/nm	270	280	290	300	310	320	330
$E_t(\lambda)$/W m^{-2}	2.32	2.22	4.82	5.14	6.89	8.30	10.59
$E_{50}(\lambda)$/W m^{-2}	1.533	1.796	4.497	5.045	6.852	8.288	10.59
$10^5 \times \Delta E_{50}(\lambda)$/W	12.522	7.504	6.142	1.858	0.739	0.246	–

power series. For the index x this takes the form:

$$e^x = 1 + x + x^2/2! + \cdots$$

Again for sufficiently weak absorption, as will be the case in our present instance, terms above the first power may be ignored. Thus equation (3.8) becomes:

$$\frac{E_L(\lambda)}{E_h(\lambda)} = 1 - 2.303\epsilon cL \tag{3.9}$$

Hence the energy absorbed within the 1-m vertical pathlength through our cube is $\Delta E_{50}(\lambda) = E_h(\lambda) - E_1(\lambda)$, producing the identity with ϵ expressed in $dm^2\ mol^{-1}$ and c the concentration of ozone expressed in $mol\ dm^{-3}$.

$$\Delta E_{50}(\lambda) = 23.03\epsilon[O_3] \times E_{50}(\lambda) \tag{3.10}$$

which expresses the energy absorbed within the cube in unit time. This parameter is tabulated as a function of wavelength in Table 3.2, on the basis of an ozone concentration of $1.7 \times 10^{-10}\ mol\ dm^{-3}$ at 50 km altitude. Because all the energy flux quantities are averages over 10-nm intervals and we have tabulated $\Delta E_{50}(\lambda)$ for 10-nm differences in wavelengths, the sum of these across the table yields the total rate of energy absorption within the cube, $\Sigma_{210}^{320}\Delta E_{50}(\lambda) = 4.89 \times 10^{-4}$ W.

When non-zero solar angles are to be considered, two factors will reduce the energy flux incident upon a horizontal surface at a defined altitude. Firstly the $\cos Z$ factor discussed in Section 2.1(iv) must be applied. Secondly the effective thickness of the ozone layer through which the radiation has passed will be $3f.\sec Z$ rather than simply $3f$ [see equation (3.6)]. However there will also be a partially compensating factor of $\sec Z$ multiplying the pathlength through the 1-m cube, giving a stronger absorption of the reduced incident energy flux.

In considering the energy balance of atmospheric layers, it is the integrated daily energy absorption which is important. Calculation of the ultraviolet heating rate, as it is termed, demands repeating the type of calculation above for the general solar angle, Z, and integrating the result between the limiting values of Z which define the hours of daylight. The details of such calculations are beyond our present scope but we may usefully cite some typical results. Conventionally the u.v. heating rate is expressed in units of $K\ day^{-1}$; this is mainly hypothetical as a direct concept since the atmospheric temperature represents a balance between heating by u.v. absorption and reradiation of energy in the infrared (the so-called infrared cooling rate). However the conventional expression can be imagined as being where the energy equivalent of the solar radiant energy absorbed in a 24-h period by a small element of volume, is stored in an imaginary external heat reservoir from which it is suddenly added to the same volume of air under stratospheric conditions. This would produce a rise of 10 K if the heating rate was $10\ K\ day^{-1}$. Cooling processes could not operate effectively within the short time scale of this imaginary process.

Park and London (1974) have made extensive calculations of this type for altitudes between 30 and 100 km. Figure 24 shows an idealized distribution of the u.v. heating rates for January and July as functions of latitude and altitude.

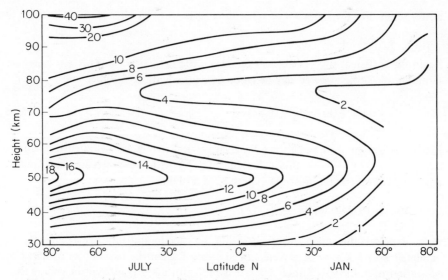

Figure 24. Model distribution of the net ultraviolet heating rate in K day^{-1} for July and January for the atmosphere between 30 and 100 km altitude. From J. H. Park and J. London, 'Ozone photochemistry and radiative heating of the middle atmosphere', *Journal of Atmospheric Sciences,* **31,** 1911 (1974). Reproduced by permission of the American Meteorological Society

We must also consider the nature of the balancing infrared emission. The principal infrared active components of the stratosphere are ozone itself, (principal band centred on 9600 nm wavelength), carbon dioxide, (principal band centred on 15,000 nm wavelength), and water vapour (principal bands centred on 6300 nm and above 30,000 nm). The efficiency of a molecular emission band in converting the thermal energy of its environment into radiated infrared energy depends mainly upon the concentration of the gas and the proximity of the band wavelengths to the Planck radiation law maximum [Wien's law – equation (1.9)] for these fundamental bands. For $T = 250$ K (general average for the stratosphere), it is predicted that λ_{max} will lie at 11,590 nm. Therefore the CO_2 and O_3 bands are more effective in this respect than the further removed H_2O bands. The typical stratospheric mixing ratios for these molecules are as $CO_2 : O_3 : H_2O :: 3 \times 10^{-4} : 2 \times 10^{-6} : 2 \times 10^{-6}$. Hence CO_2 has the most prominent role in the radiative cooling of the stratosphere.

When the results for the u.v. heating rate and the infrared cooling rate are combined for specified altitudes and latitudes, the typical pattern is obtained which is shown in Figure 25.

The most significant point to be made from this figure is that below a latitude of 40° to 50°, each point in the stratosphere is balanced virtually by radiation alone. No vertical transfer of energy of any major significance is required for stability. This situation is to be expected since the stratosphere, with its inverse gradient of temperature versus altitude, will have resistance to vertical adiabatic transport of matter and energy. In the lower stratosphere (below 25 km altitude) horizontal transport is known to be dominant over vertical processes and is manifested by high velocity winds.

A further important conclusion stems from consideration of the time scales of vertical atmospheric motions relative to the time scales associated with photo-chemical reactions. In this connection we define a parameter H known as the Scale Height [to be distinguished from the hour angle with the same symbol in section 2.1(iv)]. The scale height appears in a variation of the barometric formula [equation (3.1)], written as:

$$dP/P = -dh/H \qquad (3.11)$$

where P is pressure and h is altitude. It then follows that H is defined more fundamentally by:

$$H = RT/Mg = kT/mg \qquad (3.12)$$

where R is the gas constant, k is Boltzmann's constant, M is the mass of 1 mol and m is the mass of 1 molecule. H is equal to twice the distance through which a mole-cule (or atom) having the equipartition energy of $\frac{1}{2}kT$ in the vertical direction can rise against the force of gravity. The mean lifetime of a species with respect to vertical transport is equal to the time required for the species to move vertically through a distance equal to its own scale height under the prevailing conditions. This leads to

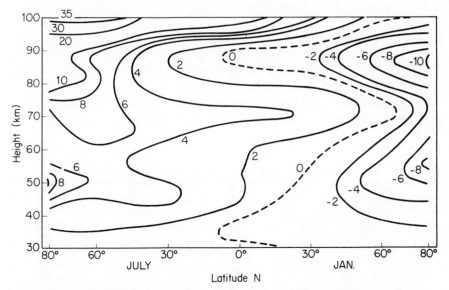

Figure 25. Model distribution of radiative heating and cooling in K day^{-1} for July and January for the atmosphere between 30 and 100 km altitude, representing the balance between ultraviolet and infrared heating and infrared cooling. From J. H. Park and J. London, 'Ozone photochemistry and radiative heating of the middle atmosphere', *Journal of Atmospheric Sciences*, **31**, 1911 (1974). Reproduced by permission of the American Meteorological Society

the equation:

$$\text{Characteristic Transport Time} = H/\dot{v} = H^2/D \qquad (3.13)$$

where $\dot{v}$ is the vertical diffusion velocity and D the effective diffusion coefficient. The predominant mode of diffusion in the stratosphere is Eddy Diffusion, which describes the transport of molecules of gases resulting from turbulent mixing in the presence of a concentration gradient. The characteristic transport time in the stratosphere averages around 10^6 s. In general this is much longer than the time scales for chemical reactions in the stratosphere, and consequently it is a very good approximation to consider that the concentration distributions of reacting species below an altitude of 60 km are determined purely by photochemical equilibrium-rate equations, upon which transport processes have a negligible effect. Only rather long-lived molecules like HNO_3, and to some extent ozone itself below 25 km, need the incorporation of vertical diffusion into their equilibrium balances. We shall be considering the chemistry of the stratosphere in detail in Chapter 9.

3.4 The Mesosphere

At altitudes above 50 km there is insufficient density of ozone for the absorption of solar radiation to maintain the temperature of around 271 K which is encountered at the stratopause. Therefore the temperature drops progressively towards the mesopause at 85 km.

The Standard Atmosphere parameters for the mesosphere are shown in Table 3.3.

In the upper mesosphere, the composition cannot in general be calculated solely on the basis of photochemical equilibrium. Figure 26 illustrates this point, showing the results of some concentration-versus-altitude profile computations, both ignoring and taking into account eddy diffusion.

One further aspect of interest in the mesosphere is the onset of ionization above about 70 km altitude, marking the overlap with the lowest (D) region of the

Table 3.3. Standard atmosphere parameters for the mesosphere

Altitude/km	T/K	Pressure/N m^{-2}	Total concentration/mol dm^{-3}	$[O_2]/[M]$
50	271	79.80	3.545×10^{-5}	0.209
55	266	42.75	1.936×10^{-5}	0.209
60	256	22.46	1.056×10^{-5}	0.209
65	239	11.44	0.575×10^{-5}	0.209
70	220	5.52	0.302×10^{-5}	0.209
75	200	2.49	0.150×10^{-5}	0.209
80	181	1.04	0.07×10^{-5}	0.209
85	181	0.41	0.03×10^{-5}	0.195

Source: *U.S. Standard Atmosphere Tables*, 1962, U.S. Government Printing Office, Washington, D.C.

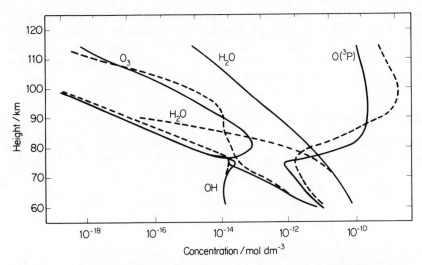

Figure 26. Altitude distributions of $O(^3P)$, O_3, OH and H_2O computed taking account of eddy and molecular diffusion (full curves) and ignoring diffusion (broken curves). From M. R. Bowman, L. Thomas, and J. E. Geisler, 'The effects of diffusion processes on the hydrogen and oxygen constituents in the mesosphere and lower thermosphere', *Journal of Atmospheric and Terrestrial Physics*, **32**, 1669 (1970). Reproduced by permission of Pergamon Press, Ltd

ionosphere. It is the result of penetration of very short wavelength ultraviolet radiation to these altitudes, with wavelengths below 135 nm having a photon energy at this limit sufficient to ionize nitric oxide, the most easily ionizable species present in significant concentrations. Table 3.4 continues the $E_t(\lambda)$ data of Table 3.2 to shorter wavelengths.

The outstanding feature in Table 3.4 is the large enhancement of the irradiance between wavelengths of 117.5 and 122.5 nm compared to that in neighbouring intervals. This is due to irradiance in the atomic hydrogen line at 121.57 nm, usually termed the Lyman α line, reflecting the overwhelming hydrogen content of

Table 3.4. Solar irradiances at the top of the atmosphere $(E_t(\lambda))$ in the wavelength range 110 to 200 nm

λ/nm	200	190	180	170	160	150
$E_t(\lambda) \times 10^3$ /W m^{-2} (10 nm)$^{-1}$	107	27.1	12.5	6.3	2.3	0.7

λ/nm	140	130	125	120	115	110
$E_t(\lambda) \times 10^3$ /W m^{-2} (10 nm)$^{-1}$	0.3	0.07	0.07[a]	9.01[a]	0.07[a]	0.03[a]

[a]Wavelength interval of integration for these data is 5 nm bandwidth centred on the wavelength given — other data have 10 nm bandwidth of integration.

From M. P. Thekaekara, 'Extraterrestrial solar spectrum, 3000–6100 Å, at 1-Å intervals', *Applied Optics*, **13**, 518–522 (1974). Reproduced by permission of the Optical Society of America.

the Sun. Accordingly it is the photoionization of nitric oxide by Lyman α radiation which is important in creating this region of the ionosphere. We shall consider the ionization processes as chemical reactions in Chapter 10, but it is worth noting at this stage that the ions in the mesosphere are subject to relatively rapid removal processes so that their concentrations diminish sharply overnight.

3.5 The Thermosphere

The apparent solar brightness temperature increases dramatically at wavelengths below 100 nm; for example at 80 nm it is 6823 K, at 50 nm it is 10,068 K, at

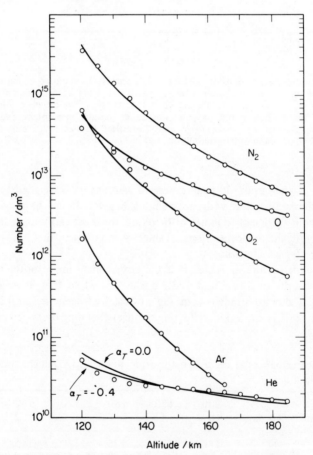

Figure 27. Average number densities for He, O (assuming no loss). N_2, O_2, Ar in part of the thermosphere. The circles represent measured values: the solid lines are calculated on the basis of static diffusive equilibrium [according to equation (3.14)], normalized to the measured values at 150 km altitude. From D. R. Hickman and A. O. Nier, *Journal of Geophysics Research*, **77**, 2884 (1972). Copyrighted by the American Geophysical Union and reproduced with permission

20 nm it is 23,288 K, and at 5 nm it reaches the colossal value of 73,225 K. As a result there is, at increasing altitudes within the thermosphere, progressively more and more high-energy radiation available which encourages the dissociation and ionization of the principal constituents N_2 and O_2 at the mesopause. Hence the thin thermosphere commences the atmospheric attenuation of solar radiation by removing X-ray and extreme ultraviolet components. The strong absorption of these high-energy photons by highly rarified gases with low volume heat capacity inevitably produces a steeply rising profile of temperature with altitude through the thermosphere.

Mass spectrometers have been carried into the lower thermosphere on rockets and have provided compositional data. One such launch was made from Fort Churchill (59° N) in Canada; the account of the procedures and results is given in the paper by Hickman and Nier cited at the end of this chapter. The compositional profiles of the neutral (i.e. uncharged) gases as a function of altitude were found to fit closely the static diffusive equilibrium model defined by the equation:

$$[X]_h = [X]_{185} \cdot \left(\frac{900}{T_h}\right)^{1+\alpha_T} \cdot \exp\left\{-\int_{6400+185}^{6400+h} \frac{M \cdot g(r)}{R \cdot T(r)}\, dr\right\} \qquad (3.14)$$

where $[X]_h$ is the concentration of species X at altitude h with ambient temperature T_h. Thermal diffusion arises as a result of the presence of large temperature gradients in gas mixtures and the thermal diffusion coefficient, α_T, is proportional to the difference between the molecular weight of the diffusing species and the average molecular weight of the medium through which it is diffusing. As a consequence, thermal diffusion will be significant only for helium and hydrogen in the thermosphere. Also in equation (3.14), M is the molecular weight of X, r is the distance from the centre of the Earth, and the temperature, T_{185}, at the altitude of 185 km of 900 K is used as the reference level.

The success of equation (3.14) in fitting the measured concentration profiles is evident in Figure 27.

3.6 Conclusion

Reasonable working approximations have been developed in this chapter to give simple physical interpretations of the phenomena of atmospheric layers and the dynamics within them. The stratosphere can be treated largely as a photochemically equilibrated layer, independent of diffusional influence from the point of view of its chemistry. The mesosphere is a photoequilibrium layer which is increasingly affected by eddy diffusion with increasing altitude. The lowest part of the thermosphere (85 to 120 km altitude) is more complicated, with the turbopause at 105 km marking the changeover from eddy to molecular diffusion as the dominant influence. However, between 120 and 200 km altitude the concentration profiles of major species without charge can be predicted by a comparatively simple static diffusive equilibrium model.

When the chemical aspects of the upper atmosphere are dealt with in Chapters 9 and 10, we shall consider them as minor perturbations of the physical situations developed here.

58

BIBLIOGRAPHY

1. 'The Structure of the Atmosphere up to 150 km', by G. V. Groves, *Contemporary Physics*, **14**, 1–24 (1973).
2. 'Remote Sounding from Artificial Satellites and Space Probes of the Atmospheres of the Earth and the Planets', by J. T. Houghton and F. W. Taylor, *Reports on Progress in Physics*, **36**, 827–919 (1973).
3. 'Ozone Photochemistry and Radiative Heating of the Middle Atmosphere', by J. H. Park and J. London, *Journal of Atmospheric Sciences*, **31**, 1898–1916 (1974).
4. 'The Effect of Diffusion Processes on the Hydrogen and Oxygen Constituents of the Mesosphere and Lower Thermosphere', by M. R. Bowman, L. Thomas, and J. E. Geisler, *Journal of Atmospheric and Terrestrial Physics*, **32**, 1661–1674 (1970).
5. 'Measurement of the Neutral Composition of the Lower Thermosphere above Fort Churchill by Rocket-borne Mass Spectrometer', by D. R. Hickman and A. O. Nier, *Journal of Geophysical Research*, **77**, 2880–2887 (1972).

Chapter 4

The Photosynthetic Origin of Fuels

At present man derives the overwhelming bulk of his energy needs from the combustion of fossil fuels, principally coal, oil, and natural gas. In this operation man is availing himself of stored prehistoric solar irradiance, fixed initially by photosynthesis. In this chapter we seek to gain an understanding of the energy fixing mechanisms of photosynthesis which have produced our fuels in the first place. Further we might ask whether present day solar irradiance could be applied to the production of fuel substances, thus stemming to some extent the increasingly rapid depletion of the Earth's comparatively meagre stored energy reserves (Table 2.6). Here we shall concentrate upon consideration of the direct harvesting of solar energy through the agency of photosynthesis, and whether such an approach to man's energy needs is practicable.

4.1 The Overall Energy Efficiency of Photosynthesis

Basically the process of photosynthesis represents a localized reversal of the tendency for the degree of disorder of matter and energy to increase with the passage of time. Solar radiant energy provides the driving force required to overcome the apparent preclusions of the Second Law of Thermodynamics in this sense. In its simplest form the chemical change induced can be represented as

$$n\,CO_2 + n\,H_2O + \text{photons} \longrightarrow (CH_2O)_n + n\,O_2$$

The general formula $(CH_2O)_n$ represents carbohydrate; for example, $n = 6$ would correspond to a hexose, $C_6H_{12}O_6$, perhaps glucose or fructose. However the carbohydrate which forms the main product of photosynthesis has the basic units polymerized into the long chain structures of cellulose and starch. The ordering achieved in this process is apparent from Figure 28, where parts of the chain structures are represented in the conventional way.

Photosynthesis utilizes light with wavelengths of up to about 800 nm in the solar irradiance. It is therefore clear that photons with energies of $150\,kJ\,mol^{-1}$ or larger may induce the photosynthetic chemical change. However, this immediately raises the problem of how the lower–energy photons above this limit can be effective since the C–O bond strength in CO_2 is $531\,kJ\,mol^{-1}$ and the O–H bond strength in H_2O is $498\,kJ\,mol^{-1}$, and both of these bonds must be broken in the course of photosynthesis. It follows that there must be available some mechanism which accumulates the energies of individual photons in order eventually to overcome the

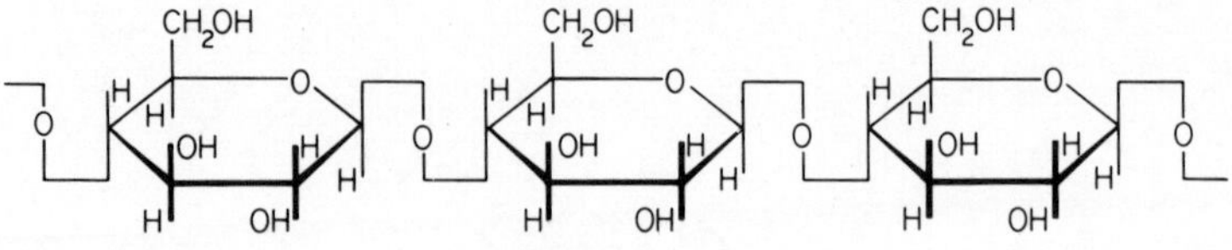

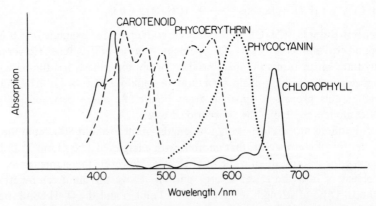

CELLULOSE

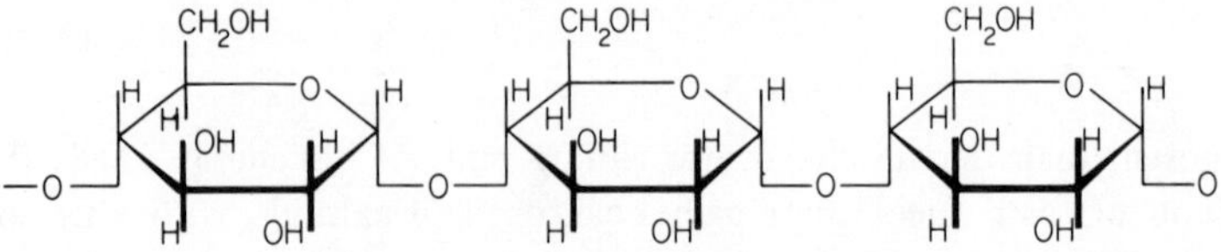

AMYLOSE (STARCH)

Figure 28. Conventional representations of parts of the chain structures of cellulose and amylose (starch).

apparent energy barriers. This cannot occur as a simple single-step process of a purely physical nature since it would contravene the basic rules of quantum theory discussed for the photoelectric effect in Section 1.1. Thus we are led to an inkling of the essential multi-step complexity of the overall photosynthetic process.

The plant pigments, dominated by chlorophyll both in terms of concentration (around 5% of a typical leaf) and, as we shall see, functional importance, are responsible for absorption of the light in the first place. Figure 29 shows the visible absorption spectra of some separated pigments. Of course there are many more of these pigments than we have space available to illustrate their absorption spectra. In almost all plants the combination of the individual pigment spectra results in

Figure 29. The absorption spectra of some chloroplast pigments. From E. Rabinowitch and Govindjee, *Photosynthesis*, John Wiley & Sons Inc., New York, 1969 p. 113. Reproduced with permission of John Wiley & Sons Inc.

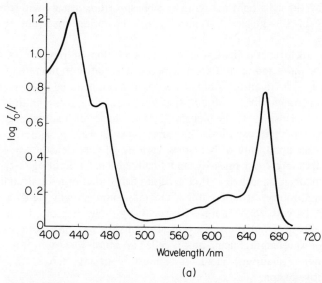

(a)

Figure 30(a). Typical absorption spectrum of plant pigments, in this instance of ethanol solution of pigments extracted quantitatively from *Chlorella* cells. From R. Emerson and C. M. Lewis, 'Chlorella photosynthesis', *American Journal of Botany*, **30**, 169 (1943). Reproduced by permission of the American Journal of Botany.

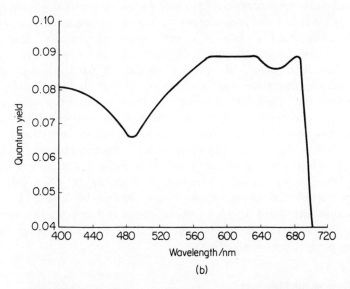

(b)

Figure 30(b). The action spectrum (the quantum yield of photosynthesis as a function of wavelength) for *Chlorella*. From R. Emmerson and C. M. Lewis, 'Chlorella photosynthesis', *American Journal of Botany*, **30**, 171 (1943). Reproduced by permission of the American Journal of Botany.

absorption by the cells right across the visible spectrum, usually with something of a minimum in the green. A typical plant absorption spectrum is shown in Figure 30(a).

Using monochromatic light, the rate of evolution of oxygen by plant cells, $P(O_2)$, can be measured in laboratory experiments relative to the rate of absorption of photons, I_a. The quantum yield of photosynthesis, denoted by Φ, is then simply $P(O_2)/I_a$. Experimentally it is found that Φ increases towards a limiting value with decreasing flux densities of the incident light and hence decreasing I_a. In very weak light the rate of photosynthesis increases linearly with I_a and the conversion efficiency is maximal, but with further increase of flux density a plot of $P(O_2)$ versus I_a falls away from linearity and finally reaches a limiting-rate plateau. In general the plateau is reached at flux densities below that of natural sunlight, which places an important restriction upon the maximum growth yields attainable in attempts to harvest solar energy. Figure 30(b) shows a plot of the maximum quantum yields for irradiation of a typical plant with weak monochromatic light as a function of wavelength. Such a plot is called an *action spectrum*. It is evident that this maximum quantum yield is almost constant across the photosyn-thetically-active region.

Generally the limiting efficiency of photosynthesis is one molecule of O_2 evolved per eight photons absorbed. Since eight photons of light of the middle wavelength of 600 nm correspond to a total energy equivalent of about $1600 \text{ kJ(mol } O_2)^{-1}$ and the stored energy in the corresponding (CH_2O) unit is 469 kJ, the limiting energy storage efficiency is around 30%.

Further laboratory experiments have used a single intense flash of light to excite the chromophores and the oxygen produced has been measured. It was found that the single flash could at best yield one oxygen molecule for around 2500 chlorophyll molecules in the sample of algae, etc., all of which absorbed one photon in the flash. Since eight photon energies are fixed in the evolution of one O_2 molecule, this means that the energy of only one photon is actually used in photosynthesis for around 300 absorbed under high flux-density conditions. Thus we come to the concept of a *photosynthetic unit* of approximately 300 chlorophyll molecules possessing one *active centre*. A photosynthetic unit in a leaf in full sunlight would be expected to absorb rather fewer than 300 photons to gain one effective photon: the typical number turns out to be around 30, so that the actual photosynthetic energy storage efficiency is $469/(30 \times 200 \times 8) \simeq 0.01$ or 1% (taking the average visible photon energy as equivalent to 200 kJ mol^{-1}).

Having thus established a photosynthetic energy storage efficiency, let us now backtrack somewhat to consider in more detail the elementary processes which combine to produce the overall effect.

4.2 The Role of the Pigments

It is immediately clear from Figures 29 and 30 that light absorbed by pigments other than chlorophyll must be effective in inducing photosynthesis. We also see that, from Figure 30(b), the larger energy content per photon of blue light does not lead to an increased value of Φ compared to the smaller energy content per photon of red light; on an energetic basis, blue light energy is stored at a lower efficiency than red light energy.

The critical laboratory experiments in this connection have been fluorescence studies upon solutions of the pigments. With chlorophyll present and irradiation with monochromatic light, the observed fluorescence appeared almost entirely in the chlorophyll fluorescence band near a wavelength of 700 nm. The yields of such fluorescence as functions of the wavelengths of the exciting light virtually parallelled the action spectra of plants. Hence there can be no doubt that the light energy absorbed by the mixture of pigments migrates by intermolecular energy transfer to chlorophyll with high efficiency. As such, these energy-transfer processes must be able to operate with substantial distances of separation between pigment molecules. The radiative lifetime of an electronically-excited pigment molecule is of the order of 10^{-8} s before it fluoresces back to the ground state. Calculations of the range factor of energy transfer have suggested a probability of around 50% for molecules 5 nm apart, so that there is no question of a collisional energy-transfer condition.

The accepted picture of the energy-transfer process is represented in Figure 31. Consider two pigment molecules with electronic energy levels E_0 (ground states) and E_1 and E_2 (respective energy levels produced by excitation of π electrons). In this figure the arrow labelled 1 corresponds to the absorption of the photon with energy sufficient to produce a vibrationally-excited state within the assembly of E_1 of the donor pigment molecule. The degradation of this vibrational energy to thermal energy of the medium is exceedingly rapid and is represented by arrow 2. Arrow 3 represents the intermolecular transfer of the energy $E_1 - E_0$ from the donor pigment 1 to the acceptor pigment 2, which proceeds as a near resonant process and accordingly populates a vibrationally-excited level of the electronic

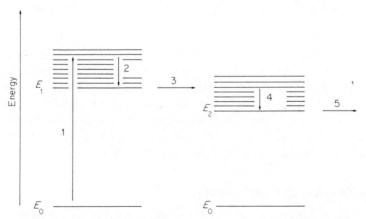

Figure 31. Schematic diagram representing part of the energy transfer mechanism in plant pigment mixtures in which chlorophyll is the final acceptor.

level E_2. Again there is degradation of the vibrational energy, arrow 4, leaving the acceptor pigment molecule with only the electronic energy $E_2 - E_0$. In the presence of a third pigment with a lower electronic energy gap, there will be another intermolecular energy transfer represented by arrow 5, analogous to process 3, passing on the energy again and producing further degradation of vibrational energy. In the range of plant pigments, it is chlorophyll which has the lowest electronic energy gap, say $E_c - E_0$, and this will be the final 'sink' of the energy-transfer processes. Chlorophyll will have received an effective energy of $E_c - E_0$ from the initially-absorbed photon, irrespective of its original energy. Therefore a greater fraction of the energy of blue light will be rejected as thermal energy in the transfer processes than will be the case for red light; the quantum efficiency will depend simply upon the number of photons absorbed rather than their energy content above the threshold energy of $E_c - E_0$. For the solution of the pigments themselves, the fluorescence will come from the 'sink' species at the end of the energy-transfer chain, i.e. chlorophyll. However, chlorophyll has a further essential distinction amongst pigments in that its presence is required for photosynthesis; in its absence no combinations of the other plant pigments has been found capable of bringing about photosynthesis.

So the plant pigments serve to collect the solar energy and pass it on into the photosynthetic units of chlorophyll molecules. In order to take in the solar radiant energy effectively, plant cells must have high absorption coefficients. In fact a leaf with a thickness of only a few layers of green cells does absorb red and violet light almost completely. Even a single cell of the algae *Chlorella* spp. with a thickness of 5 μm will absorb up to 60% of the incident light in the maximum of the red absorption band.

The highest decadic absorption coefficients of pigments are of the order of 10^6 dm^2 mol^{-1} (compare with those of O_3 in Table 2.1) while the typical thickness of a leaf may be taken as 10^{-4} dm. Let us now use the Lambert–Beer law (equation 2.2) to calculate that concentration of pigment within the leaf structure which is required for 50% absorption of the incident light. We have then:

$$\log_{10}(I_0/I_x) = \log_{10} 2 = 0.3 = \epsilon.c.x = 10^6 . 10^{-4} . c$$

from which c, the concentration of pigment averaged along the pathlength, x, is 3×10^{-3} mol dm^{-3}. Actual concentrations of chlorophyll within green leaves are more like 10^{-2} mol dm^{-3}. Hence we may expect a forest or crop to absorb all of the light incident upon it, bar that reflected, within the photosynthetically-active region of the solar spectrum. In Table 2.2 albedos of 10 to 20% were listed for photosynthesizing surfaces, so that we may apply an average factor of 0.85 to the incident irradiance to produce the typical absorbed irradiance.

4.3 The Post-Photochemical Reactions

(i) The Primary Enzyme Reactions

Only the initial step (and probably one intermediate step) are driven directly by the energy gained from the light. Some 20 to 30 'dark' reactions follow, in the course of which the initial light energy becomes trapped in the bonds of a few specific compounds. These energy-carrying molecules then serve to transport the energy from one site to another within the cell.

It is the enzymic nature of the initial reaction of the electronically-excited chlorophyll molecule which explains much of the overall kinetic behaviour and leads to an understanding of the reason for the appearance of the photosynthetic units.

Enzymes are protein molecules with molecular weights in the range of 10^5 to 10^6, while chlorophyll has a molecular weight of the order of only 10^3. Since the basic building atoms are of much the same atomic weight, we may deduce that each enzyme molecule has 100 to 1000 times the bulk of each chlorophyll molecule. Hence many chlorophyll molecules, perhaps on average about 300, will be in the environs of one bulky enzyme molecule.

Now the outstanding characteristic of enzyme reactivity is that it is associated with perhaps only one very specific part of the molecule, the active site. Therefore if a chlorophyll molecule is far from an active site, it may be excited by light absorption a large number of times without being able to utilize the energy for photosynthesis. Only that chlorophyll molecule which is excited (often by efficient energy transfer from more remote chlorophyll molecules) in the immediate vicinity of the active site and can gain an attachment of sorts thereon will be able to apply its energy content usefully.

Let us now examine the kinetics at the active site by way of a simple mechanism. Let C represent the chlorophyll ground state molecule and C* the electronically excited state. Let E represent the enzyme molecule and E:C* the complex formed by binding of C* at the active site, which goes on in non-photochemical reactions to generate the series of intermediates which result eventually in the photosynthesized carbohydrate. The kinetic scheme is:

$$E + C^* \underset{k_{-1}}{\overset{k_1}{\rightleftharpoons}} E:C^* \underset{\substack{\text{dark} \\ \text{reaction}}}{\overset{k_2}{\longrightarrow}} \text{Products}$$

Under normal conditions, E:C* will be the type of reactive intermediate to which the Stationary State Approximation can be applied, i.e. $d[E:C^*]/dt = 0$. Further, we may expect C* to be maintained in an absorption—deactivation equilibrium:

$$C \underset{k^*}{\overset{I_a}{\rightleftharpoons}} C^*$$

where I_a is the rate of absorption of photons and k^* is a first-order rate constant for the deactivation of C* molecules.

If $[E]_0$ is the 'dark' concentration of enzyme, then under irradiated conditions the actual concentration of unbound enzyme, $[E]$, is equal to $[E]_0 - [E:C^*]$. Because of the large excess of $[C]$ over $[E]_0$, $[C^*]$ will not need to be corrected for the formation of the complex.

Application of the Stationary State Approximation to $[E:C^*]$ gives:

$$d[E:C^*]/dt = k_1[E][C^*] - (k_{-1} + k_2)[E:C^*] = 0 \tag{4.1}$$

and since $I_a = k^*[C^*]$, we substitute to produce:

$$\frac{k_1 \cdot I_a}{k^*}([E]_0 - [E:C^*]) - (k_{-1} + k_2)[E:C^*] = 0 \tag{4.2}$$

Collecting the terms with [E:C*] together and dividing through by the coefficient yields:

$$[E:C^*] = \frac{k_1 \cdot I_a[E]_0/k^*}{(k_1 \cdot I_a/k^* + k_{-1} + k_2)} \tag{4.3}$$

The 'dark' reaction will be the rate-determining step so the rate of photosynthesis, R, is expressed by:

$$R = k_2[E:C^*] = \frac{k_1 \cdot k_2 \cdot I_a[E]_0/k^*}{(k_1 \cdot I_a/k^* + k_{-1} + k_2)} \tag{4.4}$$

which is of the form:

$$R = A \cdot I_a/(B \cdot I_a + 1) \tag{4.5}$$

where $A = k_1.k_2.[E]_0/k^*(k_{-1} + k_2)$ and $B = k_1/k^*(k_{-1} + k_2)$ are constants for a given system.

The plot of R, the rate of photosynthesis, versus the rate of absorption of photons, I_a, has the form shown in Figure 32, which is typical for a photosynthesizing system.

In the low irradiance region $0 - X$, $B.I_a \ll 1$ so that R is directly proportional to I_a and the photosynthesis proceeds at maximum efficiency. On the molecular basis all C* molecules produced find their way to active sites on E, either directly or by energy migration from one chlorophyll molecule to another.

The point S indicates the location expected for natural sunlight irradiance, well

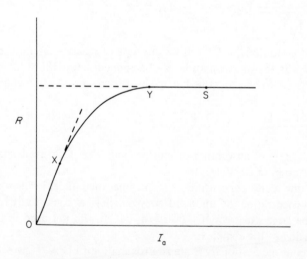

Figure 32. Schematic plot of the rate of photosynthesis (R) versus incident irradiance (I_a).

onto the plateau of saturated rate. Under these conditions $B.I_a \gg 1$ and $R = A/B = $ Constant. Here virtually all of the enzyme is bound into the complex E:C*, leaving no E itself available to take up the copious supply of C*. The wasted photons under these conditions contribute to another plant function termed transpiration, discussed in Section 7.1.

In fact the photosynthetic rate measured for a plant is a net rate; another plant function which goes on continuously is *respiration*. This is effectively a slow combustion of photosynthesized carbohydrate back to CO_2 and H_2O, which supplies energy needed for the plant's vital functions. Respiration proceeds at a near invariant rate irrespective of conditions of light and dark. It is represented by the chemical equation:

$$(CH_2O)_n + n\, O_2 \longrightarrow n\, CO_2 + n\, H_2O + heat$$

Thus the plot shown in Figure 32 should be shifted downwards if R is to represent the net rate of photosynthesis; on $O - X$, below a certain value of I_a, there would be a negative rate, corresponding to net respiration. In fact respiration over 24 h can consume up to 50% of the carbohydrate photosynthesized during the hours of daylight. Typical rates of net photosynthesis in terms of CO_2 fixed by unit leaf area are in the range 7 to 30 mg CO_2 dm^{-2} per h, with crops like wheat tending towards the top end and trees towards the bottom end of this range. Rates of dark respiration on the same basis are usually in the range 0.5 to 4 mg CO_2 dm^{-2} per h. However there is also light-induced respiration, or photorespiration, which can proceed at rates around four-times faster than the non-photochemical dark respiration. It is now considered that respiratory losses are probably the most important factor in determining net photosynthesis rates.

Photosynthesis and respiration do not represent in detail reversals of the same biochemical reactions, although both processes have enzymes and intermediates in common. Moreover they occur at different sites in the cell assembly and are therefore separated physically.

(ii) The Secondary Cycles

As we have already suggested, molecules carrying stored energy within certain of their bonds, from which it can be released by reaction, play a substantial part in photosynthesis. It is now widely accepted that the principal energy carriers are molecules containing phosphorus atoms. Preeminent in this connection is adenosine triphosphate, hereafter referred to as ATP. This is an ester of phosphoric acid which releases around 30 kJ mol^{-1} of energy upon hydrolysis to adenosine diphosphate (ADP) under standard conditions; ATP is therefore termed a high-energy phosphate. Letting AR represent the main organic part of the ATP molecule, we concentrate upon the end-groups, writing the hydrolysis reaction as:

$$AR-O-\overset{\overset{\textstyle O}{\|}}{\underset{\underset{\textstyle OH}{|}}{P}}-O-\overset{\overset{\textstyle O}{\|}}{\underset{\underset{\textstyle OH}{|}}{P}}-O-\overset{\overset{\textstyle O}{\|}}{\underset{\underset{\textstyle OH}{|}}{P}}-OH + H_2O \longrightarrow AR-O-\overset{\overset{\textstyle O}{\|}}{\underset{\underset{\textstyle OH}{|}}{P}}-O-\overset{\overset{\textstyle O}{\|}}{\underset{\underset{\textstyle OH}{|}}{P}}-OH + H_3PO_4$$

$$\text{(ATP)} \qquad\qquad\qquad\qquad\qquad\qquad \text{(ADP)}$$

The free energy released from this reaction is applied in concert with other reactions to create an energy-pumping situation. In contrast to ATP, and for that matter ADP, ordinary or low-energy phosphate esters, such as adenosine monophosphate, give slightly endothermic hydrolyses.

Another important element in the overall mechanism is the movement up and down electrochemical potential gradients resultant upon the involvement of electron-transfer and transport steps. The energy for boosting electrons against these electrochemical gradients is derived from the initially-absorbed photons.

We shall be concerned with *redox* couples, based upon different oxidation states of a species. The original meaning of the term 'oxidation' was addition of oxygen (e.g. for iron, $4\,Fe + 3\,O_2 \longrightarrow 2\,Fe_2O_3$) but the term has now the wider definitions of loss of hydrogen (e.g. for chlorine, $2\,HCl \longrightarrow H_2 + Cl_2$) or loss of negative charge, i.e. electrons (e.g. the ferrous ion is oxidized to the ferric ion in the process $Fe^{2+} \longrightarrow Fe^{3+} + e^-$). Reduction is then the reverse of these. Thus we could have a generalized redox couple AH/A based upon the equilibrium.

$$AH \rightleftharpoons A + \tfrac{1}{2}\,H_2$$

where A would be the oxidized form and AH the reduced form. In the left-to-right direction we would speak of an oxidation process, and conversely from right to left we would speak of a reduction process. The Gibbs free energy change for the oxidation process is given by:

$$\Delta G = \Delta G^{\ominus} + RT\ln([A]/[AH]) \tag{4.6}$$

where the standard free energy change, $\Delta G^{\ominus}$, is produced for equal concentrations of the two species, taking the activity of hydrogen to be unity.

In discussing redox systems, it is more usual to talk in terms of electrochemical potential differences, E, where the relationship to the free-energy change in the electrochemical reaction is:

$$\Delta G = -z\,E\,F \tag{4.7}$$

In this equation F is the Faraday constant, the quantity of electric charge always associated with one equivalent of electrochemical reaction (i.e. the passage of 6.023×10^{23} electrons) and is equal to 96,494 coulombs. In combination, E and F have units of volt faradays, simply equivalent to joules. The parameter z is the number of electron equivalents passed per mol of reaction.

The convention now adopted for a single redox couple is to express E as a *reduction potential*. This means that the conventional redox couple is always to be looked at according to the equation:

$$A + \tfrac{1}{2}\,H_2 \rightleftharpoons AH$$

even if it is actually proceeding as an oxidation in the reverse direction under particular circumstances. On this conventional basis, the free-energy change for the couple would be expressed by:

$$\Delta G = \Delta G^{\ominus} + RT\ln([AH]/[A]) \tag{4.8}$$

where the free-energy quantities are now opposite in sign to those used in equation (4.6) since they refer to the reduction process in equation (4.8). If we now substitute into this equation the reduction potentials defined by equation (4.7), we obtain:

$$E = E^{\ominus}_{AH/A} - \frac{RT}{zF} \ln([AH]/[A])$$ (4.9)

or for $T = 298$ K and putting in the values of the constants:

$$E = E^{\ominus}_{AH/A} - \frac{0.059}{z} \log_{10}([AH]/[A])$$ (4.10)

From this equation it is evident that a ten-fold change in the ratio of the concentrations results in a change of $0.059/z$ volts in the potential, E.

For practical purposes we may say that a species is 'completely' oxidized when the ratio $[AH]/[A]$ is less than 10^{-3}, corresponding to a change of $0.18/z$ volts to the positive side of standard potential, $E^{\ominus}$.

Now suppose that we have two redox couples, AH/A and BH/B, present in solution and also that a mechanism exists whereby they can interact with one another. Thus an equilibrium will be established:

$$AH + B \rightleftharpoons A + BH$$

The equilibrium state is defined by $\Delta G = 0$, i.e. $E = 0$, for the overall process. In other words, the potential for the AH/A couple under these conditions must be equal to that for the BH/B couple, both expressed as conventional reduction potentials. Hence the right-hand side of equation (4.9) is equated to the right hand side of the analogous equation for the BH/B couple, taking $z = 1$. Upon recasting this equality we obtain:

$$E^{\ominus}_{AH/A} - E^{\ominus}_{BH/B} = \frac{RT}{F} \ln([AH][B]/[BH][A])$$

Suppose that $E^{\ominus}_{AH/A} = -0.3$ V and $E^{\ominus}_{BH/B} = +0.5$ V in a particular instance. Then we have:

$$E^{\ominus}_{AH/A} - E^{\ominus}_{BH/B} = -0.8.V = 0.059.\log_{10}([AH][B]/[BH][A]),$$

so that $([AH][B]/[BH][A]) =$ antilog$(-0.8/0.059) = 2.7 \times 10^{-14}$ and the equilibrium may be considered as being completely displaced to the right for practical purposes. i.e. B has 'completely' oxidized AH to A, being itself 'completely' reduced to BH in the process.

The general point to be made here is that if two redox couples have a mechanism for interacting together and have standard reduction potentials which differ by more than $0.18/z$ volts, then we expect the final equilibrium to involve only the oxidized form of the couple with the more negative standard reduction potential and the reduced form of the couple with the more positive standard reduction potential as significant concentrations. Conversely if the standard potentials do not differ by $0.18/z$ volts, then the reaction will not go to completion effectively and we should say that the couple with the more negative standard potential had no significant oxidizing action upon the other. So in general we need only examine the standard potentials of redox couples in order to assess oxidation capabilities, even though the actual conditions in the system are not the standard ones.

If now a third redox couple, CH/C, is added into the above system and $E^{\ominus}_{CH/C}$ is, say, +1.1 V, then BH, produced in the oxidation of AH, will go on to be oxidized to B while C will become reduced to CH. Thus we come to the concept of a redox cascade or chain. A series of redox couples which have mechanisms for interaction will pass reduced nature or oxidation ability down from the couple with the most-negative standard reduction potential (e.g. AH/A in the system above) finally to stop in the couple with the most positive standard potential (e.g. CH/C). Accordingly if AH were produced (say through photochemical interaction) in a system containing B and C, then the reduced nature would be passed stepwise down the redox cascade to leave CH as its final residence (always assuming that the AH/A and CH/C do not have a mechanism for direct interaction to eliminate the involvement of BH/B).

Suppose now that there exists a mechanism whereby a non-electrochemical reaction, such as the formation of ATP from ADP and inorganic phosphate (P_i), can be coupled into the redox reaction in the AH/A–BH/B system. Such couplings are in fact common phenomena in enzyme-catalysed biochemical processes. The overall reaction now becomes:

$$AH + B + ADP + P_i \rightleftharpoons A + BH + ATP$$

The standard free energy change for the formation of ATP from ADP + P_i is close to +33 kJ mol^{-1}. We can convert this into an effective electrochemical potential through equation (4.7) to produce $E^{\ominus}_{ADP/ATP} = -0.346$ .V. If we now presume that the standard conditions are maintained for the ADP–ATP system as the overall reaction proceeds, then the equilibrium condition becomes:

$$E^{\ominus}_{AH/A} - E^{\ominus}_{BH/B} + 0.346 = 0.059 \log_{10}([AH][B]/[BH][A])$$

Hence applying the standard potentials used above we obtain:

$$-0.3 - 0.5 + 0.346 = -0.454 \text{ V}$$

so that

$$([AH][B]/[BH][A]) = \text{antilog}(-0.454/0.059) = 2 \times 10^{-8}.$$

In practical terms, this hardly represents less 'complete' reaction than before. Thus with no real penalty in terms of the oxidation of AH to A, some of the energy released by the redox process has been stored in the high-energy bond of ATP. The limiting factor, which in reality prevents us achieving something for nothing, is that in a redox cascade when the standard potential of a particular couple approaches $-(0.346 + 0.18)/z$ volts, then the equilibrium no longer goes 'completely' over to the right hand side and the capacity to generate ATP drops accordingly.

We shall now proceed to examine the currently-accepted Hill–Bendall scheme

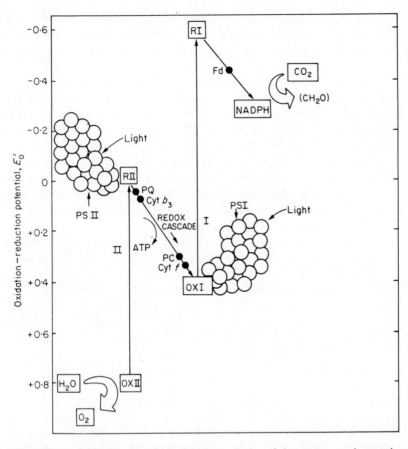

Figure 33. The Hill–Bendall scheme for the generation of the energy-carrying species in photosynthesis, ATP and NADPH. Some of the intermediate species are identified: PQ = plastoquinone, PC = plastocyanim; Cyt b_3 = cytochrome b_3; Cyt f = cytochrome f; Fd = ferrodoxin. From E. Rabinowitch and Govindjee, *Photosynthesis*, John Wiley & Sons Inc., New York, 1969, p. 190. Reproduced with permission of John Wiley & Sons Inc.

for the electron-transport chain of photosynthesis. This is represented diagrammatically in Figure 33. There are two photosynthetic centres (labelled PS I and PS II), through which the energy of the excited chlorophyll molecules (C*) is applied to achieve the creation of redox cascades. The precise identity of many of the couples involved is uncertain as yet. The system PS I generates a strong reductant (labelled R I) and a weak oxidant (labelled OX I) using the energy of one quantum transferred from an excited chlorophyll molecule. The system PS II generates a second redox pair at an earlier stage in the overall process, R II, a weak reductant, and OX II, a strong oxidant, again by electron transfer using a quantum derived from an excited chlorophyll molecule. Thus photosynthesis involves this essentially two-quantum process.

Experiments using isotopically-labelled water, $H_2^{18}O$, have shown conclusively that the O_2 evolved in photosynthesis comes solely from water and not from CO_2. It is considered that the strong oxidant OX II accomplishes the oxidation of water to liberate O_2.

The reductant R II may be considered to be analogous to AH in our example above and initiates a redox cascade which, through the agency of a series of species with progressively lower (negative) standard reduction potentials, eventually produces the species which interacts with PS I. In the course of this redox cascade there is interaction with ADP + P_i to store part of the energy release as ATP.

The strong reductant, R I, generated at PS I, starts a second redox chain, starting at a standard reducing potential of around -0.6 V, eventually producing the important energy carrying species denoted NADPH, reduced Nicotinamide Adenine Dinucleotide Phosphate, with a standard redox potential of about -0.3 V for the couple having the oxidized form denoted by NADP.

In concert, NADPH and ATP carry energy from their sites of generation to the other sites where they release their energy content in driving forward the reduction of CO_2 to carbohydrate unit (CH_2O).

We may therefore summarize the overall process of photosynthesis by the equation:

$$2\,H_2O \longrightarrow O_2 + 4\,e^- + 4\,H^+$$

$$2\,NADP + 4\,e^- + 2\,H^+ \longrightarrow 2\,NADPH$$

$$2\,H^+ + 2\,NADPH + CO_2 \longrightarrow 2\,NADP + H_2O + (CH_2O)$$

$$NET \qquad CO_2 + H_2O \longrightarrow (CH_2O) + O_2$$

The detailed mechanism of the reduction of CO_2 is termed the Calvin cycle, after its original proposer. It is now also referred to as the C_3 mechanism to distinguish it from the much rarer C_4 mechanism recently recognized as occurring in some tropical plants such as sugar cane. The C_3 designation is given on the basis of the intermediate species containing three carbon atoms involved in the Calvin cycle. Figure 34 represents this cycle in a simplified form, using numbers in ellipses to denote the total number of carbon atoms in each molecule named.

The general points of entry for ATP molecules, and for electrons and protons in the form of NADPH, are indicated. Two molecules of NADPH and three of ATP are required to fix one molecule of CO_2, shown entering at the top of the figure. Enzymes drive the reactions forward. The CO_2 molecule forms a complex of sorts with the C_5 species ribulose diphosphate and this splits in half to form two molecules of the C_3 phosphoglyceric acid. Conversion of this into the other C_3 molecule, a triose phosphate, demands a considerable input of energy, derived from one ATP molecule and one NADPH molecule per C_3 conversion. There then follows a complex set of steps, represented in the figure by a series of arrows, in the course of which branching occurs; one-sixth of the triose phosphate dimerizes and thus produces a hexose, while five-sixths goes to form ribulose phosphate and thence to reform ribulose diphosphate, thus closing the cycle. It is this last step which is

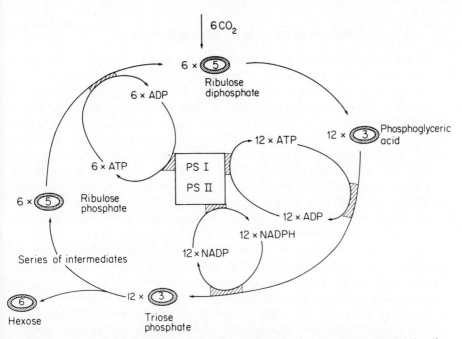

Figure 34. Schematic representation of the Calvin cycle for C_3 photosynthesis, showing the route by which CO_2 is incorporated into hexose carbohydrates and the general points at which the energy carried by ATP and NADPH is applied. The numbers in ellipses denote the number of carbon atoms in the molecule, the numbers outside indicate number of molecules.

considered to consume a further ATP molecule of energy per ribulose diphosphate reformed.

This clarifies the energy accumulation mechanism of photosynthesis. The absorbed photons are converted into the available energy content of ATP and NADPH molecules, which then transport the energy to sites where enzymes create conditions under which ATP and NADPH can couple in an energy-demanding reaction and thus revert to ADP and NADP. These 'spent' molecules then migrate back to the sites of the redox cascade generation for reenergization. O_2 is an early product at the centre PS II and it is to be noted that the hydrogen part of the water molecule does not appear as H_2 but is, rather, held in the redox system as reducing power and eventually reappears in part as water, with the oxygen atom now derived from the CO_2 molecule having been reduced.

The hexose molecule shown as the product in Figure 34 then goes on to undergo condensation polymerization to form cellulose and/or starch.

It is now relevant for us to consider respiration at this stage. The main function of plant respiration is the production of ATP, using some of the energy stored as carbohydrate during daylight hours. In laboratory experimental studies where dinitrophenol has been added to uncouple oxidative phosphorylation in respiring plants, the additional release of heat energy showed that under normal conditions, around 60% of the respiratory energy can be trapped in the form of ATP.

4.4 The Actual Efficiency of Solar Energy Storage by Plants

We now approach the question of what fraction of the incident irradiance upon a 'crop' under natural conditions can be incorporated into the plant structure? At the outset it is obvious that many factors other than irradiance control the rate of growth of a plant. For example the fine summer of 1975 in Britain produced a decrease in crop yield of around 30% because of the associated lack of rainfall, even though the integrated solar irradiance throughout the growing season was one of the highest ever. Adequate supplies of essential minerals are also required, amongst other factors. Also most plants using C_3 photosynthesis are limited in growth rate by the rate at which carbon dioxide can enter the stomata; this is clearly evidenced by enhanced growth rates in artificially-enriched air with respect to CO_2. It is worth noting that plants based upon the C_4 mechanism have an effective CO_2 priming mechanism so that this factor is not limiting for such plants.

Under typical conditions at the maximum point of the growing season, an agricultural crop such as barley can approach a 7% storage efficiency for total incident solar energy, or 14% storage efficiency for the photosynthetically-active radiant energy of wavelengths between 400 and 700 nm. At first sight this may appear to be contradictory to the storage efficiency of around 1% developed in Section 4.1, but that figure was applicable to a leaf considered to be in full sunlight i.e. in the canopy. The so-called 'sun plants' are light saturated [Section 4.3(i)] at irradiance levels of about 20% of that of full sunlight; 'shade plants' are light saturated by irradiances lower than 10% of that of full sunlight. Now leaves below the canopy will receive a much lower irradiance, but examination of the form of Figure 32 makes it clear that the plant will be able to incorporate the weakened irradiance with a much greater efficiency. In fact for leaves well down towards the ground, the efficiency may approach its limiting value of 30% (Section 4.1) if the respiration rates are sufficiently small. Figure 35 shows some light penetration characteristics for two types of bush plants at solar noon.

On an annual basis the figures are less impressive. The average British wheat crop is some 5×10^9 kg of grain from an area of around 10^{10} m^2 From the energy storage point of view we would also wish to consider the straw, chaff, roots, etc., and a reasonable guess might be that this residue was of equal weight to the grain. If we take the average combustion energy of this carbohydrate to be 17.5 kJ g^{-1} (Table 5.1), then the energy stored in the British wheat crop amounts to about 2×10^{14} kJ per annum (cf. the annual British energy consumption is of the order of 10^{16} kJ, Table 2.6), corresponding to a yield of 2×10^7 J m^{-2}. The annual insolation for Britain is about 3×10^9 J m^{-2}, so that the wheat crop represents less than 1% storage efficiency of energy in total. Taking proper account of a 4-month growing season, weather damage, etc., the true figure would still hardly reach 2%.

Agricultural crops like wheat and barley are perhaps atypical in that they have been selected for fast growth rate and are not suitable for cultivation on rough land. Moreover, under our present agricultural techniques, something of the order of 30% of the energy content of the grain has been expended in the production of the fertilizers applied. With a view to 'current account' production of fuel, perhaps the

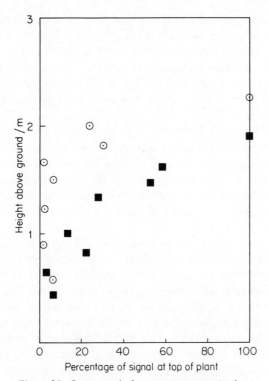

Figure 35. Some typical percentage penetrations of solar irradiance into bush plants obtained using a quantum sensor responsive to photosynthetically active wavelengths. Large-leafed bush (rhododendron) ○, small-leafed bush (hawthorn) ▪.

energy-storage efficiencies of trees are more relevant, even if much lower. For Britain the annual average energy-storage efficiency of a pine forest is of the order of 0.5%, and that of a deciduous forest of the order of 0.3%. Such considerations would lead to the concept of growing forests as energy plantations. Conifers can be grown at densities of 1500 trees per 1000 m² and could be harvested every decade or so. However conifers have the disadvantage of requiring replanting after felling whereas with some deciduous trees new shoots develop from the stumps. In the sense that it is difficult to imagine that the wood itself would find direct fuel application on a wide scale, we shall defer further discussion of this type of energy plantation to the next chapter, when we shall be in a position to assess the most viable conversion processes.

4.5 Potential Annual Photosynthesis Rates

There are few parts of the world where the annual rate of photosynthesis is limited solely by the integrated irradiance to ground and where other factors, such as water supply and soil fertility, do not provide more severe limitations. However, if the agricultural system was perfected, then the integrated solar irradiance throughout the growing season would be the true limiting factor for crop yields. Accordingly the potential annual photosynthesis rate in this sense does have significance for the so-called 'green revolution'; there is no way that a farmer in the tropics, say, can improve his crop yield (viewed as complete plants) beyond this limit.

The length of the day and the local noon solar angle during the growing season will be prime variables in this connection. However, it is necessary to consider whether the ambient temperature at a given location will make that location a realistic proposition for agricultural activity; for example it would be hopeless to consider land above 80° N, despite the 24-h daylight in summer, since ambient temperatures there are low enough to make it permanently icebound. With many plants, although the true rate of photosynthesis doubles for a 10-K rise in temperature, because of positive temperature coefficients for the rate constants of some of the elementary reactions involved, the net rate of photosynthesis is almost independent of temperature. This is particularly the case for plants with high rates of photorespiration such as wheat; increasing rates of photorespiration as the leaf warms probably almost compensate for the increased rate of true photosynthesis. However with plants such as maize, which have low rates of photorespiration, net photosynthesis rates do almost double with a 10-K rise in temperature. Another phenomenon has been observed for blue-green algae in that the saturation level of irradiance increases with rising temperature so that these unicellular organisms grow exceptionally well at around 315 K compared to temperatures below 300 K.

During the 4-month northern summer period, the lowest potential photosynthesis in the northern hemisphere is in the tropics (close to 12-h day) while Iceland and Southern Scandanavia (close to 24-h day) have the highest potential photosynthetic rates, exceeding 37.5 g of carbohydrate m^{-2} day^{-1}. These northern regions have temperatures high enough to maintain photosynthetic productivity. Western Europe between 50 and 60° N has also a high potential photosynthetic rate for plants with a 4-month growing season and in fact leads the world in wheat production efficiency (approaching 2% as total plant — see preceding section). In the tropics sugarcane grows all year round and produces carbohydrate (sucrose + cellulose) at an average rate of 5 g of carbohydrate m^{-2} day^{-1}, corresponding to an efficiency of solar energy conversion of about 0.5%. It thus becomes apparent that the harvesting of solar energy through plants is favoured in northern latitudes using a short-growing-season crop as compared to the tropical latitudes. This may be exemplified in a highly simplified manner.

Suppose that there is a plant which has an average net photosynthesis rate of $1 hour^{-1}$ in daylight and an average dark respiration rate of $0.5 hour^{-1}$. For a northern latitude situation of say 18 hours of daylight, the net daily photosynthesis rate is 18 x 1 − 6 x 0.5 = 15. For the tropics situation with a 12-hour day, the net photosynthesis rate is 12 x 1 − 12 x 0.5 = 6. Thus the northern latitude daily growth rate is 2.5-times greater, all other factors being equal.

The topic of harvesting the solar energy through photosynthesis is developed much further in the book edited by San Pietro, Greer, and Army which is cited in the bibliography list below.

Bibliography

1. *Photosynthesis*, by E. I. Rabinowitch and Govindjee, Wiley, New York, 1969.
2. *Harvesting the Sun*, (Eds A. San Pietro, F. A. Greer, and T. J. Army), Academic Press, New York, 1967.
3. *Biochemistry of Photosynthesis*, by R. P. F. Gregory, Wiley-Interscience, 1971.
4. 'Chemistry in the Environment', Chapters 7 and 8. A *Scientific American* collection (Ed. C. L. Hamilton), Freeman, San Francisco, 1973.
5. *Photosynthesis, Photorespiration and Plant Productivity*, by I. Zelitch, Academic Press, New York, 1971.
6. 'Solar Energy by Photosynthesis', by M. Calvin, *Science* **184**, 375–381 (1974).

Chapter 5

Combustion and Fuels

In this chapter we consider substances currently used as fuels and go on to discuss the scope for production and application of synthetic fuel substances with a view to the conservation of diminishing natural (i.e. fossil) reserves. Then we proceed to a detailed assessment of the thermodynamic and kinetic factors which govern the amount and rate of heat release from a fuel in combustion. Associated with the combustion of fuels is the production of pollutant gases such as nitric oxide, carbon monoxide, partially oxidized hydrocarbon molecules, and sulphur dioxide. Accordingly we must consider the origins of these. Man's activities are termed *anthropogenic*; for example, the carbon monoxide produced by man's combustion of fuels or other activities is referred to as anthropogenic carbon monoxide.

The major problem caused by combustion is the localization of both thermal and pollutant emissions. For example, nitrogen oxides as a whole are produced by biological natural sources at a rate some 15-times greater than the present rate of production from combustion. However in rural areas the total concentrations of nitrogen oxides are only around 0.02 to 0.4 parts per hundred million (p.p.h.m.) whereas levels of the order of 10 p.p.h.m. or greater are common in urban environments. Moreover there is a change from a preponderance of the comparatively harmless nitrous oxide (N_2O) to a preponderance of the combustion-generated nitric oxide (NO) and harmful nitrogen dioxide (NO_2) from rural to urban areas. In the case of carbon monoxide, over 80% of the total anthropogenic production is from internal combustion engines. The source of 40% of sulphur dioxide at present released to the atmosphere in the Northern Hemisphere is anthropogenic and projections suggests that this may rise to 70% by the end of the century. Of this anthropogenic sulphur dioxide some 83% arises from the combustion of fossil fuels. Localization of this sulphur dioxide release gives levels of 1 to 100 p.p.h.m. in urban areas as opposed to the natural background level of 0.02 p.p.h.m.

5.1 Fuels

A fuel may be defined as a substance (solid, liquid, or gas) which is exoergic when it reacts with the oxygen of the air. Usually we are concerned with straightforward combustion to yield heat but in the special case of fuel cells the energy may appear directly as electrical energy.

Table 5.1. Heats of combustion in kJ g^{-1} for 298 K
('High' values for the production of liquid water are listed here, 'low' values can be calculated from the value of 44.013 kJ mol^{-1} for the latent heat of vaporization of water at 298 K.)

Methane(g)	55.12	Hydrogen(g)	144.79	Ethylene(g)	50.30
Ethane(g)	51.38	Carbon monoxide(g)	10.11	Benzene(g)	42.23
Propane(g)	50.05	Methanol(l)	22.35	Benzene(l)	42.02
n-Butane(g)	49.53	Ethanol(l)	29.83	Acetylene(g)	49.92
n-Pentane(g)	48.71	Propanol(l)	33.73	Coal(s)	32–37
n-Octane(l)	47.90	Cyclohexanol(l)	37.28	Cellulose(s)	17.5
Isooctane(l)	47.86	Glycerol(l)	18.05	Wood(s)	up to 21
				Starch(s)	17.5

Source: Data are generally available and have been collected from a variety of sources.

For the present we shall consider only the direct, high temperature combustion of fuels in air. There are limitations to the chemical elements which are acceptable constituents of a fuel for general use: obviously elements like sulphur and nitrogen which would give rise to major productions of obnoxious gaseous oxides in combustion are excluded. Equally undesirable components are elements like boron and aluminium which would produce solid oxides. Hence a practical fuel is restricted to the elements carbon and hydrogen, which produce gaseous and immediately harmless oxides, perhaps with some oxygen incorporated also.

As pointed out in Chapter 4, all naturally occurring fuels, e.g. wood, coal, oil, and natural gas, have their origins in photosynthesis. The heats of combustion of most organic fuels do not vary markedly on a weight basis so that the choice of fuel is hardly governed by this parameter. Table 5.1 lists a selection of the heat contents per unit weight on the basis of the products being gaseous carbon dioxide and liquid water as appropriate under standard conditions of 298 K and 1 atmosphere pressure.

A list of criteria for general acceptability of fuel substances can be compiled:

(i)   Non-toxic gaseous or vapour products are required.

(ii)   For static power generation, a solid fuel like coal offers the advantages of easy and inexpensive storage over comparatively long periods of time in contrast to gases or liquid fuels.

(iii)   For mobile power sources, liquids offer easy transfer and handling characteristics.

(iv)   For mobile power sources, the fuel should have a high heat content on both a weight and volume basis.

(v)   For power sources in intermittent use, refrigerated storage is not acceptable.

(vi)   For mobile power sources, highly-compressed gaseous fuels are precluded by the weight of a sufficiently strong storage tank.

(vii)   Energy can be most efficiently transmitted from one location to another at a substantial distance by pumping gas or liquid fuel through a wide bore pipeline. The larger volume of gas can reduce the energy capacity of the volume of the pipeline itself, which must be filled prior to transmission.

Table 5.2. Comparison of fuel characteristics

Fuel	Relative heat contents		Storage
	Weight basis	Volume basis	
Petrol(l)	1.0	1.0	Simple tank
$H_2(l)(T_c = 33$ K$)^a$	2.53	0.25	High degree of refrigeration
$CH_4(l)(T_c = 190$ K$)^a$	1.16	0.68	Effective refrigeration
$CH_3OH(l)$	0.47	0.53	Simple tank
$H_2(g)(160$ atm$)$	2.53	0.05	Weighty pressure tank
$CH_4(g)(200$ atm$)$	1.16	0.23	Weighty pressure tank

^aT_c is the critical temperature, i.e. the temperature above which the gas cannot be liquified. Therefore it is necessary to refrigerate to below T_c.

In Table 5.2 several fuels in common use, or with that potential, are compared on the basis of these criteria with heat contents expressed relative to those of petrol (gasoline).

We may view these criteria of acceptability against the record and projection of fuel usage in the non-Communist world, data given in Table 5.3.

It is then interesting to compare the consumption data in Table 5.3 with the estimated reserves of fossil fuels on an energy content basis, as set out in Table 5.4.

To complete the present array of tables, Table 5.5 shows the pattern of the significant end-uses of energy (above 0.5%) for the United Kingdom in 1972. The data may be taken as applicable in a general sense to any Western industrialized country.

In Table 5.3 we see that the rate of usage of the fossil fuels is determined far more by their convenience for the major purposes shown in Table 5.5 than from any consideration of the reserves position given in Table 5.4. Transport, for example, accounts for one seventh of an industrial nation's fuel budget and is totally based on oil for obvious reasons. On the other hand, space heating is divided

Table 5.3. World^a energy demand record and forecast (in units of 10^{17} J per annum)

	1950	1960	1970	Year 1972	1977	1982	1985
Total	654	935	1610	1730	2240	2900	3391
Oil	220	410	852	980	1330	1750	2020
Natural gas	79	165	330	354	430	550	650
Water power ⎫	12	24	37	⎧ 37	49	61	73
Nuclear power ⎭				⎩ 6	31	67	104
Solid fuels	342	336	366	350	400	472	544

^{a}Excluding U.S.S.R., E. Europe, and China.

From E. Drake, *Philosophical Transactions of the Royal Society,* **A276**, 454, 455 (1974). Reproduced by permission of the Royal Society.

Table 5.4. Energy content of the Earth's reserves of fossil fuels (in units of 10^{17} J)

Fuel	Known (1972)	Forecast
Crude oil	29,000	116,000
Natural gas	20,000	73,000
Coal	141,000	2,000,000
Tar sands and shale oil	30,000	220,000 (mainly shale oil)

Source: Abstracted from the series of articles, 'Energy in the 1980's' in *Philosophical Transactions of the Royal Society*, **A 276**, 405–615 (1974).

between solid fuel (mainly by electrical generation), natural gas, and oil (direct and through electricity generation). In point of fact it can be argued that natural gas is the best proposition for most space heating applications in view of its potential for efficient transportation by pipeline, clean burning (much less sulphur content than oil) and its unsuitability for most transportation applications as such (Table 5.2). However these virtues of natural gas have hardly prevented electric power being used for a large fraction of space heating requirements, even if electricity represents only half the overall efficiency in this connection for usage of the energy content of the fuel (see Chapter 6).

The data of Table 5.3 enable us to estimate the anthropogenic output of CO_2 into the atmosphere per annum. From Table 5.1 we see that the heats of combustion of most alkanes are near constant on a weight basis. Accordingly we will not be seriously in error if we presume oil to be high molecular weight alkane in total, and to release $14 \times 48 = 672$ kJ mol^{-1} of CH_2 backbone units to yield 1 mol of CO_2, i.e. 44 g. Hence for the 1977 figures, the 1.33×10^{17} kJ of oil energy consumed (assuming that all plastics, etc., are eventually incinerated) releases in turn about 9×10^{12} kg of CO_2 to the atmosphere. Natural gas produces around 2×10^{12} kg of CO_2 and solid fuel (presumed coal) adds around 4×10^{12} kg. The total annual emission of CO_2 from the non-Communist world is therefore close to 1.5×10^{13} kg, and increasing this by one-third as a reasonable estimate for the contribution from the Communist world gives a world total of around 2×10^{13} kg of CO_2 per annum, based on the projections for 1977. This estimate for the anthropogenic output is about 10% of the annual rate of natural fixation of CO_2 by

Table 5.5. Significant end-use pattern for U.K. in 1972 (% of total)

Space heating	22	Domestic hot water	3.8	Cement	1.6
Road transport	12	Food, drink, tobacco	3.0	Agriculture	1.0
Steelmaking	8.3	Air transport	2.2	Glass, china, etc.	1.0
Chemicals	7.0	Textiles, clothing	2.1	Bricks, tiles	0.8
Engineering	6.4	Paper, printing	2.0	Rail transport	0.8
Public services	4.0	Cooking	2.0	Lighting	0.8

Data source: Abstracted from *U.K. Energy Statistics* (H.M.S.O.), 1972.

photosynthesis. As the CO_2 level in the atmosphere is fairly constant, we may therefore say that the anthropogenic output probably approximates to 10% of the natural output through bacterial oxidation, fermentation, etc., of plant materials. Put in another way, we may say that the worldwide energy budget of man, derived almost exclusively from the Earth's 'savings account' of fossil fuels, is equivalent to about 10% of the Earth's 'current account' annual energy turnover of photosynthetically-stored energy.

5.2 Synthetic Fuels

We may distinguish between the naturally occurring fuels (coal, oil, and natural gas) and other fuel substances which may be synthesized by man but which do not occur naturally in any significant quantities (e.g. hydrogen and methanol).

It is apparent from the preceding section that Nature has not produced the most convenient distribution of natural fuels – too much solid fuel, not enough liquid and gaseous fuels. Is it then feasible to convert the energy derived from solid fuel into synthesized fluid fuels? From the data of Table 5.2 it can be seen that methanol has about half the energy content of petrol on both weight and volume bases; moreover the normal boiling point of methanol is 338 K which makes storage and handling easy for transportation purposes. Hydrogen is obviously less suitable for straightforward use as transportation fuel because of its storage problems. At first sight we should envisage the principal role of synthesized hydrogen as a supplement or replacement for natural gas supplies; however, we shall see that the existence of some easily decomposed hydrides does allow consideration of hydrogen as a transportation fuel. Our discussion of this section will be limited to these two front-running possibilities.

Methanol (CH_3OH) can be made from almost any other carbon-containing fuel, and also from wood, and farm and municipal wastes. Thus the so-called 'Methanol Economy' would have considerable flexibility. It is perhaps worth noting at this point that up to 15% of methanol can be added to commercial gasoline without the need to modify the engines in current use, the major problem being the holding of the two components together as a solution or suspension. Moreover the auto-ignition temperature of methanol is high (740 K) compared to that of gasoline (595 K), which confers the high effective octane number of 106 on methanol, to be compared with 90 to 100 for gasoline. Hence the addition of methanol has the added advantage from the pollution point of view of decreasing the required additions of tetraethyl lead, the antiknock agent, and consequent emission of lead to the environment. In one investigation it was found that addition of 10% of methanol to 90 octane number gasoline raised the rating to 94, equivalent to the addition of 0.13 g of tetraethyl lead per litre of gasoline. In the absence of North Sea oil reserves comparable to those of Britain, this methanol extension of gasoline is very much in current West German thinking on the 'Energy Crisis'.

The starting point of any efficient synthetic production of methanol is the production of a mixture of carbon monoxide and hydrogen by partial oxidation of a carbonaceous fuel with O_2 and/or water vapour. Solid catalysts, in general oxides of metals (e.g. chromium, copper), are used to promote the kinetics of the achievement of the equilibrium represented as:

$$CO + 2\,H_2 \rightleftharpoons CH_3OH$$

for which the standard Gibbs energy change as a function of temperature is given by $\Delta G^\ominus = -90.800 + 0.229\ T/\text{kJ mol}^{-1}$.

Let us examine the effect of increased total pressure on this equilibrium. Suppose that we have initially 1 mol of CO and 2 mol of H_2 in a closed reactor before any methanol is formed. At equilibrium, let F be the fraction of CO reacted so that $2F$ mol of H_2 have reacted and F mol of CH_3OH have been formed. Thus at equilibrium we have:

$$CO\ +\ 2\,H_2\ \rightleftharpoons\ CH_3OH$$

molecules $1 - F$ $2 - 2F$ F

and converting into partial pressures on the basis of $3 - 2F$ mol in total and a total pressure at equilibrium of P, the partial pressures are:

$$\frac{(1-F)\cdot P}{(3-2F)} \quad \frac{(2-2F)\cdot P}{(3-2F)} \quad \frac{F\cdot P}{(3-2F)}$$

Thus the pressure equilibrium constant, $K_{p/p^\ominus}$, becomes (on the basis of pressure expressed in atmospheres and the standard pressure $p^\ominus = 1$ atm):

$$K_p = P_{CH_3OH}/P_{CO}\cdot P_{H_2}^2 = \frac{F\cdot P}{(3-2F)}\cdot\frac{(3-2F)}{(1-F)\cdot P}\cdot\frac{(3-2F)^2}{(2-2F)^2\cdot P^2}$$

$$= \frac{F\cdot(3-2F)^2}{4\cdot(1-F)^3\cdot P^2} \tag{5.1}$$

On the basis of Le Chatelier's principle the qualitative prediction can be made that increased pressure will move the equilibrium to the right, i.e. towards fewer molecules. This is predicted more quantitatively by equation (5.1). On the other hand since ΔS is negative from left to right in the equation as written above, it is clear that the yield of methanol on the equilibrium basis is favoured by lower temperatures; however a compromise with the need to obtain a suitable rate for the conversion, i.e. the kinetic factor, must be made for the synthesis to become a practical proposition. It turns out that the typical conditions under which methanol is synthesized are 50 atm total pressure at a temperature of 500 K.

This brief discussion of methanol production may be completed by mention of two promising recent developments. In the first a coal gasifier has been operated at moderate pressures to produce the so-called 'synthesis gas' on the basis of the equilibrium written as:

$$C(s) + H_2O \rightleftharpoons CO + H_2, \Delta H^\ominus = +131.4\ \text{kJ mol}^{-1}$$

Higher pressure of operation are avoided since these favour methane production through the equilibrium:

$$3\,C(s) + 2\,H_2O \rightleftharpoons CH_4 + 2\,CO$$

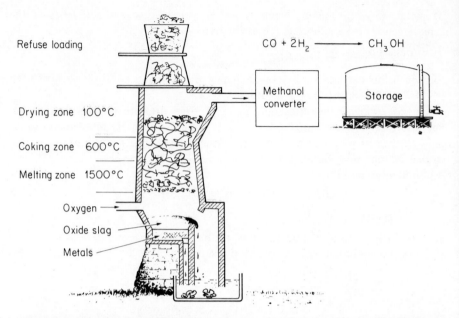

Figure 36. Schematic representation of oxygen refuse convertor for the conversion of municipal waste, etc., to CO and H_2 or methanol. From T. B. Reed and R. M. Lerner, 'Methanol: a versatile fuel for immediate use', *Science,* **182**, (28 December), 1303 (1973). Copyright 1973 by the American Association for the Advancement of Science and reproduced with permission.

Since this second equilibrium corresponds to a lesser increase in the number of gas-phase molecules from left to right as compared to the 'synthesis gas' process, Le Chatelier's principle again provides the qualitative explanation. The subsequent passage of the 'synthesis gas' over methanol-producing catalysts then yields what is called 'methyl fuel', primarily methanol but containing controlled proportions of C_2 to C_4 alcohols.

The second promising source of reactants for synthesizing methanol is what is known as the 'Oxygen Refuse Converter'. The outline design which has been proposed for this is shown as Figure 36. Municipal and industrial refuse is added at the top of the column and undergoes controlled combustion in the presence of a limited supply of oxygen as it passes down. The metals and glass are melted in the hot zone, just above the oxygen addition pipe, and are drawn off as slag with around 2% of the original volume of the refuse. All other products are gaseous or volatile. The carbon content of the refuse burns in this high temperature zone and, with the limited oxygen supply, produces carbon monoxide which rises through the column. This rising hot gas creates a zone of intermediate temperature where down-coming carbohydrates, plastics, etc., are coked and release their hydrogen to provide a gas containing, typically, near 50% CO, near 25% H_2, around 15% CO_2, and a trace of methane on a volume basis. Finally, at the top of the column the gas mixture is led out through a catalytic converter and methanol is produced as the major product. The attraction of this system is that it offers the prospect of reclamation of the energy of organic materials after primary use as fabric, packaging, etc. Data available currently suggest that the gas mixture emerging at the top of the column contains about 75% of the energy content of the original refuse, while for a typical 60% conversion of $CO + 2H_2$ to methanol in the converter the storage as methanol is about 18% of the original energy content of the refuse. Calculations suggest that in industrialized countries conversion of refuse to methanol in this way could cater for nearly 10% of the transportation energy budget. Alternatively, mixed with gasoline, it could stretch that by 10%.

Perhaps the more important aspect of the oxygen refuse converter is that it is a practical means whereby forests, etc., could be operated as energy plantations, producing an immediately-usable product. In Section 4.4 it was said that a pine forest stored incident solar energy with an efficiency of around 0.5% on an annual basis. British pine forests receive an incident flux density of $3000 \, MJ \, m^{-2}$ per annum and therefore store of the order of $15 \, MJ \, m^{-2}$ each year. Hence with an 18% efficiency of conversion to methanol, an annual energy yield of around $3 \, MJ \, m^{-2}$ in this form could be expected. If 10% of Britain's area of $2.3 \times 10^{11} \, m^2$ was managed as such an energy plantation, the energy content of the annual production of methanol alone could be of the order of $10^{17} \, J$. This represents a not inconsiderable fraction of Britain's current total energy usage of $70 \times 10^{17} \, J$ and even more of the $10 \times 10^{17} \, J$ or so applied to transportation. Moreover these comparisons take no account of the near three-times larger amount of energy present in the gases at the top of the converter, which is not converted to methanol but could be used to fire a power station. The figures for the U.S.A. are even more impressive. The land area is $9.1 \times 10^{13} \, m^2$ and currently some 23% is covered by commercial forests. If those forests were applied to methanol production as detailed above, the energy yield would be approximately equal to the present U.S.A. energy budget of the order of $10^{20} \, J$.

Thus the cycling of some 0.1% of the incident solar energy through photosynthetic fuel production before ultimate degradation to heat (which in principle does nothing to affect the Earth's heat balance) could satisfy the energy needs of almost all industrial societies. Then the energy source would be continually renewed in contrast to the current depletion of fossil fuel reserves.

Another route through ethanol rather than methanol also offers considerable promise. There are at least 13,000 microorganisms which can live on cellulose by producing the cellulase enzymes and thus digesting the long chains down to glucose units. Certain of these microorganisms can be used to obtain a cellulase extract capable of dealing rapidly with a variety of forms of cellulose including wood. In one reported instance, the extract was mixed with milled newspaper at around 320 K to provide a crude glucose syrup. It is then a simple matter to ferment the glucose to give ethanol. Pilot plants have been run at the U.S. Army Natick Laboratories which used 2000 kg of cellulose per month, with a projected capacity of nearly 10,000 kg per month.

The cycling of solar energy by these methods avoids thermal pollution (i.e.

Table 5.6. Global and localized power densities

Nature or source type	Power density/W m^{-2}
Winter solar irradiance in Britain	35
Summer solar irradiance in Britain	190
Sensible heat transfer from ground to atmosphere	10
Latent heat released by precipitation over Britain	40
Typical maximum rate of kinetic energy dissipation in American hurricane	25
Annual mean atmospheric kinetic energy in northern hemisphere	0.06
Anthropogenic power densities	
Britain (1973)	1.2
Rural area (Norfolk)	0.7
Industrial area (Lancashire)	5.3
Large chemical works in northern England	500
2000 MW power station with 1 km^2 sink	4000
Proposed nuclear power parks	10,000

evolution of heat from a 'fixed' energy resource such as fossil or nuclear fuels as opposed to 'mobile' energy sources such as contemporarily photosynthesized materials) on an overall basis. In other words it does not introduce any new term into the Earth's heat balance as represented in Figure 18. Table 5.6 makes it clear that present levels of thermal pollution are insignificant on a global basis but can be dominant on a local basis.

It is evident that even on a national scale there is a developing thermal pollution problem. It is quite feasible for local high density power outputs to modify local weather patterns by the induction of substantial upward convective motion. The national problem would be avoided to a much larger extent if our energy needs were derived within the natural heat-balance cycle, such as through the agency of photosynthesis.

There is unlikely to be a thermal pollution problem of any importance on a global basis within the forseeable future. The 1977 energy usage figure of about 3×10^{20} J per annum (Table 5.3 with a 33% increase for U.S.S.R. etc.) is to be compared in this connection with the annual receipt of solar irradiance, corresponding to around 5×10^{24} J on a global basis (Table 2.6), with around 40% of this reaching Earth's surface (Figure 15). Even if fuel consumption doubles by the year 2000, thermal pollution resulting from the mobilization of the Earth's fixed energy reserves will hardly be of significance, but the associated mobilization of fixed carbon and its release as carbon dioxide may be of more significance. The resultant enhanced greenhouse effect (see Section 1.3) could bring about a temperature rise of up to 1 K by the end of the century in terms of a worldwide average. Although small, this is by no means insignificant with regard to melting of the polar icecaps and the ensuing rise in sea level, which could amount to as much as 50.m. From this point of view, a photosynthetically-based energy economy offers the advantage of a closed cycle with respect to atmospheric CO_2, i.e. mobile carbon.

Our second synthetic fuel possibility, hydrogen, has received much attention of late. The attractions of the so-called hydrogen economy are that the raw material from which the gas would be derived, i.e. water, is in copious supply; there is in principle a closed chemical cycle with water as the end-product; and the gas can be transmitted efficiently over long distances without the resistive losses associated with the similar transmission of electric current and, again unlike electricity, the gas can be stored to meet peak demands. On the last point it is perhaps remarkable that at present a largish fraction of our expensive generating capacity is only brought into service for a few hours each winter day to meet peak electricity demand.

However hydrogen does not in fact represent the reversal in energy source concepts that methanol produced from photosynthesized materials would; the energy for the initial production of the hydrogen will be nuclear or fossil fuel derived as envisaged at present. The current opinion is that hydrogen will become a viable fuel when natural gas supplies begin to run out, predicted as 30 years from now for Britain.

In all probability, it is the electrolytic route for production of hydrogen which comes to mind first. The direct electrolysis of seawater poses some severe difficulties in connection with the formation of insoluble calcium deposits on the electrodes, and the environmental unacceptability of the necessary evolution of chlorine from the salt content. However the apparent necessity to distill the seawater prior to electrolysis is not an overwhelming disadvantage: if it is a nuclear reactor, say, built out to sea which generates the electric current, then the waste heat, around 60%, (Chapter 6) could usefully be applied to this distillation. It has been suggested that a 35% efficiency in terms of the energy stored as hydrogen compared with the original fuel energy could be achieved in such a process. The electrolysis process must be considered at present as the front runner because of its well-established technology.

A longer-term possibility is the basing of hydrogen production on the photosynthetic generation of reducing power. The biophotolysis of water has been reported using dyes coupled into the photosynthetic apparatus of blue-green algae, in the presence of a hydrogenase preparation which serves to catalyse the reduction of protons to hydrogen gas. It has been conjectured that something approaching a 10% efficiency of photosynthetically-active solar irradiance might be obtained, a high figure compared with most of the situations discussed in Chapter 4. However it would appear to be very difficult for this biophotolysis to be scaled up to practical proportions from the point of view of contributing significantly to national energy budgets.

The alternative to the electrochemical production of hydrogen, which is currently attracting a great deal of interest, is the so-called thermochemical cycle, where the thermal decomposition of water is effected in closed operations.

On the simplest Hess law basis, and since enthalpy is a function of state, the enthalpy which can be released in any single step of a cyclic operation must be equal to the amount of enthalpy taken in in the course of all the other steps. Thus for the cycle represented as:

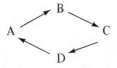

the enthalpy released in the step $D \rightarrow A$ should be equal to the enthalpy input in the course of going from A to D via B and C. In the case of the thermochemical production of hydrogen, we could identify stage D as corresponding to the appearance of 1 mol of H_2 with 0.5 mol of O_2 (separated gases) when stage A corresponds to the appearance of 1 mol of H_2O as liquid. In principle, therefore, on this simplified basis the enthalpy stored as H_2 would represent a 100% efficiency in terms of thermal energy supplied at other stages in the cycle.

Now of course the straightforward thermal decomposition of water by itself only occurs at temperatures far too high to have any practical significance. Besides which it would yield highly explosive stoichiometric mixtures of H_2 and O_2, even if the gases could be quenched sufficiently rapidly from the thermal decomposition vessel. Hence any practical cycle must use catalysts with a dual function; not only must the thermal decomposition of the water be effected at comparatively low temperatures, but the H_2 and $\frac{1}{2}O_2$ products must appear at different stages. The catalysts will be recovered so that ideally the cycle uses only water and heat to produce separated hydrogen and oxygen.

So far, over 30 possible thermochemical cycles for the production of hydrogen have been proposed but no obvious winner is discernible and none has been shown to be capable of working upon the large scale required for significant contribution to national energy budgets.

Let us focus upon one fairly simple example using readily available materials, solid carbon and iron oxides. Kinetic factors enter into the cycle in the sense that all elementary steps must take place within a practical time scale and must go essentially to completion; hence a minimum operating temperature must be given for each reaction.

$$C(s) + H_2O(l) \longrightarrow CO(g) + H_2(g), \, 970 \, K \qquad (A)$$
$$CO(g) + 2 \, Fe_3O_4(s) \longrightarrow C(s) + 3 \, Fe_2O_3(s), \, 520 \, K \qquad (B)$$
$$3 \, Fe_2O_3(s) \longrightarrow 2 \, Fe_3O_4(s) + \tfrac{1}{2} \, O_2(g), \, 1670 \, K \quad (C)$$

It is anticipated that temperatures of the order of 1500 K will be within the reach of nuclear reactors in the course of the next decade, so that the rather high temperature demanded by step (C) need not be regarded as impractical. Figure 37 shows the enthalpy changes for each process in the cycle and the movements on the temperature scale.

The enthalpy change when 1 mol of gaseous hydrogen undergoes combustion to yield liquid water is $-286.7 \, kJ$. Summation around the figure produces the anticipated equality in terms of the heat energy input on the assumption of 100%-efficient heat exchange so that no enthalpy is lost.

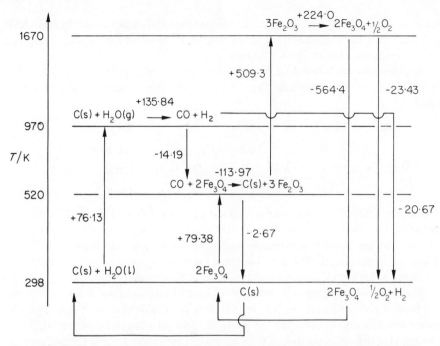

Figure 37. The carbon–iron oxide thermochemical cycle for the generation of hydrogen, with enthalpy changes/kJ and feasible temperatures inserted. Incorporating data from *The Oxide Handbook* (Ed. G. V. Samsonov), I.F.I./Plenum Press, New York (1973).

However there is an efficiency limitation deriving from thermodynamic considerations, even in the ideal limit of any thermochemical cycle involving gases. It is a fundamental deduction from the Second Law of thermodynamics that the maximum efficiency attainable in any closed cycle operation is given by the Carnot cycle efficiency, $\eta_c = (T_2 - T_1)/T_2$, where T_2 is the upper operating temperature and T_1 is the lower operating temperature (Chapter 6). Hence taking $T_2 = 1670$ K and $T_1 = 298$ K for the above cycle leads to a Carnot efficiency of 82%, still highly attractive in comparison with the efficiency of around 35% for the electrolysis process. However the efficiency of the thermochemical cycle must drop further; not only will it be impossible to prevent heat losses and achieve a 100% heat-exchange efficiency, but of necessity the processes cannot be carried out reversibly in the thermodynamic sense. In addition there is the practical need to transport large amounts of solid material from one part of the plant to another in order to ensure the separation of the gaseous products. More specifically the cycle in Figure 37 suffers from a practical disadvantage in that an endothermic reaction is operating at the highest temperature with an exothermic reaction proceeding at the relatively low temperature of 520 K. For maximum practical efficiency the reactions involved in the high-temperature stages should be exothermic with endothermic reactions being involved in the low-temperature stages; this situation would considerably ease heat-transfer problems. All of these factors combine to make it uncertain at present whether thermochemical cycles could, in large scale production, achieve a significant improvement in efficiency over the electrolytic route, which is a tried and tested operation.

Another thermochemical cycle which has received wide publicity is the so-called 'Mark I process' (de Beni and Marchetti, 1972). This is based on the following reactions:

$$CaBr_2 + 2\,H_2O \longrightarrow Ca(OH)_2 + 2\,HBr, \qquad 1000\,K$$

$$Hg + 2\,HBr \longrightarrow HgBr_2 + H_2, \qquad 520\,K$$

$$HgBr_2 + Ca(OH)_2 \longrightarrow CaBr_2 + HgO + H_2O, \qquad 470\,K$$

$$HgO \longrightarrow Hg + \tfrac{1}{2}\,O_2, \qquad 870\,K.$$

This cycle offers a considerable advantage over the one in Figure 37 in that the temperatures involved are much more realistic for a practical process. Assessments of many of the possible thermochemical cycles have been made (see, for example, Eisenstadt and Cox, 1975) and this Mark I process is expected to be capable of achieving a thermal efficiency perhaps as high as 50 to 60%.

The storage of hydrogen in the form of metal hydrides makes it a potential fuel for mobile power sources, such as the automobile. The formation of these hydrides often depends simply upon the gas being brought into contact with the surface of the metal which takes up the gas until equilibrium is reached at the applied pressure for a given temperature, provided sufficient time is allowed. The release of hydrogen when required is accomplished by reducing the applied pressure of the gas to below the equilibrium pressure when decomposition of the hydride occurs to restore the dynamic balance across the metal—gas interface. Such systems are therefore rechargeable and effectively offer a means of storing the gas without involving the high pressures (and weighty pressure tank) needed for the compressed gas, or the high degree of refrigeration needed for liquid hydrogen. In fact the metal hydrides often achieve a stored density of hydrogen greater than does the liquified gas; for example titanium hydride has a factor of 1.2 in its favour compared with liquid hydrogen. Table 5.7 compares the three modes of storing hydrogen, taking account of the weight of the minimum container required: it is obvious that

Table 5.7. Comparison of hydrogen storage systems (Basis: Energy equivalent to 54.43 kg of gasoline.)

Storage system	Weight of container and fuel/kg	Volume/m^3
Gas at 136 atm	1021	1.869
Liquid at 20 K	160	0.289
Magnesium hydride (40% voids)	314	0.306

Table 5.8. Properties of certain metal hydrides

Composition		ΔH_f/kJ per g of H_2	Available H_2/wt-%	Equilibrium temperature at 1 atm of H_2/K
Initial	Final			
Li	LiH	−90.02	12.7	1073
Mg	MgH_2	−36.98	7.7	563
Ca	CaH_2	−86.76	4.8	1193
Na	NaH	−50.01	4.2	798
$Mg_2 NiH_{0.3}$	$Mg_2 NiH_{4.2}$	−32.10	3.5	523
K	KH	−56.52	2.5	988
$VH_{0.95}$	$VH_{2.0}$	−20.03	2.0	285

magnesium hydride is a considerably better storage system than the compressed gas, whilst it has the edge over the liquid gas in its stability at ambient temperature.

Not all of the metals actually form distinct chemical compound hydrides; in many cases what is formed would be better described as a solution of hydrogen in the metal, e.g. for hydrogen in palladium. As a result, the empirical formulae do not always involve integral numbers. Potentially useful hydrides are those which contain relatively high percentages of hydrogen and which, at the same time, attain an equilibrium dissociation pressure of hydrogen of one atmosphere at a moderate temperature. Table 5.8 lists the properties of a few metal hydrides that match these criteria with magnesium hydride representing perhaps the best compromise between the opposite extremes of lithium hydride and vanadium hydride. ΔH_f is the enthalpy of formation of the hydride and with a positive sign would correspond to the amount of heat evolved in fixing 1 g of H_2 in the form of hydride.

The kinetics of recharge–discharge operations, discussed in the paper by Cummings and Powers cited in the Bibliography at the end of this chapter, appear to offer hope for the development of a practical mobile power system.

5.3 Thermodynamics of Combustion Systems

Combustion may be regarded as a process which takes a mixture of fuel and air from a state of metastable equilibrium to a final and true equilibrium state. A fuel–air mixture has a large, negative, Gibbs free energy change available in going to carbon dioxide, water, and any other products, and the reaction is therefore highly feasible in the thermodynamic sense. But the rates of the combustion reactions are zero at normal ambient temperatures; hence it is these kinetic factors which create the initial metastability of fuel–air mixtures. *Ignition* may be regarded as the establishment of a condition within the system, or part thereof, whereby a route of reactions of finite rates is provided for the system to seek its true final equilibrium state.

It will be seen that the thermodynamic and kinetic aspects of combustion can be treated largely in isolation; in this section we consider only the properties of the initial and final states, i.e. the thermodynamic view of combustion. In the next section we shall go on to consider the kinetics of the reactions which effect the release of energy.

In setting the background to the thermodynamic approach, it must be recognized that when ignition does occur, the chemical reactions proceed at exceedingly high rates. This allows us to apply the conditions of adiabatic change between the initial and final states: in other words the exothermic complex reaction occurs so rapidly that the time scale does not permit significant heat exchange with the surroundings. It follows that a high-temperature equilibrium state will result, possessing the same total energy content as the initial fuel–air mixture.

After the attainment of the high temperature equilibrium state, energy may be extracted from the system in a variety of ways. The hot gas may be brought into contact with a heat exchanger and part of its heat content transferred to another medium, as in the case of the steam production in a power station boiler. Alternatively the combustion may be effected with restricted change in volume of the gases and the resultant high-pressure, high-temperature, product-gas mixture is made to do work by expanding against a piston or turbine. These will be areas of interest for Chapter 6. For the present, interest will be confined to the composition, pressure, and temperature attained as the initial result of combustion; these parameters will clearly govern the amount of useful work which may be extracted subsequently.

A starting point might be to consider the thermodynamics of combustion based upon the simple reaction scheme for a general hydrocarbon fuel and sufficient air:

$$C_x H_y + \text{Air} \longrightarrow x\, CO_2 + \tfrac{1}{2}y H_2 O + \text{Air residue}$$

But molecules like CO_2 and H_2O, considered to be completely stable at ambient temperatures, will be subject to dissociative equilibria at high temperatures:

$$CO_2 \rightleftharpoons CO + O$$
$$H_2O \rightleftharpoons H + OH$$

Let us investigate this situation by considering a general gaseous diatomic molecule, represented as M_2. In this case the dissociative equilibrium will be:

$$M_2 \rightleftharpoons M + M$$

and the enthalpy change from left to right will be positive, reflecting the strength of the M–M bond. Under conditions of constant applied pressure, the pressure equilibrium constant (K_p referred to the standard pressure of $p^{\ominus} = 1$ atm) as a function of temperature will vary according to the relationship:

$$d(\ln K_p)/d(1/T) = -\Delta H/R \qquad (5.2)$$

Here ΔH is the average value of the enthalpy change across the temperature range concerned and R is the gas constant. In the case of a positive ΔH for the dissociation of M_2, the equilibrium, as written above, will move to the right as the temperature rises. In this way heat energy is withdrawn from the system to provide the energy of formation of the atoms, M, so that for a given input of energy into the system, the final temperature is reduced compared with that of a hypothetical system where no dissociation has occurred. This is but a further reflection of Le Chatelier's principle, with the real system reacting (dissociating) to resist the applied change (increasing temperature).

In Figure 38(a) the total enthalpy (H_T) of two examples of M_2–M systems, namely M = Cl and M = H, is represented as a function of temperature for a total pressure of 1 atm in each case. At low temperatures the composition is overwhelmingly M_2 so that H_T increases rather slowly with rising temperature reflecting the heat capacity of M_2 itself. On the other hand, at very high temperatures the system is almost entirely composed of M atoms and the increase of H_T with further rise in T reflects the heat capacity of the M atom. Between the extremes of temperature in the figure, however, the molecules will show a progressively greater degree of dissociation as the temperature increases; not only will heat be entering the heat capacity modes but it will also be withdrawn into the endothermic formation of atoms. Accordingly the total enthalpy of the system would be expected to rise much more rapidly with temperature across the range in which dissociation becomes effective in view of the fact that the ΔH quantity of equation (5.2) is large in comparison with H_T. Since the effective heat capacity of the system at constant pressure is given by $C_p = (\partial H_T/\partial T)_p$, the effect of the sharply rising H_T in the dissociation range is shown in Figure 38(b) as a high heat-capacity barrier opposing further rise in temperature.

The particular range of temperatures over which the steep rise in H_T occurs for a given diatomic molecule will depend upon the strength of the molecular bond. The species shown in Figure 38 have very different bond strengths of $D_0^{\ominus}$ (Cl–Cl) = 238.9 kJ mol^{-1} and $D_0^{\ominus}$(H–H) = 432.0 kJ mol^{-1}, and are chosen to illustrate this point. Cl_2 systems show their highest effective heat capacities near 2000 K and the gas is almost completely in the atomic form by 3000 K. However the stronger bond of H_2 raises the temperature of highest effective heat capacity to near 3800 K, and the degree of dissociation is still increasing with rising temperature at 4500 K.

In Table 5.9 are listed the primary dissociation energies or bond strengths of those molecules most frequently appearing as products of combustion.

It can be seen that all of the bond strengths in Table 5.9 are greater than that for hydrogen. Therefore within the temperature range of 2000 K to 3000 K encountered in normal combustion systems, it is expected that degrees of dissociation will be small. Nevertheless significant withdrawal of energy occurs in producing these small degrees of dissociation, and the actual temperature attained is much lower as a consequence than in the hypothetical situation where no dissociation occurred. However we can be sure than CO_2, H_2O, and N_2 will be the major species in the high-temperature equilibrium produced by the combustion of a hydrocarbon fuel in a stoichiometric amount of air.

In order to appreciate the adiabatic nature of combustion processes, let us

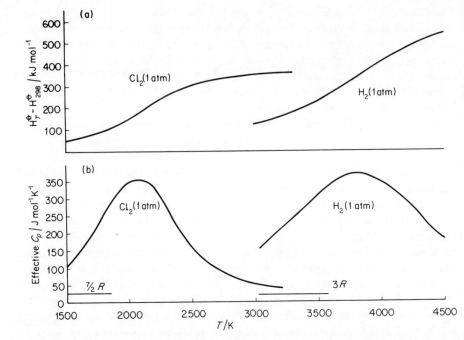

Figure 38. (a) Plots of total effective enthalpies per mol of gas of equilibrated systems of Cl_2 and H_2 with total pressures of 1 standard atmosphere, as functions of temperature. (b) Plots of effective heat capacities (C_p) of equilibrated Cl_2 and H_2 systems at total pressures of 1 standard atmosphere, as functions of temperature [differentials of the corresponding curves of (a)]. Based upon data in *JANAF Thermochemical Tables*, issued by the Thermal Research Laboratory, Dow Chemical Company, Midland, Michigan, under U.S. Air Force Sponsorship.

consider the simple, hydrogen-free system of carbon monoxide and oxygen in the stoichiometric ratio of two parts by volume of carbon monoxide to one part of oxygen. The exothermic reaction will be:

$$CO + \tfrac{1}{2} O_2 \longrightarrow CO_2 \quad \Delta H_0^{\ominus} = -285.0 \text{ kJ mol}^{-1}$$

In a preliminary exercise we can calculate the hypothetical temperature (T_h) which would be reached by this adiabatic combustion if the CO_2 product were not subject to dissociation. The overall process can be divided hypothetically into two stages, invoking Hess's law, as shown in Figure 39.

We imagine the combustion reaction as proceeding at 298 K in the presence of a heat reservoir which absorbs all of the heat released. Thus in the first stage it is imagined that the CO_2 product is produced at 298 K with a release of $-\Delta H_{298}^{\ominus}$ to the heat reservoir. $\Delta H_{298}^{\ominus}$ is obtained from *JANAF Thermochemical Tables* as -282.99 kJ mol^{-1} CO. In the imaginary second stage the heat is withdrawn from the heat reservoir and applied to heat the CO_2 up to temperature T_h. In this connection *JANAF Thermochemical Tables* list molar heat content differences as excesses over the molar heat content at 298 K. Alternatively, expressions for the

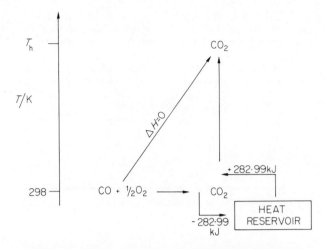

Figure 39. Hess's law cycle concept of adiabatic flame temperature (T_h) for a stoichiometric $CO + \frac{1}{2}O_2$ flame, ignoring dissociative equilibria. The numbers indicate enthalpy changes.

molar heat capacities at constant pressure as functions of temperature may be used in an integrated form between the limits of 298 K and the upper temperature being considered.

The problem posed here then simply reduces to finding the temperature at which the $H_T^{\ominus} - H_{298}^{\ominus}$ value for the 1 mol of CO_2 produced is equal to 282.99 kJ Inspection of the *JANAF* table for CO_2 shows that the required T_h is between 5000 K and 5100 K (100-K tabulation interval) and is 5058 K on a linear interpolation between the two molar heat contents. One glance at the plot for H_2 in Figure 38(a), taking into account that the bond strength in CO_2 is only some 25% higher than that in H_2, suggests that the non-dissociative view of this $CO-O_2$ combustion system is totally unrealistic.

In the real $CO-O_2$ combustion situation, the following dissociative equilibria will be established:

$$CO + O \rightleftharpoons CO_2 \quad K_p(1)$$
$$O_2 \rightleftharpoons O + O \quad K_p(2)$$

where K_p is the equilibrium constant for the reaction in terms of the partial pressure expressed in atmospheres. On the basis of the very high dissociation energy of CO (Table 5.9) and anticipating the final result of such a calculation, the equilibrium involving CO dissociation:

$$CO \rightleftharpoons C + O$$

may be ignored and carbon monoxide can be considered as effectively undissociated at the temperatures encountered in typical combustion systems; it follows that atomic carbon concentrations are negligible on the equilibrium basis.

Table 5.9. Primary bond dissociation energies
of molecules involved in combustion systems

Molecule	Bond	Dissociation energy $(D_0^\ominus)$/kJ mol^{-1}
CO_2	OC–O	531.4
H_2O	HO–H	497.9
N_2	N–N	941.8
CO	C–O	1069.4
O_2	O–O	493.6
NO	N–O	627.2
SO_2	OS–O	565.0

Source: Generally available data.

Values of $K_p(1)$ and $K_p(2)$ at 100 K intervals are obtainable from the *JANAF* tables. The *JANAF* tabulations take the form of listings of K_p parameters for the formation of a species from the standard states of its elementary components. Hence for CO_2 the equilibrium constant quoted refers to the equilibrium:

$$C(\text{graphite}) + O_2 \rightleftharpoons CO_2$$

and is defined as $K_p(3) = p_{CO_2}/p_C \cdot p_{O_2}$ where p_C is the pressure of atomic carbon which represents equilibrium with a graphite surface at the temperature in question. For the oxygen atom the *JANAF* listed equilibrium constant refers to the equation:

$$\tfrac{1}{2} O_2 \rightleftharpoons O$$

and is therefore equal to $(K_p(2))^{\frac{1}{2}}$ as written above. For carbon monoxide the equilibrium referred to is evidently:

$$C(\text{graphite}) + \tfrac{1}{2} O_2 \rightleftharpoons CO$$

with the listed equilibrium constant $K_p(4) = p_{CO}/p_C \cdot p_{O_2}^{\frac{1}{2}}$. By combination of the K_p parameters taken directly from the tables, we obtain:

$$K_p(1) = p_{CO_2}/p_{CO} \cdot p_O = K_p(3)/(K_p(4) \cdot K_p(2)^{\frac{1}{2}}),$$

where p_{CO_2}, p_{CO} and p_O are partial pressures expressed in atmospheres.

The calculation of the actual or adiabatic flame temperature (T_a) attained by the combustion of a stoichiometric $CO–O_2$ mixture at atmospheric pressure is reiterative in nature and is therefore best performed on a computer. However it will bring home the general method if we attempt to follow through the various stages.

At the outset of the calculation, three parameters will be known:

(i)   The initial ratio of the total number of carbon atoms to the total number of oxygen atoms (0.5 for the stoichiometric mixture),
(ii) The initial temperature of the gases (say 298 K),
(iii) The total pressure (say presumed constant at 1 atm).

Table 5.10. Reiterative approach to mass balance at the guessed temperature of 3000 K in the stoichiometric $CO-O_2$ combustion system

Trial p_{CO_2}/p_{CO}	p_O	p_{O_2}	p_{CO}	p_{CO_2}	Atomic ratio $\dfrac{p_{CO} + p_{CO_2}}{p_O + p_{CO} + 2p_{CO_2} + 2p_{O_2}}$
5	0.184	2.679			
2	0.0737	0.429	0.168	0.336	0.284
1	0.0368	0.107	0.428	0.428	0.558
1.2	0.0442	0.154	0.364	0.437	0.5044
1.21	0.0446	0.157	0.362	0.437	0.5013
1.215	0.04475	0.1582	0.3599	0.4372	0.4996
1.214	0.04471	0.1579	0.3602	0.4372	0.49991

The first step is to make a reasonable guess for T_a, say 3000 K. Values of $K_p(1) = 27.151$ and $K_p(2) = 0.01266$ are derived from the *JANAF* tables for 3000 K.

Four species, namely CO_2, CO, O, and O_2 may exist in significant concentrations or partial pressures in the high temperature equilibrium gas. We may find it convenient to use p_{CO_2}/p_{CO} as our prime variable for the next stage. Table 5.10 then summarizes the approach to the final composition at 3000 K satisfying the condition that the total number of carbon atoms within CO_2 and CO must be exactly half the total number of oxygen atoms within CO_2, CO, O, and O_2, according to the initial ratio and thus conserving mass balance. In all cases listed in Table 5.10, the additional constraint is imposed that the sum of the partial pressures, expressed in atmospheres, must be unity. This emphasizes the value of choosing p_{CO_2}/p_{CO} as the trial parameter; p_O then follows directly using $K_p(1)$; p_{O_2} is then calculated using $K_p(2)$ and $p_{CO} + p_{CO_2}$ is given by $1 - (p_O + p_{O_2})$. Combination of the ratio and sum then yields the explicit partial pressures.

The last set of partial pressures in Table 5.10 may be regarded as the equilibrium composition if the guess of 3000 K for the final temperature happens to be correct.

Now, however, we must test whether $T_a = 3000$ K is indeed correct by checking the heat balance. In other words we must ask whether this composition at 3000 K possesses a total excess heat content over the heat content at 298 K which is exactly equal to the heat released by the effective reaction considered as taking place at 298 K and having the stoichiometry:

$$CO + \tfrac{1}{2} O_2 \longrightarrow 0.5485\ CO_2 + 0.4515\ CO + 0.1977\ O_2$$
$$+ 0.05602\ O$$

This is simply a second way of expressing the mass balance and relative concentrations or partial pressures. The heat released by this process is calculated from the heats of formation $[\Delta H_f^{\ominus}(298\ K)]$ given by *JANAF* tabulations to be

141.26 kJ. Again from *JANAF* tables, the sum of the heat contents $(H_{3000}^{\ominus} - H_{298}^{\ominus})$ of the product mixture is 148.64 kJ. Hence our guess of T_a = 3000 K was close but slightly too high, since not enough heat is released by the hypothetical combustion at 298 K to raise the equilibrium composition mixture to 3000 K.

At this stage we must return to the beginning of the reiterative cycle and make a new and reasoned lower guess of T_a, say 2800 K The set of equilibrium partial pressures which then emerges is p_{CO_2} = 0.6015, p_{CO} = 0.2593, p_{O_2} = 0.1204 and p_O = 0.0187. The heat released in proceeding to this composition at 298 K is 192.35 kJ on the same basis as before. The sum of the heat content of the products is 137.79 kJ. Thus more heat is released than can be taken up by the equilibrium composition at T_a = 2800 K, the reverse of the situation for T_a = 3000 K.

It is therefore evident that the correct adiabatic flame temperature is between 2800 K and 3000 K. We might try 2900 K and so on.

After four or five attempts we will arrive at a stage where a sufficient collection of heat imbalance data has been accumulated for a plot to be made of the differences between the heat released and the heat content of the product mixture against temperature, whereby the correct value of T_a can be interpolated. Such plots are almost linear close to the balance point, and where the line intercepts the heat difference axis the value of T_a can be read off. In the case of the stoichiometric $CO-O_2$ mixture the final answer is T_a = 2973 K, with partial pressures of p_{CO_2} = 0.46, p_{CO} = 0.35, p_{O_2} = 0.15 and p_O = 0.04 representing the equilibrium composition.

The large difference between T_h = 5058 K and T_a = 2973 K emphasizes the importance of dissociative equilibria in determining the peak temperature reached in a homogeneous combustion mixture. In the particular case of the above example, the reason can be seen in simple terms: in effect almost half the carbon monoxide fuel is not consumed so that the actual heat released is only around half of that which would be involved in the hypothetical situation where all of the carbon monoxide is oxidized to CO_2.

Now of course we have picked upon a fairly impracticable combustion system from the point of view of energy generation, just in order to illustrate the basic principles. At large, the generation of energy mainly involves the elements carbon and hydrogen in the fuel, and nitrogen and oxygen in the air. This creates a considerable complexity in any calculations analogous to our simple one above, which was restricted to just two elements. The CO_2 dissociative equilibrium will still be very much involved, and the presence of CO in the high temperature equilibrium means some reduction in the energy released. Similarly H_2O is involved in equilibria with species such as H, OH, H_2, etc. Nitrogen is a strongly-bonded molecule (Table 5.9) and as such it is like carbon monoxide in being undissociated, for practical purposes, in normal combustion systems in the sense that the partial pressures of the nitrogen atom are negligible. However nitrogen molecules are involved in a nitric-oxide formation equilibrium represented as:

$$N_2 + O_2 \rightleftharpoons 2 \text{ NO}, \Delta H_0^{\ominus} = +181 \text{ kJ mol } N_2^{-1}$$

Therefore we expect the equilibrium for this endothermic formation of nitric oxide, expressed as $p_{NO}^2/p_{N_2} \cdot p_{O_2}$, to be governed by the exponential term, $\exp(-181{,}000/RT)$, as a function of temperature. Values of this exponential function for temperatures of 2000 K, 2500 K and 3000 K are 1.9×10^{-5}, 1.7×10^{-4} and 7.1×10^{-4} respectively. Hence it is expected that small but not insignificant levels of nitric oxide will be formed in typical combustion systems using air, these levels rising fairly sharply with the flame temperature.

The equilibria which are significant in normal combustion systems may be listed as follows:

$$CO + O \rightleftharpoons CO_2 \quad \text{(i)}$$
$$O + O \rightleftharpoons O_2 \quad \text{(ii)}$$
$$H + OH \rightleftharpoons H_2O \quad \text{(iii)}$$
$$H + H \rightleftharpoons H_2 \quad \text{(iv)}$$
$$O + H \rightleftharpoons OH \quad \text{(v)}$$
$$N_2 + O_2 \rightleftharpoons 2NO \quad \text{(vi)}$$

This may be regarded as a minimum set which can be recast to yield other combinations. For example the combination of (iii) and (iv) can be considered to refer to the equilibrium:

$$OH + H_2 \rightleftharpoons H_2O + H$$

and the combination of (ii), (iv), and (v) can be referred to the equilibrium represented as:

$$H_2 + O_2 \rightleftharpoons 2 OH$$

which has more the form of (vi).

Various other possible components, such as hydrocarbon radicals, NO_2, and H_2O_2, are eliminated from consideration as the predicted equilibrium levels at high temperatures are vanishingly small.

We may conclude this section by tabulating the results of such high-temperature equilibrium calculations for some typical combustion systems in Table 5.11 (p. 103).

We may note in Table 5.11 the effect of increased operating pressure on the methane–air combustion system. Dissociation at the higher pressure is discouraged on the basis of Le Chatelier's principle. Hence greater partial pressures of CO_2 and H_2O are formed at 100 atm pressure compared with the situation at 1 atm total pressure. As a result more energy is released for heating the product mixture so that T_a is 171 K higher for methane combustion at 100 atm compared to that at 1 atm. This is particularly relevant in connection with internal combustion engines, where peak pressures may be some 70-times atmospheric. We may illustrate the suppression of dissociation by looking at the ratios of less dissociated species to more dissociated species in terms of partial pressures. For example p_{CO_2}/p_{CO} is increased from 9.4 in the 1-atm system to near 27 in the 100-atm system. Other comparable increases are p_{H_2O}/p_{OH} from 57 to 124, p_{H_2O}/p_H from 433 to 3500, p_{H_2}/p_H from 10 to 23, and p_{O_2}/p_O from 20 to 67. However it should be noted

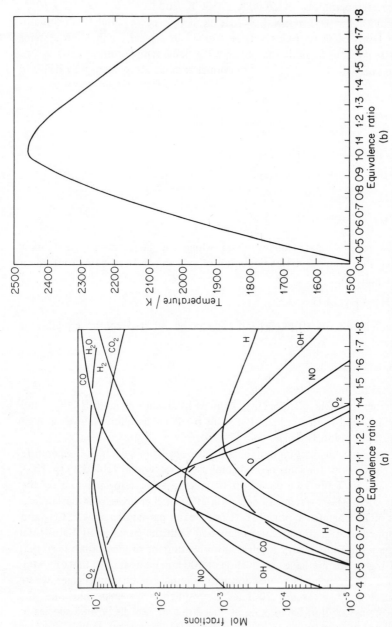

Figure 40(a). Equilibrium composition of burnt gas versus equivalence ratio for aviation kerosene ($CH_{1.76}$) fuel, for combustion air-pressure of 4 atm and initial temperature of 450 K. From A. R. Morr and J. B. Heywood, 'Partial equilibrium model for predicting concentration of CO in combustion', *Acta Astronautica*, 1, 952 (1974). Reproduced with permission of Pergamon Press Ltd.
Figure 40(b). Adiabatic flame temperature as a function of equivalence ratio corresponding to (a). Same source as Figure 40(a), p. 953. Reproduced with permission of Pergamon Press Ltd.

that the partial pressure of nitric oxide is more influenced by the change in temperature, with the ratio p_{N_2}/p_{NO} only changing from 361 in the 1-atm system to 328 in the 100-atm system, as would be expected seeing that equilibrium (vi) has the same number of molecules on each side.

It is clear that the equilibrium composition of a combustion system will depend upon the initial composition of the fuel—air mixture as will T_a. Figure 40 shows the results of computations on the equilibrium composition of the burnt gas as a function of the fuel/air equivalence ratio (unity for the stoichiometric mixture) for aviation kerosene, together with the T_a profile.

5.4 Kinetics of Combustion Systems

Books have been written on the kinetics of combustion alone, so our view here is restricted necessarily and is confined to the basic theme of the chapter as a whole, namely the release of energy by combustion under typical conditions and the formation of pollutant species. Furthermore we shall confine our attention largely to alkane vapour—air premixed systems. In the first place we shall consider what are called *ignition* phenomena therein; alternatively we may say that we are interested in the *explosion* of such mixtures. The term explosion in exact definition applies to the violent increase in pressure which is associated with autoacceleration of the rates of combustion reactions in a constant-volume system.

In alkane—air systems there exists a regime of pressure—temperature conditions under which *slow* combustion takes place at a finite rate. Variation of total pressure and/or temperature can move the system into an explosive regime where the reaction rates tend towards infinity. Because cycles of branching-chain reactions are involved with rates which are dependent upon both temperature and the concentrations of reactive atoms and radicals, and also the overall reaction is highly exothermic, it is usually a combination of the inability of the gas mixture to dissipate the energy released and the inability of the termination reactions of the active species to keep pace with the propagation reactions which leads to explosion. Hence the resultant explosion is often a mixture of a *thermal* explosion and a so-called *isothermal* explosion, the latter in the sense that the accumulation of chain-carrier concentration produced by branching is not *per se* dependent upon increasing temperature.

For the typical alkane—air system the boundary on a pressure-versus-temperature diagram which represents transition from slow combustion to explosive combustion, has the form shown in Figure 41. Our interest will be entirely in the explosion or ignition regime.

The combustion of fuel—air mixtures usually results in a *flame*. This has its origin in the non-uniform nature of the normal ignition situation. For example, if a sparking device is used to initiate the explosive reaction then the zone of ignition is at first localized around the spark before spreading progressively through the mixture. The resultant combustion front, travelling with a finite velocity, is termed the flame. The typical stationary flame established on, say, a Bunsen burner is in reality

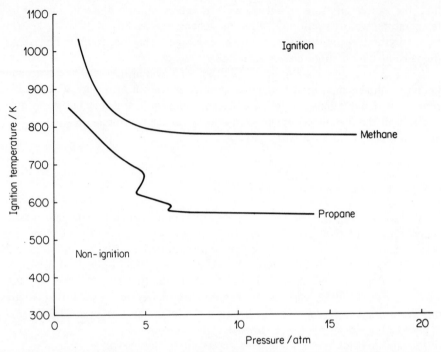

Figure 41. Representative forms of the ignition boundaries of near-stoichiometric hydro-carbon—air mixtures on a pressure-versus-temperature diagram. Cool flame and multistage ignition phenomena are ignored.

a downward travelling combustion front balanced by the upward travelling fuel—air mixture and stabilized by the heat-sink characteristics of the burner assembly.

In alkane—air combustion systems, hydrogen atoms are an important active species and, because of their low mass, they can diffuse through the mixture ahead of the actual flame front itself. Three zones may be envisaged. The first is a preheating zone which terminates when the probabilities of chain branching and chain breaking for any collision of an active centre become equal. The system is then said to have attained the effective ignition temperature. Hydrogen atoms diffusing into this preheating zone commence the reaction of the alkane, and the resultant dehydrogenation and fragmentation continue to develop into the second zone, the explosion zone. Here the net rates of propagation of active centres increase rapidly leading to increasing rates of energy release and hence rising temperature, which in turn enhances the reaction rates in an autocatalysis. When a hypothetical critical temperature, T_c, is reached, a critical concentration of active centres (atoms and radicals) is attained and ignition or explosion results. The major part of the heat release of the overall reaction takes place between T_c and T_f, the final flame temperature in the third bulk-reaction zone.

The relevant point then is that in the real ignition situation the bulk of the enthalpy release occurs very rapidly within a narrow reaction zone, across which

Table 5.11. Adiabatic flame temperatures and equilibrium compositions of combustion systems based on air (all stoichiometric)

Initial system	H_2-Air (1 atm)	Methane$-$Air		Propane$-$Air (1 atm)
		(1 atm)	(100 atm)	
T_a/K	2100	2222	2393	2219
Partial pressures/atm				
CO_2	–	0.082	8.76	0.1004
CO	–	0.009	0.33	0.0099
O	0.00	0.0002	0.004	0.0002
O_2	0.00007	0.004	0.282	0.0047
H	0.00	0.0004	0.005	0.0003
H_2	0.005	0.004	0.116	0.0027
OH	0.0097	0.003	0.143	0.0024
NO	0.00018	0.002	0.221	0.0020
N_2	0.66	0.724	72.37	0.7344
H_2O	0.325	0.171	17.77	0.1430

there is a very steep temperature gradient and also steep gradients of concentrations of species. This situation encourages the transport of heat and of active species so that the flame is essentially self-propagating.

The high temperatures achieved in the explosion and bulk-reaction zones mean that relatively high concentrations of atoms and radicals can exist therein in chemical equilibrium. If we refer to Table 5.11, it can be seen that in the propane flame the species O, H, and OH account for nearly 0.3% of the total concentration of all species.

The kinetic route of approach to the final equilibrium composition is through branching-chain propagation steps. Because of the dehydrogenation of the fuel in the preheating zone, the major sequence is usually closely related to that which occurs in the hydrogen$-$oxygen system:

$$H + O_2 \longrightarrow OH + O \tag{1}$$

$$O + H_2 \longrightarrow OH + H \tag{2}$$

$$OH + H_2 \longrightarrow H_2O + H \tag{3}$$

It is at this point that we commence our kinetic numbering scheme, where each elementary reaction in this and subsequent chapters retains a single identification number. The important elementary reactions are then collected in the Appendix at the end of the book for easy reference.

The branching nature of the propagation cycle expressed by reactions (1), (2), and (3) can be seen by summing left and right hand sides to obtain the net reaction for one cycle (noting that two OH radicals must be consumed in (3) per cycle):

$$H + O_2 + 3\,H_2 \longrightarrow 2\,H_2O + 3\,H \tag{5.3}$$

Thus one hydrogen atom becomes three each time that the cycle proceeds. Such a cycle will contribute relatively little to the energy release as a whole: in fact at 2000 K the overall cycle represented by equation (5.3) results in the release of 49.53 kJ per mol O_2. It is only in the subsequent recombination or chain termination reactions such as:

$$H + H + M \longrightarrow H_2 + M \tag{4}$$

for which $\Delta H^{\ominus}_{2000} = -453.8$ kJ per mol of H_2 that the major heat release takes place. In reaction (4) M is used to denote any molecule in the system which acts as the inert third body.

Now the rate of reaction (1) is usually rate determining for the cycle of reactions represented by equation (5.3), and therefore determines the rate of accumulation of hydrogen atoms in the system. The rate of generation is represented by the differential form:

$$d[H]/dt = 2\,k_1\,[H]\,[O_2] \tag{5.4}$$

with the coefficient 2 appearing because there is a net gain of two hydrogen atoms per cycle. The rate constant k_1 has been measured many times in laboratory experiments and its variation with temperature can be summarized by the Arrhenius form:

$$k_1 = 2.2 \times 10^{11} \exp(-8455K/T)\,dm^3\,mol^{-1}\,s^{-1}$$

Around 2000 K, a 100-K rise in temperature causes k_1 to increase by a factor of 1.22, and the rate constant corresponds to a reaction having an efficiency of one collision in approximately 60 of H with O_2. We shall see that this collisional-efficiency viewpoint is quite helpful in giving meaning to rate-constant numbers.

On the other hand three body-recombination rate constants like k_4 are almost independent of temperature at around 2000 K. An appropriate value of k_4 for a combustion system would be of the order of 10^9 dm^6 mol^{-2} s^{-1}, the exact value depending upon the composition of M. For a total pressure of 1 atm and for a temperature of 2000 K, k_4 corresponds to two hydrogen atoms recombining for every 70,000 collisions between hydrogen atoms.

The evident disparities in the respective probabilities per collision of the reactant species between bimolecular reactions of type (1) and termolecular reactions of type (4) emphasizes an important point. It is likely that the path to the full equilibrium composition will pass through an 'overshoot' position, particularly at lower total pressures. At this stage the propagation steps, depending upon k_1, will be proceeding so fast compared with the recombination steps that supra-equilibrium concentrations of atoms and radicals may well exist transiently. This may occur despite the fact that the temperature will rise following the overshoot position due to the substantial heat release caused by the three-body recombination processes which take the concentrations down towards the final equilibrium values.

The 'overshoot' phenomenon and the generally high rates of the reactions leading to the concentrations of CO_2, H_2O, CO, etc., in the final high-temperature mixture mean that the final state is invariably established within a very short time scale. Even the three-body processes will have reached equilibrium within the time scales involved in internal combustion engines and furnaces in general. Hence the kinetics of the actual combustion of the alkane with oxygen turn out to be largely irrelevant from our point of view since it is clear that the high temperature equilibrium state will be produced in the combustion chamber.

However there are two important minor exceptions to this general postulation, concerning the formation of nitric oxide, and combustion occurring near to the walls of the combustion chamber. These are of major importance in the sense that they produce the main pollutant species encountered in the exhaust gases.

The Formation Kinetics of Nitric Oxide

Nitric oxide arises from the reaction of atmospheric nitrogen and oxygen put into combustion but the mechanism is complex. A key to a possible mechanism is provided by the kinetics of the formation of nitric oxide in shock-heated air by itself. The shock-tube method is very relevant from our point of view since it enables the reaction to be studied under the conditions of high pressure and temperature which correspond closely to the conditions of operation of many practical combustion systems. In the shock tube these conditions are produced by the rapid adiabatic compression of air or N_2-O_2 mixtures across the shock front created by the sudden release of high-pressure driver gas at one end of the tube. The temperature gradient across the front is exceedingly steep; a time scale of the order of only 10 molecular collisions is required for the reactant mixture to achieve the high temperature of the gases reacting behind the shock front. Hence pressure and temperature rise virtually instantaneously in what, from the time scale of reactions occurring in the heated air, is a discontinuity in the profiles.

When the rate of formation of nitric oxide is followed using variable mixtures of N_2 and O_2 as reactants, the experimental rate law has the form:

$$d[NO]/dt = k[O_2]^{1/2}[N_2] \qquad (5.5)$$

As such this immediately eliminates the possibility that nitric oxide could be formed by an elementary step represented as:

$$N_2 + O_2 \longrightarrow 2NO$$

when the rate law would be first order in each reactant.

The experimental rate law is interpreted in terms of a chain mechanism, often referred to as the Zeldovich mechanism. This is based on the elementary reactions

$$O_2 \rightleftharpoons O + O \qquad (5.6)$$
$$O + N_2 \longrightarrow NO + N \qquad (5)$$
$$N + NO \longrightarrow N_2 + O \qquad (-5)$$
$$N + O_2 \longrightarrow NO + O \qquad (6)$$
$$O + NO \longrightarrow N + O_2 \qquad (-6)$$

The two minus signs in the elementary reaction numbers are used to denote reactions which are the reverse of those with the positive numbers.

The equilibrium between oxygen atoms and molecules represented by equation (5.6) is always established well before any significant amount of nitric oxide has been formed. This will be even more the case in a combustion system where oxygen atoms are coupled into the overshoot phenomenon.

The rate-determining step of the Zeldovich cycle is reaction (5). The Arrhenius expression for its rate constant may be taken as:

$$k_5 = 1.4 \times 10^{11} \exp(-37947 \, K/T) \, dm^3 \, mol^{-1} \, s^{-1}$$

and for $T = 2000$ K, this corresponds to reaction at only one collision in around 100 million of O and N_2. Comparison with the reaction probability even of three-body recombination reactions on a collisional basis (about 1 in 70,000 as mentioned above) leaves little doubt that the formation of nitric oxide in heated air is an extremely slow process. Hence we can appreciate that the nitric oxide equilibrium level can only be reached through the Zeldovich chain mechanism in a combustion mixture long after the passage of the flame front, if indeed it can be reached fully within the millisecond time scale of the typical internal combustion engine cycle.

However it appears that nitric oxide can be assumed to have reached the high-temperature equilibrium partial pressure within the millisecond time scale where a combustion system using hydrocarbon fuel is concerned. This is because of the intervention of an additional formation mechanism in such systems, often referred to as the 'prompt' NO mechanism. It appears probable that this depends upon the attack of radicals like CH upon N_2 in a near thermoneutral elementary reaction:

$$CH + N_2 \longrightarrow HCN + N \tag{7}$$

In fact HCN has been detected in the burnt gases of hydrocarbon flames. Possible subsequent elementary reactions could be:

$$H + HCN \longrightarrow CN + H_2 \tag{8}$$

$$CN + O_2 \longrightarrow CO + NO \tag{9}$$

$$CN + OH \longrightarrow CO + NH \tag{10}$$

$$O + NH \longrightarrow NO + H \tag{11}$$

However the definitive 'prompt' NO mechanism cannot be given as yet and must await further experimental investigations.

A second source of 'prompt' nitric oxide is the nitrogenous content of a fuel itself. Typical coals contain 1 to 2% of nitrogen by weight and in thermal decomposition yield ammonia, hydrogen cyanide, and free nitrogen as gases, while heavier organic nitrogen compounds such as pyridine and aniline are incorporated into the tars formed. It is certain that a significant fraction of this nitrogen content will be available to form nitric oxide rapidly in the combustion of coal. With fuel oils it is the highest boiling distillates which tend to have the highest nitrogen content; the weight fraction can be 0.01 to 0.02. Frequently the nitrogen in the oil exists as amine side chains or as 5- or 6-membered ring compounds such as pyridines and carbazoles. Again we may expect these to give rise to nitric oxide in combustion.

However the final conclusion of real importance for our present purposes is that in combustion systems using typical organic fuels, there is every reason to believe that nitric oxide reaches its equilibrium level at the peak temperature. It is worth noting that this need not be the case where hydrogen gas is used as fuel, because of the lack of any 'prompt' NO mechanism.

Combustion near the Walls of a Chamber

We now turn to the fact that a vast spectrum of hydrocarbon and partially-oxidized hydrocarbon molecules is detected at the exhaust of internal combustion engines in particular. Figure 42 shows a portion of the range. The species detected at exhaust are largely unrelated to the original nature of the fuel and represent a considerable degree of fragmentation thereof. Table 5.12 shows the distribution between the various classes of compounds found at the exhaust of a typical automobile engine.

The normal fuel for the automobile engine is petrol (gasoline), which is roughly the 310 to 470 K fraction from crude petroleum. In general, there are something of the order of 2000 different molecular species, mostly C_4 to C_9 paraffins, olefins, and aromatics, in gasoline, the proportions varying markedly with source. However examination of the major 'unburned' hydrocarbon species in Table 5.12 shows that a large degree of degradation has occurred within the engine chamber, with C_1 to C_3 species by far the most prominent at exhaust. Further there is a much enhanced proportion of olefins and aromatics.

Now as we have seen in our calculations of the equilibrium compositions of combusted systems, there should be essentially zero concentrations of any such hydrocarbons. Moreover it is impossible to conceive that these could form in the course of the exhaust process.

In fact the 'hydrocarbons' (including aldehydes, etc.) originate within the boundary layers which must exist near the walls of any combustion chamber, so that the system is not completely homogeneous as we have assumed in our equilibrium calculations. The high heat capacity of the walls means that there must be layers of gas in their vicinity which are at much lower temperatures than the bulk of the combusting mixture. The heat transfer to the relatively cold walls suppresses and alters the kinetics so that only partial degradation and oxidation of the fuel molecules is achieved, i.e. the combustion is wall-quenched. Convincing evidence on this point has been obtained in experiments where, using propane as fuel in an internal combustion engine, gas samples drawn off from the boundary layers contained ethylene, acetylene, propylene, and methane (Daniels, 1967). As we shall mention later, it is possible that a considerable formation of carbon monoxide also takes place in the boundary layer.

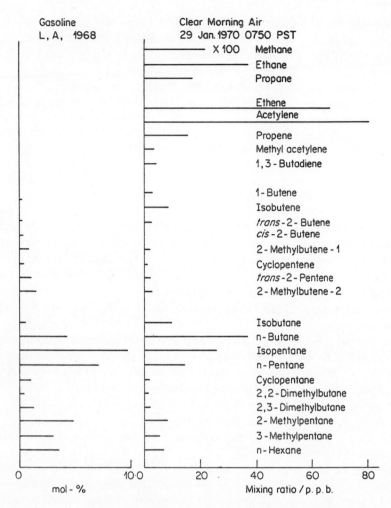

Figure 42. Spectrum of hydrocarbon species in gasoline and in unreacted morning air at Riverside, California, the large bulk of which originate in automobile engines. From E. R. Stephens and M. A. Price, 'Comparison of synthetic and smog aerosols', *Journal of Colloid and Interface Science,* **39,** 279 (1972). Reproduced by permission of Academic Press, Inc., New York.

Knock and the Role of Lead

Autoignition (also referred to as preignition, knock, or pinking) arises in an internal combustion engine largely on account of the finite velocity of flame propagation from the spark zone. A compression shock wave precedes the passage of the flame front through the mixture; the resulting adiabatic compression and heating can lead to the localized crossing of ignition boundaries of the type represented in Figure 41, with subsequent spontaneous ignition and irregular

Table 5.12. Typical exhaust composition for automobile engine

Gross composition (% by volume)		'Unburned hydrocarbons' (parts per million by volume)	
CO_2	9.0	Methane	178
O_2	4.0	Ethylene	231
H_2	2.0	Acetylene	104
CO	1.0	Propylene	89
NO	0.06	1-Butene + isobutene	80
SO_2	0.006	1,3-Butadiene	40
		Benzene	25
		Aldehydes (mainly HCHO) + many others in trace amounts	400

burning rather than smooth flame propagation. The external effects are unpleasant noise and loss of power, together with potential mechanical damage.

We shall see in Chapter 6 that a high compression ratio is desirable, at least from the point of view of maximizing power output and the efficiency of fuel energy utilization. Unfortunately there is a greater tendency to induce knock with higher compression ratios. Table 5.13 lists the compression ratios at which knock occurs typically for some fuels and also the *octane number* which serves as a measure of the knocking tendency for a given fuel. The *octane number* is defined as the percentage by volume of isooctane in its mixtures with n-heptane which has the same knock resistance as the fuel in question.

It is evident that currently-used compression ratios of around 9 for typical automobile engines could hardly be used but for the use of agents where the addition raises the octane number. Of such *antiknocks*, the potential use of methanol with octane number 106 has been mentioned in Section 5.2, but by far the most commonly used today is tetraethyl lead, $Pb(C_2H_5)_4$.

Table 5.13. Limiting compression ratios for knock resistance and octane numbers of some fuels

Fuel	Limiting compression ratio (at 600 r.p.m.)	Octane number
Isooctane (2,2,4-Trimethylpentane)	6.6	100
n-Hexane	3.0	26.0
2,2-Dimethylpentane	5.55	95.6
2-Methyl-2-butene	6.45	84.7
Cyclohexane	4.7	77.2
Toluene	11.35	100.3
Ethylbenzene	8.2	97.9

From Research Project 45, *18th Annual Report of the American Institute of Petroleum*, 1956. Reproduced by permission of the American Petroleum Institute.

Although the detailed mode of action of lead as an antiknocking agent is not fully understood as yet, some important aspects are reasonably clear. Studies of the thermal decomposition rates of the closely-related tetramethyl lead in shock tubes (Homer and Hurle, 1972) have shown that lead atoms must be generated in leaded combustion mixtures well before the possibility of spontaneous ignition within the adiabatic compression front arises. Moreover it was found that when air was present in the shock-heated gases, a lead monoxide smoke or aerosol formed very rapidly indeed; when air was used as the diluent with 0.5% by volume tetramethyl lead, the half life for the formation of the smoke was of the order of 10 μs for temperatures around 1000 K. This is a much shorter time scale than the induction periods for hydrocarbon—air mixtures to explode spontaneously.

It is considered that the action of the lead monoxide aerosol as the effective antiknocking agent arises from the ability of its surface to destroy chain-propagating atoms and radicals such as H and OH heterogeneously. With the introduction of this additional termination step to compete with chain-branching propagation, the mixture lacks the ability to explode until the arrival of the flame front itself overwhelms the resistance to propagation created by the presence of the aerosol.

Radiant Energy from Combustion Systems

One of the most notable visual characteristics of a typical flame is the emission of light. For example the normal Bunsen flame has a bluish-green inner cone surrounded by a pale bluish outer zone.

When the flame emission for a hydrocarbon fuel is analysed spectroscopically, it is found that species such as CH, C_2 and OH are principal emitters, and are identified by their characteristic spectral bands. Figure 43 shows the typical spectrum obtained from a methane—air flame.

The major components of the flame gases, such as N_2, CO_2, and H_2O are transparent across the visible region; therefore they have vanishing absorption coefficients and by Kirchhoff's law (Sections 1.3 and 3.1), do not emit thermal radiation in the visible. This is emphasized when a luminescent flame is produced where the yellowish emission of soot particles is reasonably intense. Solid carbon absorbs strongly across the visible region of the spectrum (black surface) and hence by Kirchhoff's law will emit near black-body radiation with a high emissivity. Thence the typical non-sooting flame has no effective mode of visible radiant-energy loss from the major component gases. Only in the infrared spectral region, within the vibrational bands of CO_2 and H_2O, is there a significant output of radiant energy. For a small Bunsen flame with normal full aeration, about 18% of the heat of combustion is lost by radiation, mainly in the infrared; in the visible and ultraviolet regions the heat loss by radiation is only around 0.4%. However under the high pressure conditions within a combustion chamber, much of this radiant energy is trapped so that the radiative cooling tends towards insignificance and the true temperature is not far short of the adiabatic flame temperature.

The spectacular visible emissions from flames with organic fuels under non-

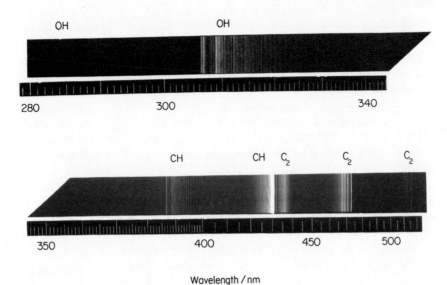

Wavelength / nm

Figure 43. Typical spectrum of the inner core of a premixed methane–air flame, with the major emitting species identified. (Courtesy of Dr. E. Collinson, University of Leeds.)

sooting conditions have their origins mainly in chemical excitation and departure from full chemical equilibrium. Chemical excitation means that the emitting state (electronically excited) is produced directly by a chemical reaction, rather than by the ground state acquiring the energy of excitation from its environment. Frequently bimolecular reactions are concerned, such as:

$$C_2 + OH \longrightarrow CO + CH^*$$
$$CH + O_2 \longrightarrow CO + OH^*$$

The asterisks in these equations denote electronic excitation, generally in the range 200 to 400 kJ mol^{-1}. These excited species have short radiative life times, of the order of 10^{-6} to 10^{-9} s, and will radiate photons in transitions usually to their ground electronic states, e.g.

$$CH^* \longrightarrow CH + h\nu$$

However in flame gases at atmospheric pressure, the electronically-excited species may suffer some 50 or more collisions with other molecules during their radiative lifetime and there is usually a finite probability for non-radiative deactivation, represented as:

$$CH^* + M \longrightarrow CH + M$$

and the electronic energy of CH* is degraded to thermal energy of M, representing any other molecule. Hence it would be expected that a greater proportion of the total rate of production of electronically-excited species would give rise to what is termed *chemiluminescence* in the visible region under lower total pressure conditions.

Our principal interest in the visible and ultraviolet chemiluminescence lies in whether or not this leads to effective radiative cooling. From what was said above, it is anticipated that the answer would be negative, but this can be put on a more quantitative basis.

In a stoichiometric propane–O_2 flame at atmospheric pressure, absolute measurements of the total emission intensity from the electronically-excited state of C_2, denoted by C_2^*, produced between wavelengths of 400 nm and 600 nm, have shown that 4×10^{-7} photons are produced for each O_2 molecule consumed (Gaydon and Wolfhard, 1950). Taking the average wavelength as 500 nm and converting to a molar energy basis through the Planck equation (1.1) (incorporating the Avogadro number 6.023×10^{23}), we obtain a radiative energy output through C_2^* of 0.096 J per mol of O_2 consumed in the flame. From the data of Table 5.1, it is found that the molar heat of combustion of propane is 2215.4 kJ, which corresponds to 5 mol of O_2 consumed on the basis of the stoichiometric equation:

$$C_3H_8 + 5\,O_2 \longrightarrow 3\,CO_2 + 4\,H_2O$$

Hence the total energy released in the flame on this basis is 443.1 kJ for one mol of O_2 consumed, of which the 0.096 J radiated in the C_2 bands is only some $2 \times 10^{-5}\%$.

In an acetylene–O_2 stoichiometric flame run at 3 atm total pressure, the yield of light in the C_2 band system is 1.35×10^{-4} photons per O_2 molecule consumed (Gaydon and Wolfhard, 1950), which corresponds to 0.032 kJ per mol of O_2. The total energy released per mol O_2 consumed on the basis of the stoichiometric equation is 519 kJ. Hence the C_2 radiation represents a radiative loss of energy of only $6 \times 10^{-3}\%$ of the total released. Similar considerations apply for CH and OH emissions in this flame. In general we need pay little attention to such chemiluminescence with regard to energy production from a fuel.

It is also possible to demonstrate that the intensities of such chemiluminescences represent departures from full equilibrium in the flame, which is however totally insignificant for our present purposes. In the stoichiometric acetylene–O_2 flame at atmospheric pressure, the light yield in the C_2 band system has been measured as 3.6×10^{-5} photons emitted per molecule of O_2 consumed (Gaydon and Wolfhard, 1950). Now the fundamental relationship between the rate of emission of C_2^*-derived photons (I) and the concentration of C_2^* within the flame front is:

$$I = k_r[C_2^*] \times V \tag{5.7}$$

where k_r is the reciprocal of the radiative lifetime (i.e. k_r is the radiative rate constant) and V is the volume of the flame front. The volume of the flame front in this case was estimated at 2×10^{-7} dm^3. The number of O_2 molecules entering the flame was 5.38×10^{20} s^{-1} and, using a radiative lifetime of 8×10^{-9} s for C_2^*, the concentration of C_2^* was obtained as 1.29×10^{-9} mol dm^3. The temperature of the flame was measured as 3320 K, so that the corresponding partial pressure of C_2^* was

3.4×10^{-7} atmospheres. If this is to correspond to a full equilibrium situation, the partial pressures of C_2^* and the ground state C_2 must be related through the Boltzmann equation:

$$p_{C_2^*}/p_{C_2} = \exp(-230000/RT) = 2.43 \times 10^{-4},$$

where the excitation energy of C_2^* with respect to C_2 is $230\ kJ\ mol^{-1}$. Hence we predict a partial pressure for C_2 of 1.4×10^{-3} atm and if we are correct in the assumption of full equilibrium, this should agree with the partial pressure of C_2 obtained from the full equilibrium calculation according to the methods outlined in Section 5.3. However, that turns out to be 5×10^{-6} atm. Hence we are forced to conclude that the concentration and partial pressure of the excited state C_2^* are well in excess of what they should be for a full equilibrium situation.

However in comparison with the total energy of the flame situation, the extent of departure from full equilibrium and the energy involved therein which produces the chemiluminescent emission, is insignificant, as is the radiative cooling so produced.

To summarize the conclusions of this section, it may be said that kinetic factors impose few limitations within a typical combustion chamber. In the general case therefore, the full equilibrium composition is applicable at the threshold of the exhaust expansion processes, which is the topic of the next section. As we shall see, kinetic factors are more significant there, in contrast to the essentially thermo-dynamic situation within the chamber.

5.5 The Exhaust Expansion of Combustion Products

The final stage of any combustion process is the expansion of the high temperature and pressure combustion mixture to atmosphere. It is during this stage that the energy is extracted as useful work, which may be in terms of work done against a piston or turbine, or heat exchange as in steam generation for work to be extracted subsequently. Alternatively propulsion may be gained through conservation of momentum considerations as in a jet engine or rocket.

The exhaust process invariably involves a very rapid decrease in total pressure, so rapid that the expansion is adiabatic, i.e. the time scale is too short for effective heat exchange with the atmospheric gases. Therefore it will be the adiabatic gas laws which will govern the concurrent fall in temperature of the combustion products. For the high temperatures concerned we may apply the ideal gas laws with confidence.

The form of the adiabatic gas law which will be appropriate is:

$$P_1^{1-\gamma} \cdot T_1^{\gamma} = P_2^{1-\gamma} \cdot T_2^{\gamma} \tag{5.8}$$

where P_1 and T_1 identify one state in the expansion process, and P_2 and T_2 identify another state, γ is the ratio of the molar heat capacities for constant pressure (C_p) and constant volume (C_v), i.e. $\gamma = C_p/C_v$. For the assumption of ideal gas behaviour we have the relationship:

$$C_p - C_v = R \tag{5.9}$$

γ is dependent upon the nature of the molecule in question, the number of atoms, the vibrational frequencies, and the rate at which energy modes can be activated under changing conditions. Since combustion products are a highly complex mixture of different small species, it will be well nigh impossible to consider the adiabatic expansion quantitatively without access to extensive computer facilities. Moreover, the chemical adjustments to the changing conditions may release or abstract energy and increase or decrease the total number of species present.

Our approach to the exhaust expansion will therefore be simplified of necessity, but our objective remains the gaining of an insight into the critical features which produce the composition finally emitted to atmosphere.

Nitric Oxide

The experimental observations of the nitric oxide level as a fraction of the total exhaust gas on a volume basis show that this is emitted to the atmosphere substantially at the level achieved within the combustion chamber, i.e. the exhaust expansion produces no effective change.

We may gain an understanding of the factors which produce this situation for nitric oxide by considering an adiabatic expansion of air itself from an initial high-temperature equilibrium composition. The initial conditions are taken to be a temperature of 2500 K and a total pressure of 70 atm, which are close to the conditions in the real internal combustion engine.

First we must calculate the equilibrium composition of the high-temperature and high-pressure air, involving species N_2, O_2, NO, O, and N. We ignore the argon and other minor constituents, and for simplicity take the molar fractions of 0.2 for O_2 and 0.8 for N_2 at the outset.

From *JANAF* thermochemical tables we obtain the values of the K_p equilibrium constants for 2500 K appropriate to the equilibria:

$$N_2 + O_2 \rightleftharpoons 2\,NO \qquad K_p(5)$$
$$O_2 \rightleftharpoons O + O \qquad K_p(2)$$
$$N_2 \rightleftharpoons N + N \qquad K_p(6)$$

These have the values $K_p(5) = 3.51 \times 10^{-3}$, $K_p(2) = 2.07 \times 10^{-4}$ and $K_p(6) = 8.51 \times 10^{-14}$ for 2500 K, all on the basis of a standard state of 1 atm pressure. Table 5.14 shows the partial pressures and corresponding concentrations at the commencement of the expansion.

In the real exhaust expansion processes the rate of fall of temperature is extremely large: something of the order of 1 K per μs is typical. So we shall consider our present high-temperature air to be subject to such a cooling rate.

The appropriate value of γ which should be applied must be considered. The values of C_p given in *JANAF* tables, combined with equation (5.9), yield for N_2, the major component, $\gamma = 1.29$ at 2500 K, $\gamma = 1.31$ at 1500 K, and $\gamma = 1.4$ at 298 K. The variation reflects the increasing activation of vibration as an effective heat-capacity mode as temperature increases. Because of the high vibrational

Table 5.14. Equilibrium parameters for air (assumed 20% O_2, 80% N_2) for total pressure of 70 atm and temperature of 2500 K

	Species				
	N_2	O_2	NO	O	N
Partial pressure/atm	55.19	13.19	1.61	0.05	2×10^{-6}
Concentration/mol dm^{-3}	0.268	0.064	0.00784	2.43×10^{-4}	9.7×10^{-9}

frequency of the N≡N bond (7.08×10^{13} Hz), even at 2500 K the vibrator contributes only about 91% of its full equipartition value. The result of measurements on the vibrational relaxation rate of N_2 in shock tubes (Millikan and White, 1963) suggest relaxation times of the order of 60 μs at 70 atm and 2500 K. Reference to the real exhaust process for combustion products, where there are species such as CO_2 and H_2O which should considerably reduce the vibrational relaxation time for N_2, suggests that in our approach with air, in order to be compatible with the real expansion process, we should consider N_2 to be fully vibrationally relaxed under all conditions. We shall use $\gamma = 1.29$ as an average for the high temperature range but it should be noted that the appearance of $1 - \gamma$ as a power in equation (5.8) makes the choice of the value more critical than might first be apparent.

If the expanding mixture is to remain in full equilibrium in passing from P_1, T_1 to P_2, T_2, then the rates of all chemical reactions concerned must be high enough to effect the required changes in equilibrium partial pressures within the time available. If this is not the case, then a non-equilibrium composition for the state P_2, T_2 will be produced. The reactions which have rates too low to effect the projection of full equilibrium states through the expansion are termed 'frozen'.

Two general types of reactions will be of importance in our simplified system.

The first of these is the three-body reaction of the type mentioned in connection with the 'overshoot' phenomenon within a flame front in the preceding section. Of specific interest to us here will be the reaction:

$$O + O + M \longrightarrow O_2 + M \tag{12}$$

With temperature decreasing throughout the expansion, this reaction will be responsible for maintaining the equilibrium between oxygen atoms and molecules. If it should happen that the rate of reaction (12) required to keep pace with the changes in partial pressures demanded by $K_p(2)$ is too high on the basis of the value of the rate constant k_{12}, then the $[O]/[O_2]$ ratio will become that for a supra-equilibrium for all stages of the expansion after the 'freezing' of the reaction (12). The three-body reaction rate will be governed principally by the local concentrations of the reactant species, O and M, as the expansion proceeds. Measurements of the rate constants for three-body recombinations involving atomic species have shown that these are almost independent of temperature in the high temperature range concerned here. The reverse dissociation reaction:

$$O_2 + M \longrightarrow O + O + M \qquad (-12)$$

will have a finite rate and we will therefore be considering a net rate of recombination of oxygen atoms in the course of the expansion, represented by $R_{12} - R_{-12}$, according to the equalities:

$$-\tfrac{1}{2}d[O]/dt = R_{12} - R_{-12} = k_{12}[O]^2[M] - k_{-12}[O_2][M] \qquad (5.10)$$

R_{12} and R_{-12} are then the instantaneous rates of the forward and reverse steps, while k_{12} and k_{-12} are the respective rate constants.

As the expansion proceeds, R_{12} must be of sufficient magnitude to effect the decrease of the ratio $[O]/[O_2]$ required for the maintainance of full equilibrium if freezing is not to occur. We can therefore investigate the ability of the equilibrium to project itself into the expansion process by comparing R_{12} with the required rate of oxygen atom recombination (R_r). If R_{12} becomes equal to or less than R_r at a given point in the expansion, then we may legitimately consider that reaction (12) is effectively frozen at that point and beyond.

The calculation might be performed as follows, assuming that the rate of fall of temperature is $1 \text{ K } \mu s^{-1}$ and that the adiabatic gas law (5.8) holds with $\gamma = 1.29$ near 2500 K. We then calculate time scales and corresponding pressures for a specified temperature.

It would seem logical to investigate the situations for those points in the expansion process corresponding to temperatures of 2400 K, 2300 K etc., taking 10-K intervals around these. Hence within the series of calculations, we should derive equilibrium concentrations of O and O_2 for points defined by $T = 2405$ K and $T = 2395$ K. Then we should correct the concentrations derived for 2395 K for the decrease in total gas density compared with 2405 K in order to isolate the chemical component of the change in the oxygen atom concentration between the two points, which is $R_r \times 10^{-5}$ since the time interval is 10 μs. The results of such a calculation are reproduced in Table 5.15.

It is clear from the results of Table 5.15 that the three-body recombination of

Table 5.15. Equilibrium and rate parameters for air expanding adiabatically from $P = 70$ atm and $T = 2500$ K

T/K	p_O/atm	$10^5 \times [O]/\text{mol dm}^{-3}$	$R_r/\text{mol dm}^{-3} \text{ s}^{-1}$	$R_{12}/\text{mol dm}^{-3} \text{ s}^{-1\,a}$
2405	0.028837	14.609 ⎫		
2395	0.027111	13.792 ⎭	0.28	4.03
2305	0.015857	8.380 ⎫		
2295	0.014813	7.861 ⎭	0.18	1.25
2205	0.007841	4.3326 ⎫		
2195	0.007284	4.0432 ⎭	0.104	0.29
2105	0.003645	2.1095 ⎫		
2095	0.003363	1.9559 ⎭	0.061	0.052

[a]Based upon a temperature independent value of $k_{12} = 7 \times 10^8 \text{ dm}^6 \text{ mol}^{-2} \text{ s}^{-1}$.

oxygen atoms effectively 'freezes' very early in the expansion process, at a temperature above 2100 K, when the total pressure is in excess of 30 atm. The underlying reason is that whereas R_{12} is not limited by any change in its rate constant, k_{12}, with falling temperature, the much steeper drop in total pressure is coupled with a sharp temperature dependence of the equilibrium constant $K_p(2)$. R_{12} has a cubic dependence upon concentration; even if the ratio $[O]/[M]$ did not vary through the expansion, R_{12} would decrease by a factor of 8 for a twofold decrease of total density (achieved by 2100 K). However at the same time we can see from Tables 5.14 and 5.15 that the oxygen atom equilibrium concentration has decreased by a factor of almost 12 from 2500 K to 2100 K. Hence the total factor by which R_{12} will have decreased for equilibrium compositions is approximately 12 x 12 x 2 or 300 over this range. It can be seen in Table 5.15 that the decrease of R_r is considerably less (a factor of around 6) over this range. Therefore it is hardly surprising that oxygen-atom concentrations for equilibrium are not projected far out into the expansion, and the concentrations thereafter are frozen at supra-equilibrium levels.

Although we have concentrated entirely upon the recombination of oxygen atoms through reaction (12) here, the final conclusion is of general validity: three-body reactions freeze early in the expansion of high-temperature air at rates comparable to the exhaust expansion of internal combustion engines. We may also extrapolate the present conclusion to the real combustion exhaust gas and state that there also any three-body reactions will rapidly become ineffective in modifying the composition.

The second general type of reaction occurring during the adiabatic expansion of air from a high-temperature equilibrium is the bimolecular 'shuffle' reaction. The terminology 'shuffle' is applied to these reactions since they do not lead to the removal of atoms or other reactive species in total but merely convert one into another. In the case of high-temperature air, we are concerned with the elementary reactions of the Zeldovich cycle (preceding section). For clarity, these may be listed again:

$$O + N_2 \longrightarrow NO + N \tag{5}$$

$$N + NO \longrightarrow N_2 + O \tag{-5}$$

$$N + O_2 \longrightarrow NO + O \tag{6}$$

$$O + NO \longrightarrow O_2 + N \tag{-6}$$

The rate constants are denoted by k_5, k_{-5}, k_6 and k_{-6} respectively. The 'shuffle' nature can be seen in that reactions (5) and (−6) convert an oxygen atom into a nitrogen atom, while the reverse happens in (−5) and (6). At the same time reactions (5) and (6) generate nitric oxide while reactions (−5) and (−6) remove it. Thus, if these reactions can proceed at significant rates in the course of the adiabatic expansion, they could lead to an alteration in the level of nitric oxide at exhaust compared to that in the high-temperature equilibrium at the start. As mentioned earlier, the experimental observations suggest that this does not happen, so we must seek an explanation of this.

Thermodynamic considerations predict that the level of nitric oxide for fully equilibrated air at 300 K is negligible. The relevant equilibrium constant, $K_p(5)$, decreases from a value of 3.5×10^{-3} at 2500 K to 7.6×10^{-31} at 300 K, illustrating the force of this point. Therefore we must conclude that kinetic factors are responsible for the significant amounts of nitric oxide emitted to the atmosphere by combustion engines.

The first step is to establish the temperature variations of k_5, k_{-5}, k_6 and k_{-6}. We note that the pairs k_5 and k_{-5}, and k_6 and k_{-6} are related through the equilibrium constants for the systems;

$$O + N_2 \rightleftharpoons NO + N \qquad K_5 = k_5/k_{-5} \qquad (5.11)$$

$$N + O_2 \rightleftharpoons NO + O \qquad K_6 = k_6/k_{-6} \qquad (5.12)$$

The equilibrium constants K_5 and K_6 can be obtained by combination of equilibrium constant data from *JANAF* tables. The point here is that where one of the pair of rate constants has a considerable temperature coefficient — and we would need to make a considerable extrapolation from the temperature range where it has been measured, producing some uncertainty — it may happen that the other rate constant of the pair has a much smaller temperature coefficient, so that combination with the appropriate K will yield a much more secure value for the first rate constant. Such is the case for k_5 and k_{-5}; the first has a very large temperature coefficient, indicated by an Arrhenius activation energy of near 316 kJ mol^{-1}. However k_{-5} has a value of the order of 10% of the collision frequency even at 300 K; as such it can have virtually no temperature coefficient and a value of $k_{-5} = 3.0 \times 10^{10}$ dm^3 mol^{-1} s^{-1} can be applied in our current investigation with reasonable certainty. Hence values of k_5 are derived using k_5 and k_{-5} from equation (5.11).

In the case of the second equilibrium system, both k_6 and k_{-6} have significant temperature coefficients, but the former is much less than the latter.

Table 5.16 shows the array of rate constant values for 2500 K and 2400 K; as we shall see, there is no need to consider values for lower temperatures in our present area of interest.

The values shown in this table illustrate the large degree of temperature dependence of k_5 and the lesser temperature dependence of k_{-6}.

We now proceed to test the ability of the Zeldovich-cycle reactions to match the rate of change of nitric oxide concentration required to extend the chemical equilibrium into the expansion process. The equilibrium calculation which gave rise

Table 5.16. Estimated values of the Zeldovich cycle rate constants/ dm^3 mol^{-1} s^{-1}

T/K	k_5	k_{-5}	k_6	k_{-6}
2500	3.5×10^4	3.0×10^{10}	4.8×10^9	1.6×10^6
2400	1.9×10^4	3.0×10^{10}	4.3×10^9	1.1×10^6

Table 5.17. Rate parameters $(\text{mol dm}^{-3} \text{ s}^{-1})$ for nitric oxide reactions in air expanding adiabatically from $P = 70$ atm and $T = 2500$ K

T/K	$\text{NO}/\text{mol dm}^{-3}$	R_{NO}	R_5	R_{-5}	R_6	R_{-6}
2500	7.84×10^{-3}		2.27	2.02	2.84	3.06
2405	5.6619×10^{-3} $\Big\}$	7.71	0.63	0.56	0.87	0.94
2395	5.4822×10^{-3}					

to Table 5.15 is extended to yield the nitric oxide concentrations shown in Table 5.17 above. Hence we can evaluate a parameter R_{NO} which expresses the instantaneous rate of chemical removal of nitric oxide necessary if full chemical equilibrium is to prevail at the point in the adiabatic expansion designated by a particular temperature. Again we assume a rate of fall of temperature of $1 \text{ K} \mu s^{-1}$. From the equilibrium concentrations of all the species in the expanding air and the rate constants of Table 5.16, we derive the instantaneous rates of the reactions, $R_5, R_{-5}, R_6,$ and R_{-6}, which are shown in Table 5.17.

From the data for around 2400 K it is clear that none of $R_5, R_{-5}, R_6,$ or R_{-6} are of the order of magnitude of the required R_{NO}. Therefore we conclude that the Zeldovich 'shuffle' reactions will freeze almost at the start of the exhaust process. This is emphasized by the fact that not even the equilibrium rates at 2500 K can match R_{NO} at 2400 K.

The underlying reasons for this situation can be seen by reference to preceding tables. R_5 and R_{-6} are limited largely by the low values of the rate constants k_5 and k_{-6} shown in Table 5.16 and occasioned by the relatively large Arrhenius activation energies involved. On the other hand R_{-5} and R_6 involve relatively large rate constants; rather these are limited by the common factor of the nitrogen atom being a reactant, concentrations of which are very low (Table 5.14).

Hence the overall picture which emerges for the adiabatic expansion of high-temperature, high-pressure, $N_2 - O_2$ mixtures is of all reactions, both the three-body recombinations and the bimolecular 'shuffle' reactions which could alter the nitric oxide level, freezing very early in the process. As a result we have here the kinetic explanation for the experimental observations that the nitric oxide levels measured at exhaust reflect the peak temperature reached within a combustion chamber.

Our conclusion for simple $N_2 - O_2$ mixtures turns out to apply equally to exhaust expansion of combustion products. Potential additional reactions for the generation of nitric oxide in the more-complex real systems have insignificant rate compared to even the Zeldovich reactions. For example, we could consider reactions of nitrogen atoms with carbon dioxide and water:

$$N(^4) + CO_2(^1) \longrightarrow NO(^2) + CO(^1)$$
$$N(^4) + H_2O(^1) \longrightarrow NO(^2) + OH(^2)$$

The numbers in brackets are the spin multiplicities of each species. In Chapter 8,

120

where such terminology is explained and the spin conservation rules for elementary reactions are enunciated, it will become clear that the above reactions are made improbable on this basis, i.e. they are said to be 'spin forbidden' and therefore the corresponding rate constants are very small, even at high temperatures.

Moreover the total oxygen content in the form of oxygen atoms or molecules of a real mixture of combustion products will be much lower than for the air we have considered; much of it will have been consumed by the combustion itself preceding nitric oxide formation in time, and will in this case be in the unreactive forms of carbon dioxide and water. The resultant lower concentrations of oxygen atoms and molecules in the expanding gases will cause 'shuffle' reactions to freeze to a greater extent than in our above treatment.

The experimental variation of the nitric oxide level at the exhaust of a typical combustion engine as a function of the fuel/air equivalence ratio is shown in Figure 44.

The fuel/air equivalence ratio of unity is defined as the mass of air required to effect the complete combustion of a mass of fuel, say $C_x H_y$, considering CO_2 and

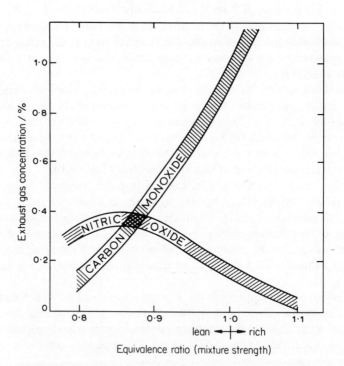

Figure 44. Typical variations of nitric oxide and carbon monoxide levels at automobile engine exhaust as functions of the fuel–air equivalence ratio. From E. S. Starkman, 'Theory, experiment and rationale in the generation of pollutants by combustion', *Proceedings of the Twelfth (International) Symposium on Combustion* (1969), p. 598. Reproduced by permission of the Combustion Institute, Pittsburgh.

H_2O as the only products, according to the equation:

$$C_xH_y + Air \longrightarrow x\ CO_2 + \tfrac{1}{2}y\ H_2O + Nitrogen.$$

For octane, C_8H_{18}, and air (20.946% O_2 by volume) the stoichiometric equation would become:

$$C_8H_{18} + 12.5\ O_2 + 47.18\ M' \longrightarrow 8\ CO_2 + 9\ H_2O + 47.18\ M'$$

where M' represents all the inert species (substantially N_2). Hence on a volume basis this stoichiometric mixture (fuel/air equivalence ratio of 1) has a volumetric air/fuel ratio of 59.68. The molecular weight of octane is 114.23, while the average molecular weight of air is 28.94. On the mass basis, the stoichiometric ratio of air to octane is therefore 15.12. If a particular mixture is rich in air, i.e. it contains more oxygen than is required to effect the complete combustion of the fuel to CO_2 and H_2O, and has, say, an air/fuel ratio by mass of 20, then the fuel/air equivalence ratio is $15.12/20 = 0.756$.

Because the exhaust expansion exerts very little influence upon the nitric oxide level produced at exhaust, we may look back to Figure 40 as representing a typical variation in molar fractions for equilibrium in the combustion chamber as a function of the fuel/air equivalence ratio. We see that this follows the same general form for nitric oxide as shown in Figure 44. The maximum molar fraction appears at an equivalence ratio of around 0.8, representing a compromise between the increased concentrations of O_2 available in air-rich systems and the decreased peak temperature resulting from the increased heat-dilution effect of the excess air. The equilibrium:

$$N_2 + O_2 \rightleftharpoons 2\ NO$$

favours nitric oxide formation with increased availability of N_2 and O_2. However from left to right it is endothermic with $\Delta H^{\ominus} = +181$ kJ $(mol\ N_2)^{-1}$ at 2500 K, producing a term $\exp(-21770\ K/T)$ in the equilibrium constant, which therefore decreases sharply with decreasing peak temperature.

On the fuel-rich side, not only is the peak temperature decreased by the dilution effect of excess fuel, but very little oxygen remains to equilibrate with nitric oxide. Accordingly the yield of nitric oxide at exhaust is less for both reasons.

Carbon Monoxide

In the case of carbon monoxide, the pollutant is emitted at exhaust from an internal combustion engine at a molar fraction lower than that which corresponds to the high-temperature equilibrium in the combustion chamber. However the emitted level vastly exceeds that corresponding to equilibrium at ambient temperature. Hence our first inclination, by analogy with nitric oxide, is to suggest that the reactions involving carbon monoxide as reactant freeze during the adiabatic expansion.

The reactions determining the change in the molar fraction of carbon monoxide during the expansion will only involve species based on the three atomic species C,

H, and O; we have already shown that nitrogen chemistry is frozen very early and there is no interaction with C or H containing species to any significant extent. Examination of Figure 40 reveals that only the species CO, CO_2, H_2O, OH, O, H, O_2, and H_2 need to be considered.

We may straightaway eliminate any three-body recombination reactions from our discussion. These may be taken as frozen at an early stage of the expansion process, as discussed in the preceding subsection. Hence we must direct our attention to 'shuffle' bimolecular reactions, eight in all, of which four are reverse steps of the other four.

$$CO + OH \underset{k_{-13}}{\overset{k_{13}}{\rightleftarrows}} CO_2 + H \tag{13}$$

$$H + O_2 \underset{k_{-1}}{\overset{k_1}{\rightleftarrows}} OH + O \tag{1}$$

$$O + H_2 \underset{k_{-2}}{\overset{k_2}{\rightleftarrows}} OH + H \tag{2}$$

$$OH + H_2 \underset{k_{-3}}{\overset{k_3}{\rightleftarrows}} H_2O + H \tag{3}$$

The rate-constant symbols appear above and below the appropriate arrow.

It will be noted that the forward steps, reactions (1), (2), and (3), are simply the branching mechanism of the hydrogen–oxygen reaction mentioned in Section 5.4.

In the case of nitric oxide, we have located the origin of the freezing of the Zeldovich reactions as being, substantially, the sharp temperature coefficient of the rate constant k_5, which can be expressed as $\exp(-37947\,K/T)$. We may use this as a basis of comparison for the temperature coefficients of the rate constants involved in the carbon monoxide 'shuffle' mechanism. Accordingly in Table 5.18 we list rate constants and the temperature coefficient parameter θ, so that the rate constants vary as $\exp(-\theta\,K/T)$ over the high temperature range in question.

Table 5.18 indicates immediately that there are no reactions in the C–H–O system which have particularly large values of θ relative to that appropriate to k_5. Consequently none of the rate constants in Table 5.18 has a particularly low value in the high-temperature regime (cf. k_5 and k_{-6} in Table 5.16). Further, there are no very sharp decreases in value of any of the rate constants between 2500 K and 2000 K. The rates of these reactions will also be dependent upon the concentrations of the reactant species; Table 5.11 shows that none of these are likely to be low enough to impose any real limitations comparable to those imposed upon R_{-5} and R_6 by the very low concentrations of nitrogen atoms (Table 5.14) in the nitric oxide reaction system.

The contents of above paragraph lay the foundations of the argument for the C–H–O system. However the increased complexity of this as compared with the nitric oxide set of reactions forces us to adopt a less quantitative approach. It is reasonable to consider that the set of bimolecular 'shuffle' reactions (13), (1), (2), and (3), and their reverse steps have large enough rates to adjust *relative* concentrations of the species involved to the changing conditions imposed by the adiabatic expansion. However, the *absolute* concentrations of the atomic and

Table 5.18. Temperature coefficient parameter (θ) and high temperature rate constants for the reactions (13), (1), (2), and (3)

	Reaction			
	(13)	(−13)	(1)	(−1)
θ	4000^a	$15,700^a$	8440	0
k for 2500 K/dm^3 mol^{-1} s^{-1}	6.5×10^8	7.3×10^8	7.5×10^9	2.5×10^{10}
k for 2000 K/dm^3 mol^{-1} s^{-1}	4.2×10^8	2.2×10^8	3.2×10^9	2.5×10^{10}
	(2)	(−2)	(3)	(−3)
θ	4750	3670	2590	10,100
k for 2500 K/dm^3 mol^{-1} s^{-1}	2.5×10^9	1.7×10^9	7.8×10^9	1.5×10^9
k for 2000 K/dm^3 mol^{-1} s^{-1}	1.6×10^9	1.2×10^9	6.0×10^9	5.3×10^8

[a] Approximate values, only applicable above 1500 K.

radical species can only be decreased to meet the full equilibrium requirements at each stage in the expansion if three-body recombination reactions have large enough rates to produce the change in composition. The argument which we developed earlier strongly suggests that this cannot be the case and that the three-body reactions freeze very early in the exhaust process.

We may now investigate the consequences of such considerations. Equilibrium constants are defined as below for the 'shuffle' reactions:

$$K_{13} = k_{13}/k_{-13} = [CO_2][H]/[CO][OH] \tag{5.13}$$

$$K_1 = k_1/k_{-1} = [OH][O]/[H][O_2] \tag{5.14}$$

$$K_2 = k_2/k_{-2} = [OH][H]/[O][H_2] \tag{5.15}$$

$$K_3 = k_3/k_{-3} = [H][H_2O]/[OH][H_2] \tag{5.16}$$

Combination of equations (5.13) and (5.16) produces another equilibrium constant, K':

$$K' = K_3/K_{13} = [H_2O][CO]/[CO_2][H_2] \tag{5.17}$$

K' is then the equilibrium constant for the water–gas shift reaction:

$$CO + H_2O \rightleftharpoons CO_2 + H_2 \tag{5.18}$$

Reference to Figure 40 shows that equation (5.18) involves the major species of a typical hydrocarbon–air combustion system.

Taking pairs of the above equilibrium-constant expressions in turn, as has been done to produce equation (5.17), it becomes apparent that a set of concentration ratios are proportional to one another:

$$\frac{[OH]}{[H]} \propto \frac{[CO_2]}{[CO]} \propto \frac{[O_2]}{[O]} \propto \frac{[H_2O]}{[H_2]}$$

This must be true both for the actual system with 'shuffle' rates able to compete with the rate at which conditions change and for the corresponding hypothetical system with full equilibrium established at each stage. The obvious conclusion, therefore, is that all species are able to maintain their full equilibrium ratios of concentrations with respect to one another in the real adiabatic-expansion situation, even if the absolute concentrations are supra-equilibrium ones on account of frozen three-body reactions. Under these circumstances, each state in the real expansion process is referred to as a *state of partial equilibrium.*

Extensive computations have been made in an attempt to predict the concentration profiles of all species throughout the expansion of combustion products. The details are beyond the scope of this book However it appears that, even with the conditions of partial equilibrium imposed (allowing a considerably higher level of carbon monoxide to appear at exhaust than would the projection of full equilibrium throughout), there is substantial uncertainty as to whether these computations can predict sufficiently high levels of carbon monoxide at exhaust to match those actually observed. We must ask accordingly what other origin could the carbon monoxide have?

In the first place it is possible that the assumption of a homogeneous mixture for the computations is significantly in error. As we have remarked earlier (Section 5.4), the presence of partially oxidized or dehydrogenated hydrocarbon molecules represents a considerable departure from a homogeneous reaction mixture and originates in the boundary layer at the walls of the combustion chamber. Of course, it is possible to regard carbon monoxide itself as a partially oxidized molecule, hence it may well be formed to some extent in the partial combustion achieved within the boundary layer.

However, there have been laboratory studies on premixed, free-standing, flames of hydrocarbons in air where, even when the flame contained sufficient oxygen in principle to convert all CO to CO_2 in practice carbon monoxide was still a final product. The principal clue as to origin was that carbon monoxide production was enhanced when the flame gases were quenched very rapidly. The water–gas shift reaction (5.18) should produce substantial conversion of CO to CO_2 if partial equilibrium is maintained throughout under these circumstances. Table 5.19 shows the variation of the equilibrium constant K' for this system as a function of temperature, bearing out this point.

We are therefore driven to the view that even partial equilibrium may break down in a sufficiently fast quenching process since the excess production of carbon monoxide in these flames must have a homogeneous origin. Reference back to Table 5.18 suggests that it is to the reactions (-13) and (-3) that we must look for any freezing during the course of adiabatic expansion, since they have the largest temperature coefficients associated with their rate constants. However both of these have major constituents, CO_2 and H_2O respectively, as reactants which should keep their actual rates high. On the other hand, reaction (1), with the lesser constituents of H and O_2 as reactants, and also with a rate constant subject to a temperature coefficient of not too much less than that applying to k_{-3}, might well be a better candidate for freezing during the exhaust expansion of combustion

Table 5.19. Variation of the water gas shift equilibrium constant, K', as a function of temperature

T/K	2500	2000	1500	1000	500	298
K'	6.09	4.52	2.56	0.69	0.00724	0.0000961

products. In this way the supply of hydroxyl radicals would become depleted below the level demanded by partial equilibrium; this would then become the limiting factor for R_{13}, the sole removal route for carbon monoxide being reaction (13) using hydroxyl radicals as the coreactant.

In conclusion to our present consideration of carbon monoxide emission levels from combustion engines, we have to admit that the detailed explanation is indeterminate. It still remains uncertain whether a homogeneous model of the gases is a satisfactory basis upon which to base quantitative interpretations.

5.6 Conclusion

In this chapter we have considered combustion from both the kinetic and thermodynamic viewpoints. Both have dominant roles at different stages. Thermodynamic criteria can be applied with confidence within the combustion chamber itself. Kinetic factors are of overwhelming importance in the subsequent exhaust process, and the emission of the principal pollutants to the atmosphere cannot be understood in thermodynamic terms. Departure from thermodynamic equilibrium need not be total under rapidly changing conditions as are encountered in the exhaust process; conditions of partial equilibrium, based upon relatively rapid bimolecular but on frozen termolecular reactions, project equilibrium ratios of concentrations of species well beyond the point at which absolute concentrations correspond to full chemical equilibrium. But there is evidence, particularly in the high levels of carbon monoxide appearing at exhaust, that not even partial equilibrium can be maintained throughout the typical exhaust expansion.

The appearance of partially oxidized and unsaturated hydrocarbon molecules at exhaust, typically about 0.2% of the gross composition of the exhausted gases, points to inhomogeneity in the boundary layers of the combustion chamber since there is no thermodynamic equilibrium basis for anything remotely of this order. It is a notable indication of the limiting kinetic factors during the exhaust expansion that these organic species are emitted to atmosphere comparatively unscathed.

Bibliography

1. 'Methanol – A Versatile Fuel for Immediate Use', by T. B. Reed and R. M. Lerner, *Science*, **182**, 1299–1304 (1973).
2. 'The Hydrogen Economy', by C. A. McAuliffe, *Chemistry in Britain* **9**, 559–563 (1973).
3. 'A Systems Approach to Energy', by W. Häfele, *American Scientist*, **62**, 438–447 (1974).

4. 'The Storage of Hydrogen as Metal Hydrides', by D. L. Cummings and G. J. Powers, *Industrial and Engineering Chemistry, Process Design and Development*, **13**, 182–192 (1974).
5. *'Flames. Their Structure, Radiation and Temperature'*, by A. G. Gaydon and H. G. Wolfhard, 3rd edition, Chapters V, VIII, IX, and XII, Chapman and Hall, London, 1970.
6. *'Flame and Combustion Phenomena'*, by J. N. Bradley, Methuen & Co. Ltd, London. 1969.
7. 'Chemistry of Pollutant Formation in Flames', by H. B. Palmer and D. J. Seery, *Annual Review of Physical Chemistry*, **24** 235–262, (1973).

Chapter 6

The Realization of Energy

We have seen in Chapter 5 how the high-pressure, high-temperature gas mixture is produced within a combustion chamber; this is the predominant present mode for the generation of energy. At the high-temperature equilibrium stage within the chamber unharnessed thermal energy has been produced. In this chapter we examine how this form of energy is converted either into work or into more useful forms of energy, and the factors which govern the efficiency of such conversion. Both static and propulsion power-generation systems are important sources of atmospheric pollutants; it will be shown that in conventional systems in current use the efficiency of fuel conversion must be compromised with the levels of pollutants produced at exhaust.

Propulsion systems in most common use are the spark-ignition piston engine of the automobile, the compression-ignition or Diesel engine of heavy transport, and the gas turbine engine used in aircraft under names of turbojet, turboprop or turbofan, or in industrial or marine settings as a turboshaft engine. These combustion cycles will be examined in some detail in Section 6.1. The cycles involved in static furnace or boiler systems are the subject matter of Section 6.2 and special attention will be given to the generation of electricity, perhaps the end-product of most importance. Apparently intractable conflict between maximizing the energy output and minimizing the hazard to the atmospheric environment arises in all of these systems. Sections 6.3 and 6.4 look a little into the future and to the possibilities of resolving this conflict by way of fluidized bed combustors and fuel cells, two particularly fruitful areas where the basic technology is fairly well understood and which are currently receiving considerable attention.

Nuclear reactors will only receive passing reference. They create a potential thermal pollution problem and there is the danger of release of radioactive gases to the atmosphere. But most of the potential pollution problems of nuclear energy production are land or water based and so are outside our present scope. It is worthwhile noting that where a reactor provides the input for a heat-exchange system raising steam, the overall principles of energy yield are much the same as for a combustive energy source.

6.1 The Internal Combustion Cycles

The Air Standard Cycle

As will become apparent in the course of this section, all of the internal combustive cycles rely upon the same general fundamental set of operations: the heat is released under either constant volume conditions or constant pressure conditions, or alternatively some combination of both in the idealized cycle. The first stage is a compression where either a fuel-air mixture or the air itself is subjected to an adiabatic increase in pressure and hence the temperature is raised. After the combustion stage (or in part during it) the gas mixture expands adiabatically, when it does work against a piston or a turbine rotor or, as in the jet engine, simply provides conserved momentum thrust. One can imagine a hypothetical final stage where, to close the cycle, the exhausted gases are returned to the starting conditions for compression in the repeated first stage.

From what was said in Chapter 5, it is evident that the detailed analysis of such real cycles will be an exceedingly complex undertaking, if only from the chemical standpoint. In practice, mechanical features and imperfections would be superimposed, e.g. friction, heat losses, gas leakage past piston rings, and incomplete scavenging. Therefore in order to gain an insight purely into the energy conversion aspects, we must simplify the modelling of the internal combustion cycles considerably. For our present purposes it would be sufficient if the model generated the basic interrelationships between power output of the cycle and the resultant chemical consequences in a semi-quantitative manner. The *air standard cycles* provide such a working approximation.

The fundamental concepts of the Air Standard Cycle (A.S.C.), corresponding to a particular real engine cycle, may be listed as follows:

(i) The working fluid is air.
(ii) All operations during the cycle are conducted *reversibly* in the thermodynamic sense.
(iii) The air remains unaltered in chemical composition throughout the course of the cycle, i.e. it is not considered to be subject to dissociation at high temperatures.
(iv) The perfect or ideal gas laws are applicable.
(v) The heat capacities, C_v and C_p, are invariant through the cycle and have their respective values for 298 K.
(vi) No heat is gained or lost by the working fluid except during the stages of the cycle prescribed for heat addition or removal, and which are imagined as being effected by an external heat reservoir with 100% efficiency of heat transfer.

Now although of course the air standard cycle does not portray at all correctly what happens in an actual combustion cycle, it does indicate the directions of changes in the energy conversion efficiency resultant upon variations in external parameters such as the compression ratio and the fuel/air equivalence ratio. It also allows the superposition of fundamental chemical consequences upon the purely physical nature of the air standard cycle.

At the outset we must define some parameters.

The *thermodynamic efficiency* (η) is defined by the equation:

$$\eta = \Sigma \Delta W / Q \tag{6.1}$$

where $\Sigma \Delta W$ is the net work done by the cycle and Q is the quantity of heat which flows *into* the working fluid in the course of the cycle, both expressed in the same unit, conventionally joules. Q may be considered as the equivalent of the heat generated by combustion in the real cycle corresponding to a particular A.S.C.

The *power output* (P) is defined by the equation:

$$P = \eta \cdot Q \cdot f \tag{6.2}$$

where f is the number of cycles per unit time. Equation (6.2) is adapted into the form of equation (6.3) for application to real engine cycles:

$$P = M_a \cdot \bar{Q} \cdot \eta' \tag{6.3}$$

where M_a is the mass of air supplied in unit time, $\bar{Q}$ is the quantity of heat generated per unit mass of air by combustion of the inmixed fuel, and η' is the actual operating efficiency of the engine, which must necessarily be less than the ideal efficiency η since actual operation is thermodynamically irreversible by definition.

The Spark-Ignition Cycle

This cycle is also referred to as the Otto cycle or the Constant Volume cycle.

The principles of operation of the spark-ignition cycle are illustrated in Figure 45.

In the actual cycle, fuel vapour and air are drawn into the cylinder on the piston downstroke and are mixed. The piston then drives upwards, compressing the gases adiabatically and effectively achieving preheating. When the piston reaches the top of its stroke, the spark is fired. Explosion in this situation takes place so rapidly that the combustion stage is essentially completed with the piston stationary, i.e. the combustion takes place under constant volume conditions for all purposes.

The *compression ratio*, given the symbol r, is defined by the ratio of the volume occupied by the gases at the bottom of the downstroke (V_1) and the volume occupied by the gases at the top of the upstroke (V_2)

$$r = V_1/V_2 \tag{6.4}$$

In typical spark ignition engines, r has a value between 8 and 10. As mentioned above the combustion takes place in a very short time compared with the time scale for piston motion; the total pressure of the gases rises sharply as a consequence of the increased temperature at constant volume and to a lesser extent, as in the case of octane fuel, as a result of the increased number of molecules in the system (see Section 5.3). Typical peak pressures are of the order of 70 atm.

The high pressures achieved in the combustion stage drive the piston downwards on the power stroke, doing the work of expansion permitted by the value of the compression ratio. Further adiabatic expansion, unharnessed, occurs when the exhaust valve opens for expansion of the spent gases to the atmosphere. In actual operation there must be a subsequent scavenging cycle where the piston is driven upwards with the exhaust valve open in order to expel a further proportion of the combustion products, before the exhaust valve closes and the inlet valve opens at

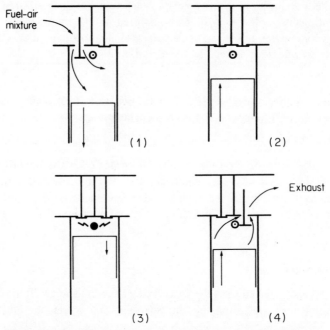

Fuel-air mixture

(1)

(2)

Exhaust

(3)

(4)

Figure 45. Simplified representation of the stages of the operating cycle of a constant volume combustion engine: (1) Downstroke. Intake of mixture; (2) compression upstroke; (3) spark ignition at top of stroke; (4) exhaust upstroke following power downstroke. Arrows indicate directions of flow or movement. Spark plug indicated by circled dot.

the top of the stroke, and a new charge of fuel—air mixture is drawn into the cylinder on the downstroke, ready to begin the next combustion cycle. It should be noted that the cylinder is never completely scavenged of the products of the previous combustion stroke, obviously a source of minor inefficiency.

In the corresponding air standard cycle, this second scavenging — recharging stroke is not considered. In the A.S.C., where air alone is the working substance, we consider that the same charge is used over and over again, with the rejection of waste heat to an external reservoir.

The A.S.C. corresponding to the spark ignition cycle is represented on a pressure—volume diagram in Figure 46. Stage 1—2 corresponds to the initial adiabatic compression induced by the piston upstroke, when work W' must be done on the air. Stage 2—3 corresponds to the combustion at constant volume achieved in the real fuel—air cycle; in the A.S.C., it is considered that an amount of heat Q, is added from an external reservoir. Stage 3—4 corresponds to the power stroke, when the air does work W in its adiabatic expansion. Stage 4—1 is the hypothetical completion of the cycle, where it is imagined that the air is restored to its initial state by the rejection of a quantity of heat Q' to the external reservoir.

Let us now analyse this cycle thermodynamically, using P, V, and T respectively to represent total pressure, volume, and temperature of the air, with number suffixes indicating the point on Figure 46. We shall use the ideal adiabatic gas laws [such as equation (5.8)] with the ratio of the heat capacities γ taken as 1.4.

Stage 1–2: Adiabatic compression. Since by definition no heat is exchanged with the surroundings, the First Law of thermodynamics states that the internal energy increase of the air, ΔU, must be equal to the work, W', done on it. Since the heat capacity at constant volume, C_v, is defined as $(\partial U/\partial T)_v$, this means that we can write the equation:

$$\Delta U_{1-2} = W' = C_v(T_2 - T_1) \tag{6.5}$$

for this stage.

Stage 2–3: Heat addition at constant volume. The quantity of heat, Q, added from the external reservoir is equated to the increase in the internal energy of the air through the First Law since no work is involved. Hence the equation is:

$$Q = \Delta U_{2-3} = C_v(T_3 - T_2) \tag{6.6}$$

Stage 3–4: Adiabatic expansion – power stroke. By analogy with state 1–2 and equation (6.5), the equation governing this stage is:

$$W = -\Delta U_{3-4} = C_v(T_3 - T_4) \tag{6.7}$$

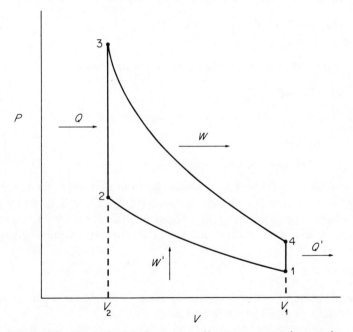

Figure 46. Air standard cycle corresponding to constant volume engine cycle.

The negative sign indicates that the loss of internal energy appears as the work done by the air.

Stage 4–1: Rejection of heat at constant volume. The quantity of heat, Q', must be rejected to the external reservoir in order to restore the air to state 1, as is made clear by Figure 46. By analogy with stage 2–3, the relevant equation is:

$$Q' = -\Delta U_{4-1} = C_v(T_4 - T_1) \tag{6.8}$$

In the set of equations (6.5) to (6.8) we have considered W, W', Q and Q' all as positive quantities. However if we are to consider the net work done *by* the cycle, $\Sigma\Delta W$, then it is clear that we must change the sign of W', so that it is effectively negative work done *by* the air. Hence the net work done by the air in the course of the cycle is:

$$\Sigma\Delta W = W - W' \tag{6.9}$$

Now the thermodynamic efficiency (η), as defined by equation (6.1), for this air standard cycle is:

$$\eta_i = \Sigma\Delta W/Q = (W - W')/Q \tag{6.10}$$

We here apply the subscript i to η in order to make it clear that this is the ideal thermodynamic efficiency, corresponding to the carrying out of all the stages in the A.S.C. reversibly, i.e. infinitely slowly. We may substitute for W, W', and Q from equations (6.7), (6.5), and (6.6) respectively as follows:

$$\eta_i = \frac{C_v(T_3 - T_4) - C_v(T_2 - T_1)}{C_v(T_3 - T_2)} \tag{6.11}$$

Cancelling out C_v and regrouping the numerator terms leads to:

$$\eta_i = 1 - (T_4 - T_1)/(T_3 - T_2) \tag{6.12}$$

From equations (6.6) and (6.8) this may be expressed as:

$$\eta_i = 1 - (Q'/Q)$$
$$= 1 - (\text{Heat rejected/Heat supplied}) \tag{6.13}$$

Such a relationship might have been postulated from first principles at the outset.

We may now develop relationships between T_1, T_2, T_3, and T_4 on the basis of the ideal adiabatic gas law. Since $P.V^\gamma = $ Constant while at the same time $P.V = R.T$ for 1 mol of an ideal gas, it follows that there is an alternative form of the adiabatic gas law:

$$T.V^{\gamma-1} = \text{Constant} \tag{6.14}$$

Therefore in the A.S.C. above, the temperatures and volumes at the ends of the adiabatic stages 1–2 and 3–4 are related as:

$$\frac{T_2}{T_1} = \left(\frac{V_1}{V_2}\right)^{\gamma-1} = \left(\frac{V_4}{V_3}\right)^{\gamma-1} = \frac{T_3}{T_4} \tag{6.15}$$

since by virtue of the constant volume stages 2–3 and 4–1, $V_2 = V_3$ and $V_1 = V_4$.
Substitution of equation (6.15) into equation (6.12) then yields:

$$\eta_i = 1 - (T_1/T_2) = 1 - (V_2/V_1)^{\gamma-1}$$
$$= 1 - (1/r)^{\gamma-1} \tag{6.16}$$

by inserting the definition of the compression ratio, r.

Thus we come to the conclusion that the ideal efficiency of the air standard cycle corresponding to the spark-ignition cycle is determined entirely by the initial compression stage and the compression ratio achieved therein. At first sight, it is a somewhat surprising to find that the efficiency with which work is developed from the heat supplied, concerning stages 3–4 and 2–3 respectively, is governed only by the work which is done on the air in stage 1–2. Moreover it is equally surprising perhaps that the efficiency is in no way related to the magnitude of the heat addition, Q, corresponding to the heat released by combustion in the real engine situation. But the key to understanding this is the fact that the compression ratio does govern the power stroke stage 3–4 in the sense that it limits the extent of the adiabatic expansion which may be applied in doing useful work. It is clear from Figure 46 that the air is still above atmospheric pressure at point 4 and is therefore, in theory, capable of yielding more work. However the mechanical arrangement cannot permit this, so that part of the heat content of the air must be rejected in the form of hot air at the exhaust.

Let us now consider the states which are explored in the course of the A.S.C. A typical engine will use $r = 8$; this may be substituted into equation (6.16) with $\gamma = 1.4$, to evaluate the ideal efficiency.

$$\eta_i = 1 - (1/8)^{0.4} = 0.565$$

Thus we see that even under these most idealized circumstances no more than 56.5% of the inherent energy of a fuel can be converted into useful work, the remainder appearing at exhaust as waste heat.

Let us specify starting conditions at point 1 of $P_1 = 1$ .atm and $T_1 = 300$ K. If we use the adiabatic gas law to follow the state of the air up stage 1–2, using $V_1/V_2 = 8$, the values of T_2 and P_2 are obtained as shown in Table 6.1 below.

In order to make Q realistic in magnitude, let us consider octane, which would

Table 6.1. State parameters for the air standard cycle of Figure 46 with $r = 8$ and $Q = 79.65$ kJ (mol air)$^{-1}$

Point	Pressure/atm	Temperature/K
1	1.0	300
2	18.4	689
3	120.7	4521
4	6.6	1968

burn stoichiometrically according to the equation:

$$C_8H_{18} + 12.5\ O_2 + 50\ N_2 \longrightarrow 8\ CO_2 + 9\ H_2O + 50\ N_2$$

taking air for simplicity as 20% O_2 and 80% N_2. Reference to Table 5.1 shows that this will produce a heat release of 47.9 kJ g^{-1} of fuel, i.e. 5461 kJ per mol of octane or per 62.5 mol of air. Hence for 1 mol of air the heat released is 87.37 kJ. However this is on the basis of the formation of liquid water so that it would perhaps be better to use $Q = 79.65$ kJ per mol of air for the water formed as vapour. We use $C_v = 20.79$ J mol^{-1} K^{-1}, the value for 298 K and, within the terms of the A.S.C. approximation, this is considered as invariant with temperature. Under these circumstances, equation (6.6) leads to the equality:

$$Q = 79650 = C_v(T_3 - T_2) = 20.79(T_3 - T_2)$$

and $T_3 - T_2 = 3832$ K. This is unrealistically large because the A.S.C. does not involve dissociative equilibria. Since T_3 and V_3 are now established, the ideal gas law yields the corresponding P_3.

The adiabatic stage 3–4 is dealt with in the same manner as was 1–2 above, with $V_4/V_3 = 8$, thus evaluating P_4 and T_4.

Table 6.1 shows the final results of this A.S.C. calculation. The value of $T_4 = 1968$ K makes very clear the point that, even at the end of the power stroke, the air entering the exhaust process retains a substantial part of the energy released by combustion, thus accounting for the thermodynamic efficiency of the cycle being only 56.5%.

At this stage we should wish to examine how the A.S.C. will compare with the corresponding real fuel–air cycle. It will be appreciated that the calculations involved in the latter are exceedingly complex but can be performed on an idealized basis, still considering the stages as being performed reversibly and ignoring any mechanical factors, but taking account of variable heat capacities and heat capacity ratios as functions of temperature and composition, and also allowing for the inevitable high-temperature dissociative equilibria. The details of such calculations are beyond our present scope but Figure 47 illustrates the final results for octene, C_8H_{16}, as fuel, with the corresponding air standard cycle shown for comparison.

As anticipated the A.S.C. peak temperature, corresponding to T_3, of 5200 K and peak pressure, corresponding to P_3, of 125.2 atm are drastically lowered to 2794 K and 71.8 atm respectively in the ideal fuel–air cycle. We see that the principal source of the decreased peak parameters is that the equivalent of $T_3 - T_2$ is reduced from 4433 K in the A.S.C. to 2139 K in the fuel–air cycle, largely reflecting the intervention of dissociative equilibria in the latter case (cf. the CO–O_2 flame temperature calculation in Section 5.3). Moreover the equivalent of $T_3 - T_4$ is reduced from 2936 K in the A.S.C. to 1094 K in the fuel–air cycle. From equation (6.7) we can realize that this will considerably reduce the work which may be extracted from the adiabatic expansion between points 3 and 4 in the fuel–air cycle. However the value of Q (in this case 92.53 kJ per mol of air) is still derived from the stoichiometric equation with CO_2 and H_2O considered as the products of combustion. Hence we expect the ideal fuel–air cycle to show a considerably reduced thermodynamic efficiency compared with the A.S.C. In the case of Figure 47 the decrease is from 0.57 for the A.S.C. to 0.355 for the ideal fuel–air cycle.

A further fall in thermodynamic efficiency is found for the actual fuel–air cycle as conducted in an engine. In this case the thermodynamic efficiency is usually

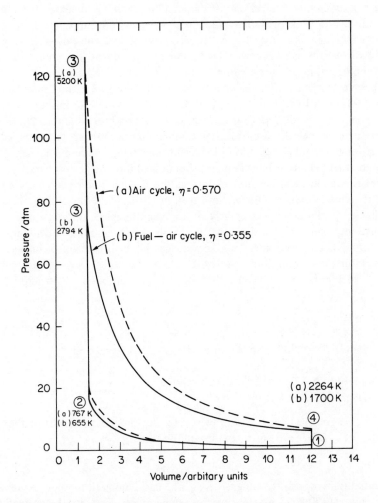

Figure 47. Comparison of air standard cycle, (a), with a calculated real fuel (octene)–air cycle, (b). The compression ratio is 8 and T_1 is 333 K with P_1 of 1 standard atmosphere. The temperatures attained at each point in the two cycles are indicated. The heat gain in stage 2–3 is 3.187 kJ per gram mass in each case. From C. F. Taylor, *The Internal Combustion Engine in Theory and Practice*, Vol. 1, 2nd ed., M.I.T. Press, 1966, p. 70. Reproduced by permission of the Massachusetts Institute of Technology.

below 0.25. Aside from mechanical factors such as friction and heat leakage, one of the main reasons for the further decrease is the irreversibility of operation at finite speed. However despite the very considerable drop in efficiency on going from the A.S.C. to the actual working cycles, the principle still holds that the higher the compression ratio used, the higher is the efficiency of fuel energy utilization.

According to the theory the efficiency of a spark-ignition engine should benefit from the use of higher and higher compression ratios. The obvious limitation in this connection is the mechanical one of the strength of the cylinder walls required to stand up to the peak pressure. In practice, for peak pressures above 70 atm, the weight of the cylinder block begins to become excessive; the power developed per unit engine weight can become unfavourable compared with that of an engine with a lighter block using a lower compression ratio.

However there are chemical penalties also in the use of higher compression ratios. In Table 5.13 it can be seen that pre-ignition in fuel–air mixtures will result if the compression ratio is increased to above 7 or so for the most available and suitable fuels. Hence the use of higher compression ratios must entail the addition of increasing amounts of antiknocking agent, inevitably at present tetraethyl lead. Hence the first penalty is increased emission of lead into the atmosphere.

The second chemical penalty for increased efficiency will be enhanced emission of nitric oxide at exhaust. As we have seen in Section 5.5, the level of nitric oxide at exhaust is determined very largely by the peak temperature reached within the combustion chamber, usually on the basis of established high-temperature equilibrium. Let us examine the consequences in this direction of increasing the compression ratio from 8 to 12 for a fixed value of Q, say corresponding to $T_3 - T_2 = 2000$ K. Let us say also that the initial conditions are $T_1 = 350$ K.

(i) $r = 8$

From equation (6.15):
$T_2/350 = (8)^{0.4} = 2.3$, i.e. $T_2 = 805$ K.

Hence $T_3 = 2805$ K and from equation (6.16) $\eta_i = 0.565$.

(ii) $r = 12$

Similarly to (i), $T_2 = (12)^{0.4} \cdot 350 = 945$ K.

Hence $T_3 = 2945$ K and $\eta_i = 0.63$.

The increase achieved in η_i is hardly spectacular, but at first sight neither is the increase in T_3. But for the nitric oxide equilibrium:

$$N_2 + O_2 \rightleftharpoons 2\,NO$$

the equilibrium constant, either K_p or K_c, has a numerical value expressed by $21.12 \exp(-21770\ K/T)$ for this range of temperatures. For 2805 K the equilibrium constant is 9.0 × 10^{-3}, rising to 1.3 × 10^{-2} for 2945 K. Thus a near 10% rise in the efficiency of heat conversion to work in the A.S.C. entails a near 50% increase in the nitric oxide equilibrium constant; ignoring any small variations in the concentrations of N_2 and O_2, this means a near 25% rise in the emission of nitric oxide as a penalty. So, even if new material for the construction of the cylinder block can be produced, having increased mechanical strength for a given weight and apparently opening up a higher range of compression ratios, the penalties in terms of the enhanced output of pollutants make this of dubious merit. In Chapter 8 we shall see that such enhanced nitric oxide emission is thoroughly undesirable on the basis of the increased rates of photochemical cycles which would ensue.

An alternative approach which might suggest itself is the use of higher compression ratios to increase the efficiency in conjuction with a limitation of the peak temperature by the use of leaner fuel—air mixtures, i.e. a lower effective value of Q. This course, however, reduces the power of the engine, as expressed by equation (6.3), as is shown by the following argument. The stoichiometric fuel—air mixture for octane as fuel is 1 mol of octane per 62.5 mol of air approximately (see the equation above). The total volume of the cylinder at any stage is constant. If we reduce the amount of octane to 0.8 mol, increasing the amount of air to 62.7 mol in compensation, then we shall have decreased the quantity of heat generated per unit mass of air, $\bar{Q}$, by 20%. At the same time we shall only have increased $\dot{M}_a$, the mass of air supplied in unit time, by 0.2 in 62.5, i.e. about 0.3%. Equation (6.3) indicates that the delivered power is dependent upon the product of $\bar{Q}$ and $\dot{M}_a$, which will therefore be close to 20% lower than for the stoichiometric mixture, a decrease unlikely to be significantly offset by any rise in the actual thermodynamic efficiency, η', achieved by the increased compression ratio, also a factor in equation (6.3). Moving to the fuel rich side is hardly a solution either. Not only will there be energy wastage from the exhaust emission of excess fuel but, on the basis of Figure 44, there will be greatly enhanced emissions of carbon monoxide and also 'unburned hydrocarbons', again undesirable pollutants.

One interesting development in the design of the spark-ignition combustion chamber deserves mention, since it resolves the pollution problem to a considerable extent, with only a comparatively minor penalty in terms of power output. The designers considered that the two combustion-chamber system shown in diagrammatic form in Figure 48 offered this partial solution.

In this concept the primary chamber (1) (65 to 85% of the total clearance volume) is separated from the main cylinder by a dividing orifice (3). The fuel injector (4) supplies fuel to the primary chamber only and the design and timing prevent the fuel itself from penetrating into the main cylinder (2). Air is drawn into the main cylinder through the intake port (5) during the downstroke of the piston. During the subsequent compression stroke, the air is compressed and flow through the orifice (3) occurs. It is during the compression stroke that the fuel is injected into the primary chamber. Near to the top of the compression stroke the spark (6), located near the orifice, ignites the mixture in the primary chamber, which is then necessarily fuel rich. As the resultant flame front propagates through the primary chamber, thermal expansion forces high-temperature, rich-mixture, combustion products (see Figure 40) into the relatively cool air in the main chamber. Under these conditions carbon monoxide, hydrogen, and partially-oxidized species can continue to burn under excess-air conditions. We may refer back to Figure 44 to gain an understanding of the pollutant-forming characteristics of this engine design. In the primary chamber, nitric oxide formation is suppressed by the fuel-rich conditions. However we would expect copious formation of species such as carbon

138

monoxide. The dilution effect of the excess air in the main chamber means that the peak temperature attained is greatly reduced compared to that for the combustion of a stoichiometric mixture, again inhibiting the formation of nitric oxide (the typical fuel/air equivalence ratio for this engine is around 0.7). Under normal spark-ignition, single-cylinder, conditions, mixtures this low in fuel would have very difficult ignition characteristics, but, because the combustion is induced in the local fuel-rich conditions in the primary chamber, it will propagate quite easily even when the air excess conditions of the main chamber are encountered. Hence the carbon monoxide and 'unburned hydrocarbons', emanating from the primary chamber at high levels, are almost completely consumed in the main chamber. Figure 49 shows the typical nitric oxide and carbon monoxide levels detected at the

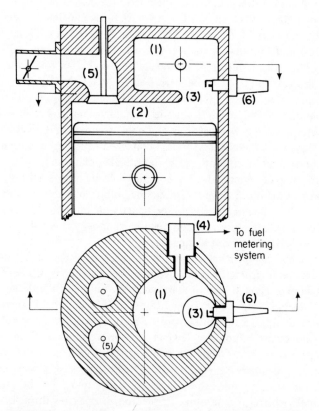

Figure 48. The design of the double combustion chamber proposed by Newhall and El-Messiri: above, sideview; below overhead view. (1) Primary combustion chamber; (2) main (secondary) combustion chamber; (3) connecting orifice; (4) fuel injector; (5) air intake; (6) spark plug. From H. K. Newhall and I. A. El-Messiri, 'A combustion chamber concept for control of engine exhaust air pollutant emissions', *Combustion and Flame,* **14**, 156 (1970). Reproduced by permission of American Elsevier Publishing Company, Inc.

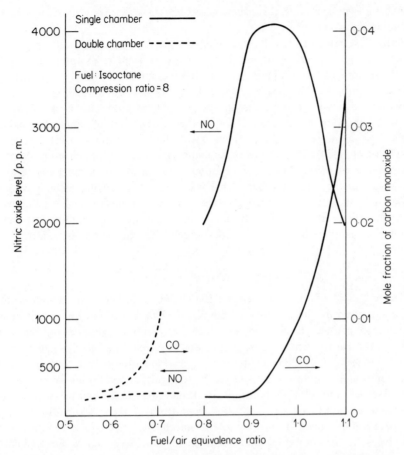

Figure 49. Comparison of the exhaust emission levels of nitric oxide and carbon monoxide from the prototype double combustion chamber engine of Figure 48 and those from a conventional single chamber engine over the operating ranges of fuel–air equivalence ratios. Based on figures in H. K. Newhall and I. A. El-Messiri, 'A combustion chamber concept for control of engine exhaust air pollutant emissions', *Combustion and Flame*, **14**, 156, 157 (1970). Reproduced by permission of American Elsevier Publishing Company, Inc.

exhaust of the two-chamber engine, with those for a single-chamber engine inserted for comparison.

Figure 49 emphasizes the operation with the much leaner mixture which is possible with the two-chamber engine. It is perhaps of significance that the extrapolation of the single-chamber results could yield similar levels for the same fuel-lean mixtures but, as mentioned above, a conventional engine cannot run on these. The two-chamber engine eases the ignition problems and could in fact be regarded as a means of allowing the engine to run on lean mixtures and achieve the pollutant emission levels expected from this operation, possibly largely independent of design. In Japan an engine has now been developed along the principles of the two-chamber concept and this has managed to satisfy the stringent 1976 American pollutant emission indices.

Shell Research in Britain have developed recently a somewhat different system to achieve lean mixture operation. This 'Vapipe' concept relies upon efficient heat exchange from the hot exhaust gases to the incoming liquid fuel to vaporize and preheat it prior to injection. It appears that this procedure enhances the burning characteristics of the lean mixture in a single-chamber conventional cylinder to the extent of allowing smooth running with the achievement of similar low exhaust levels of pollutant gases to those obtained with the two-chamber engine.

It has been reported that the two-chamber engine is capable of delivering maximum power levels of 65 to 85% of those obtained from a conventional engine of the same cylinder capacity. Moreover it has been suggested that this design possesses a higher resistance to knock than does the conventional design; higher compression ratios are therefore available without incurring significant penalties in terms of pollutant emissions which offsets to some extent the reduced power output.

The Compression-Ignition (Diesel) Cycle

The principle of the Diesel engine is that air alone is drawn into the cylinder and compressed by the upstroke of the piston. The fuel (liquid) is 'atomized' into the charge of heated air and the high temperature achieved by the initial compression is sufficient to cause spontaneous ignition. The Diesel engine therefore is based on the use of much higher compression ratios than are applied to the spark-ignition engine in order to reach the self-ignition conditions for the injected fuel. Because much higher pressures are achieved prior to ignition in the Diesel cycle, the constant-volume combustion conditions of the spark-ignition cycle are not practical in this case in view of the mechanical strength limitations of the cylinder walls. Thus the fuel injection is timed to achieve what is termed a *limited pressure cycle*.

In the idealized limited pressure cycle, the first part of the fuel combustion does occur under constant volume conditions, when the piston is at the turning point at the top of its upstroke but, as might be anticipated, the rate at which combustion proceeds under self-ignition conditions is rather slower than would result from spark-ignition; the second element of the combustion in the Diesel cycle takes place with the piston withdrawing on its power stroke. The consequent expanding volume then, in principle, allows the combustion to go to completion under constant or limited pressure conditions. On a historical note it is told that Herr Diesel took some time to perfect his timing system so that for a considerable time the neighbourhood of his workshop reverberated to enormous explosions, no doubt reflecting too constant volume combustions·

The idealized air standard cycle corresponding to the limited pressure operation is represented in Figure 50. At stage 2–3 part of the combustion occurs at constant volume, while at stage 3–4 it is envisaged as continuing under constant pressure conditions. In practice, of course, the two stages are not as clear-cut as those shown in the figure, the pathway at point 3 being much more curved in response to a more gradual transition from constant volume to constant pressure conditions. However

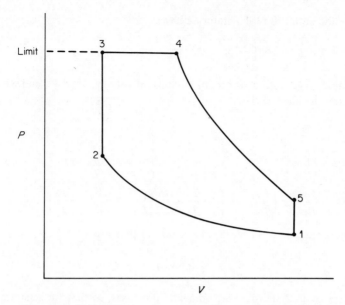

Figure 50. Air standard cycle corresponding to limited pressure engine operation.

the sharp transition pathway is more amenable to analysis so that we shall concentrate upon that, confident that points of general applicability will emerge.

At the outset of our analysis of the ideal A.S.C. thermodynamic efficiency, we shall define the compression ratio as before, $r = V_1/V_2$, and also two new parameters, $a = P_3/P_2$ and $b = V_4/V_3$, which evidently reflect the proportions of heat additions made under constant volume conditions and under constant pressure conditions respectively.

The easiest approach to the expression for η_i is to start off with the final result and prove identity:

$$\eta_i = 1 - \left\{ \left(\frac{1}{r}\right)^{\gamma-1} \cdot \frac{ab^\gamma - 1}{(a-1) + \gamma\, a(b-1)} \right\} \tag{6.17}$$

We split the complex expression inside the main brackets into more amenable components and seek to express each in terms of temperature (T) parameters.

(i) $\left(\dfrac{1}{r}\right)^{\gamma-1} = \left(\dfrac{V_2}{V_1}\right)^{\gamma-1}$

Since stage 1–2 is an adiabatic compression just as in the spark-ignition cycle, we may apply equation (6.15) so that we obtain:

$$\left(\frac{1}{r}\right)^{\gamma-1} = \frac{T_1}{T_2} \tag{6.18}$$

(ii) By the definitions of a and b we have:

$$ab^\gamma - 1 = \left(\frac{P_3}{P_2}\right) \cdot \left(\frac{V_4}{V_3}\right)^\gamma - 1$$

Now because stage 2–3 is heat input at constant volume, $V_2 = V_3$ and we may also deduce from the ideal gas law:

$$\frac{P_3}{P_2} = \frac{T_3}{T_2} \tag{6.19}$$

Thus

$$ab^\gamma = \left(\frac{T_3}{T_2}\right) \cdot \left(\frac{T_4}{T_3}\right)^\gamma = \frac{T_4}{T_3} \cdot \frac{T_3}{T_2} \cdot \left(\frac{T_4}{T_3}\right)^{\gamma-1} = \frac{T_4}{T_2} \cdot \left(\frac{T_4}{T_3}\right)^{\gamma-1}$$

Also because stage 3–4 is heat input at constant pressure:

$$\frac{V_4}{V_3} = \frac{T_4}{T_3} = \frac{V_4}{V_2} \quad \text{as} \quad V_2 = V_3 \text{ (above)} \tag{6.20}$$

Further, stages 1–2 and 4–5 are adiabatic processes, so that we generate the two forms of equation (6.14):

$$\left(\frac{V_1}{V_2}\right)^{\gamma-1} = \frac{T_2}{T_1} \tag{6.21}$$

$$\left(\frac{V_5}{V_4}\right)^{\gamma-1} = \frac{T_4}{T_5} \tag{6.22}$$

Combination of equations (6.21) and (6.22) yields:

$$\frac{T_2 \cdot T_5}{T_1 \cdot T_4} = \left(\frac{V_4}{V_2}\right)^{\gamma-1} \tag{6.23}$$

since the constant volume heat rejection stage 5–1 means that $V_5 = V_1$. Accordingly the expression developed above for ab^γ can be substituted on the basis:

$$\left(\frac{T_4}{T_3}\right)^{\gamma-1} = \left(\frac{V_4}{V_2}\right)^{\gamma-1} = \frac{T_2 \cdot T_5}{T_1 \cdot T_4}$$

to produce:

$$ab^\gamma = \frac{T_5}{T_1}$$

Hence the component of the expression for η_i is simply:

$$ab^\gamma - 1 = (T_5 - T_1)/T_1 \tag{6.24}$$

(iii) $(a - 1) = \left(\frac{P_3}{P_2}\right) - 1$ by the definition of a.

Hence from equation (6.19) we obtain:

$$(a - 1) = (T_3 - T_2)/T_2 \tag{6.25}$$

(iv) $\gamma \, a(b - 1) = \gamma \left(\dfrac{P_3}{P_2} \right) \cdot \left\{ \left(\dfrac{V_4}{V_3} \right) - 1 \right\}$ from the definitions.

On applying the identities (6.19) and (6.20) this becomes:

$$\gamma \, a(b - 1) = \gamma \left(\frac{T_3}{T_2} \right) \left\{ \left(\frac{T_4}{T_3} \right) - 1 \right\} = \gamma \, (T_4 - T_3)/T_2 \tag{6.26}$$

Now we return to equation (6.17) for η_i and substitute in the equations (6.18), (6.24), (6.25), and (6.26) developed separately above. As a result we obtain:

$$\eta_i = 1 - \left\{ \left(\frac{T_1}{T_2} \right) \cdot \frac{(T_5 - T_1)/T_1}{(T_3 - T_2)/T_2 + \gamma \, (T_4 - T_3)/T_2} \right\}$$

$$= 1 - \left\{ \frac{C_v(T_5 - T_1)}{C_v \cdot (T_3 - T_2) + C_p \cdot (T_4 - T_3)} \right\} \tag{6.27}$$

since $C_p/C_v = \gamma$.

Equation (6.27) then simply states:

$$\eta_i = 1 - \left(\frac{\text{Heat rejected (stage 5–1)}}{\text{Heat supplied (stages 2–3, 3–4)}} \right)$$

which may be compared with the general definition of the ideal A.S.C. efficiency as expressed by equation (6.13), thus proving the identity of equation (6.17). We see that in the compression-ignition cycle, although the initial compression stage 1–2 determines r and so exerts a major influence upon η_i, the following heat input stages determine a and b and so exert an effect also.

It is to be noted that the constant volume or spark-ignition cycle can be considered as but a special case of the compression-ignition cycle as far as the air standard cycles are concerned, where $b = 1$. In that event equation (6.17) reduces to equation (6.16). It is also worth noting that the term involving a and b, and multiplying $(1/r)^{\gamma - 1}$ in equation (6.17) is more than unity for the conditions $a > 1$ and $b > 1$ of the compression-ignition A.S.C.. Accordingly for the same values of r, the spark-ignition η_i exceeds that of the compression-ignition A.S.C. However as we have remarked, the compression-ignition cycle uses much larger values of r, commonly in the range 16 to 21, thus gaining compensation.

Let us investigate the compression-ignition A.S.C. by calculating both the ideal efficiency and the temperature and pressure parameters for each point in Figure 50. Typical operating parameters are $r = 16$, $a = 1.4$ and $b = 3.3$. The results are obtained in a manner similar to that used to obtain Table 6.1, and are set out in Table 6.2.

From equation (6.17) it is calculated that $\eta_i = 0.567$ under these conditions, making this cycle marginally more efficient than that of Table 6.1.

We may see on comparing the state parameters of Tables 6.1 and 6.2 that the

Table 6.2. State parameters for the air
standard cycle of Figure 50 with
$r = 16$, $a = 1.4$, and $b = 3.3$

Point	Pressure/atm	Temperature/K
1	1.0	300
2	48.5	909
3	67.9	1273
4	67.9	4200
5	7.45	1735

peak temperatures are quite similar for the spark-ignition and compression-ignition air standard cycles. This also extends to the actual fuel—air cycles. Thus it is hardly surprising that there are no marked differences between the nitric oxide emission levels of typical spark-ignition and Diesel engines. There is some evidence that the exhaust levels of carbon monoxide and 'unburnt hydrocarbon' species are considerably lower for the Diesel engine. This cannot be interpreted on any fundamental chemical basis as yet but it may reflect the fact that in the diesel engine the fuel is sprayed into the central region of the heated air and therefore may have less opportunity to reach the boundary layers close to the walls.

In principle, by virtue of the three adjustable parameters, r, a, and b in equation (6.17), η_i for the Diesel engine can be greater than that for the spark-ignition engine. For this, and other reasons beyond our present scope, it turns out that the working thermodynamic efficiencies for the real fuel—air Diesel cycle can be around 0.4 or 40%, considerably higher than the best achievable in spark-ignition engines.

The Gas Turbine or Jet Cycle

The basic design of a typical jet engine is shown in Figure 51.

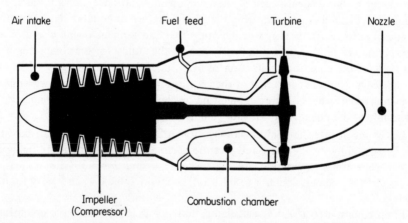

Figure 51. Basic design of a typical aircraft jet engine. Reproduced by permission of Rolls-Royce (1971) Ltd. from publication TJ 372/1, 1973.

Air is drawn in and compressed in the primary impeller stage, prior to entering the combustion chamber. The liquid fuel (e.g. kerosene of Figure 40) is 'atomized' into the compressed air and combustion takes place at constant pressure by virtue of the flowing operation. In the jet engine the subsequent adiabatic expansion of the combustion products passes through a relatively inefficient subsidiary turbine, which merely serves to power the impeller, before release into the atmosphere with a high velocity, producing an opposite momentum acting on the aircraft through its engine mountings. The industrial gas turbine is similar except that the exhaust turbine is now highly efficient and extracts most of the available gas heat content to provide shaft power, leaving a very low residual thrust.

These processes may be reduced to the simplified block diagram of Figure 52.

The air standard cycle corresponding to the gas turbine cycle is shown in Figure 53.

Typical bulk gas temperatures for the actual fuel—air cycle at the points shown in the figure are $T_1 = 330$ K, $T_2 = 560$ K, $T_3 = 1250$ K and $T_4 = 800$ K. As before the stage 4—1 represents the imaginary cooling at constant pressure required to close the A.S.C.

Analysis of the A.S.C. for gas turbine operation proceeds as follows.

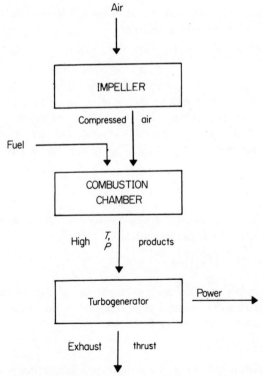

Figure 52. Block diagram representing gas turbine or jet engine operation.

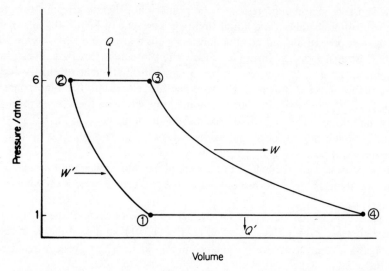

Figure 53. Air standard cycle corresponding to gas turbine or jet engine operation.

Stage 1–2: Adiabatic compression in the impeller. Here work, W', is done on the air and equation (6.5) is equally applicable:

$$\Delta U_{1-2} = W' = C_v(T_2 - T_1) \tag{6.5}$$

Stage 2–3: Heat addition at constant pressure. It is considered here that an amount of heat, Q, is transferred to the air in the A.S.C. under constant pressure conditions from the external heat reservoir. The air evidently performs work of expansion in the course of the heat addition but, since by the definition of heat capacity at constant pressure, $C_p = (\partial H/\partial T)_p$, where H is the enthalpy, we have the equation:

$$\Delta H_{2-3} = Q = C_p \, (T_3 - T_2) \tag{6.28}$$

Stage 3–4: Adiabatic expansion – Power delivery. In this stage work, W, is done by the air and equation (6.7) can be applied:

$$W = -\Delta U_{3-4} = C_v(T_3 - T_4) \tag{6.7}$$

Stage 4–1: Heat rejection at constant pressure. Let us consider that a quantity of heat, Q' is rejected to the external heat reservoir in the A.S.C. in this stage. By analogy with equation (6.28) the relevant equation is:

$$-\Delta H_{4-1} = Q' = C_p(T_4 - T_1) \tag{6.29}$$

Now from equation (6.13) the air standard cycle ideal efficiency is simply:

$$\eta_i = 1 - (Q'/Q) = 1 - \frac{(T_4 - T_1)}{(T_3 - T_2)} \tag{6.30}$$

The next task is to make use of the gas laws to obtain the relationships between these temperature parameters.

Application of the ideal gas law to the constant pressure stages 2–3 and 4–1 produces two equations:

$$T_3/T_2 = V_3/V_2 \tag{6.31}$$

$$T_4/T_1 = V_4/V_1 \tag{6.32}$$

Application of the adiabatic gas law equation (6.14) to the two adiabatic stages 1–2 and 3–4 yields two further equations:

$$\left(\frac{V_2}{V_1}\right)^{\gamma-1} = \frac{T_1}{T_2} \tag{6.33}$$

$$\left(\frac{V_3}{V_4}\right)^{\gamma-1} = \frac{T_4}{T_3} \tag{6.34}$$

The incorporation of equations (6.33) and (6.34) into equations (6.31) and (6.32) leads to the equalities:

$$\frac{T_3}{T_2} = \frac{T_4}{T_1} \cdot \left(\frac{V_4}{V_1} \cdot \frac{V_2}{V_3}\right)^{\gamma-1} = \left(\frac{T_4}{T_1}\right)^{\gamma} \left(\frac{T_2}{T_3}\right)^{\gamma-1}$$

On rearrangement this yields:

$$\left(\frac{T_3}{T_2}\right)^{\gamma} = \left(\frac{T_4}{T_1}\right)^{\gamma} \quad \text{i.e.} \quad \frac{T_3}{T_2} = \frac{T_4}{T_1} \tag{6.35}$$

Therefore on substitution of equation (6.35) back into equation (6.30), the simple expression appears:

$$\eta_i = 1 - (T_1/T_2) \tag{6.36}$$

which is identical in form to equation (6.16) for the ideal thermodynamic efficiency of the constant volume cycle.

In the case of compression by a rotary impeller it is more meaningful to define a pressure compression ratio, $r_p = P_2/P_1$, than to use the volume compression ratio, r, used before. In order to express equation (6.36) in terms of r_p, it is necessary to refer to the form of the adiabatic gas law involving P and T:

$$P_1^{1-\gamma} \cdot T_1^{\gamma} = P_2^{1-\gamma} \cdot T_2^{\gamma}$$

From this we reexpress equation (6.36) as:

$$\eta_i = 1 - \left(\frac{1}{r_p}\right)^{\gamma-1/\gamma} = 1 - \left(\frac{1}{r_p}\right)^{0.268} \tag{6.37}$$

with the latter form based upon the air standard cycle value of $\gamma = 1.4$.

Gas turbine engines normally run with fuel/air ratios of about one quarter of the stoichiometric value, i.e. with a large excess of air. This then explains the rather low bulk peak temperatures of the order of 1250 K mentioned before. Compression

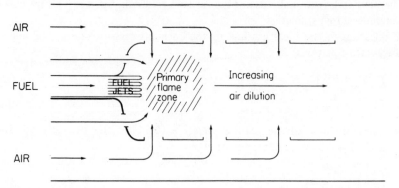

Figure 54. Diagrammatic representation of a gas turbine or jet engine flame tube showing the air injection pattern and the development of the primary flame zone.

ratios are usually relatively low, such as $r = 6$ or $r_p = 12$, under which conditions equation (6.37) predicts an ideal A.S.C. efficiency of 0.512. Actual operating efficiencies turn out to be around 0.35, thus approaching those of Diesel engines.

In view of the comparatively low bulk peak temperatures, it might be anticipated that gas turbine or jet engines would produce very low levels of nitric oxide at exhaust. Typical observations suggest that this is only realized in a minor way at best, with the nitric oxide emission indices only up to a factor of 3 lower than those for spark-ignition engines. There are however very large reductions in the carbon monoxide (factor of 10 to 100) and 'hydrocarbon' (factor of 5 to 20) emission indices, as might be expected for the very fuel lean operating conditions.

The explanation of the perhaps surprisingly high nitric oxide emissions from gas turbine engines lies in the typical design of the combustion chamber with its flame tube, shown in Figure 54. The figure shows the idealized flow patterns and lays emphasis on the fact that the compressed air enters the perforated flame tube in stages. The flame itself relies upon the turbulent mixing of the air with the 'atomized' fuel droplets. Under these circumstances both the localized burning zones around fuel droplets and the existence of a primary flame zone, which has a composition much nearer to stoichiometric than the bulk mixture, create localized temperatures approaching those of a typical combustion situation. It seems, therefore, that nitric oxide formation can proceed at least part of the way towards its equilibrium level for these localized high temperatures; once formed it cannot be significantly destroyed within the available time scale. This might then be regarded as a nitric oxide 'overshoot' phenomenon, which has a substantial degree of irreversibility in the chemical sense.

In conclusion to this section, we may say that review of the modes of generation of propulsive power has shown that at best only some 40% of the fuel energy content can be used to obtain useful work, with the Diesel engine being most efficient in this direction. Design parameters can be important in determining the pollutant emission indices, with the ability to combust fuel-lean mixtures having prime significance.

6.2 Static Power Generation

In this section we shall be interested primarily in the external combustion systems used to generate electric power. In these the heat released by combustion of the fuel in air is transferred to a different working medium, usually water/steam. Modern boilers can convert around 88% of the chemical energy of the fuel into heat content of the working medium; in our discussion we shall presume this efficiency of heat transfer and examine the subsequent processes.

The maximum theoretical thermodynamic efficiency of a water/steam working cycle is achieved in the Carnot cycle. This is represented in Figure 55 on a pressure/volume diagram.

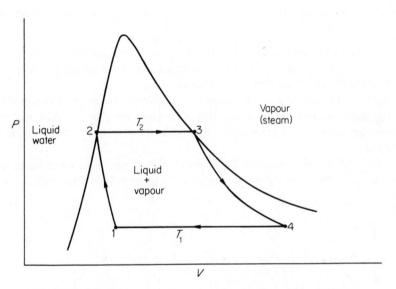

Figure 55. The Carnot cycle for water–steam as the working substance.

The general Carnot cycle consists of two reversible isothermal heat transfer processes, one at the higher temperature T_2 and the other at the lower temperature T_1. These are the processes 2–3 and 4–1 in Figure 55. The cycle is closed by two reversible adiabatic processes, 1–2 and 3–4.

It is an elementary application of the Second Law of Thermodynamics to show that the Carnot cycle represents the cycle of maximum possible efficiency for the conversion of heat into work of expansion. The Carnot efficiency, η_c, is then given by:

$$\eta_c = (T_2 - T_1)/T_2 \tag{6.38}$$

This is equally applicable to the steam cycle of Figure 55. Any deviation from the Carnot pathway will produce a lower efficiency.

150

The devices necessary to perform the Carnot cycle in Figure 55 are:

(i) A steam generator for the reversible isothermal conversion of water to steam at constant pressure, represented as 2–3. The curved boundary drawn in the figure and passing through these two points divides the phases. To the left of the left-hand limb is liquid, to the right of the right-hand limb is steam, while between the two limbs both phases coexist.

(ii) An adiabatic expander for the isoentropic expansion 3–4 during which work is extracted. Point 4 lies inside the right-hand limb and hence corresponds to partially-condensed steam.

(iii) A condenser in which reversible isothermal extraction of heat takes place, taking the system from one mixed phase at point 4 to another at point 1, which can be described as 'wet' steam.

(iv) An adiabatic compressor in which 'wet' steam at point 1 is condensed to liquid water at point 2.

It is immediately clear that a Carnot cycle will be impossible to carry out in practice, not only because it is based upon imaginary reversible processes, but because states 1 and 4 lie within the two phase region.

In fact in practical power generation it is a different cycle, the Rankine cycle, which is used, necessarily of lower efficiency than the Carnot cycle, operating between the same two temperatures. The ideal Rankine cycle is represented in Figure 56.

Process 1–2, a compression of liquid water, is performed by a small feed pump. Process 2–3 involves two stages in principle; the temperature of the water at point

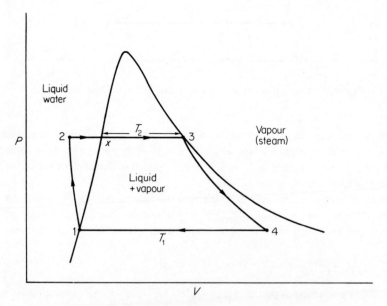

Figure 56. The ideal Rankine cycle for water–steam.

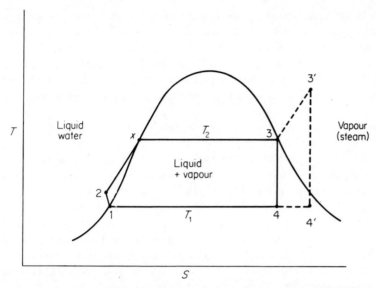

Figure 57. Rankine (full lines) and superheated Rankine (dashed lines) for water–steam on a temperature (*T*) versus entropy (*S*) diagram.

2 is well below the upper working temperature, T_2. The water is first heated at constant pressure to reach T_2 (point X) and then steam is generated at the same pressure to reach state 3. Process 3–4 is the work-yielding adiabatic expansion which produces part condensed low pressure steam through the turbine. Process 4–1 is the constant pressure heat extraction from the 'wet' steam accomplished in cooling towers, to produce liquid water at point 1.

The superheated Rankine cycle involves the superheating of the steam beyond state 3 at constant pressure and produces an improvement in the efficiency by increasing the upper working temperature. The difference between the normal and superheated Rankine cycles is shown in the temperature versus entropy pathways in Figure 57.

The normal cycle (1–2–3–4) corresponds to that represented in pressure-versus-volume form in Figure 56. The superheated version deviates around the pathway 3–3'–4'–4; it increases the mean temperature of heat input along the pathway 2–3–3', so raising the efficiency above that of the normal cycle. Superheating is carried out in practice by ducting the vaporized steam (at point 3) back through the boiler heat-exchange system.

Unlike the Carnot cycle, the expression for the thermodynamic efficiency of the Rankine cycle is not a simple one, and its evaluation involves consulting steam tables and charts for each specific case. Accordingly it is beyond our present scope, except that we may look at one specific comparison with the corresponding Carnot cycle.

A steam plant used steam produced by the superheated Rankine cycle at a pressure of $3000 \ \text{kN m}^{-2}$ with a temperature of 598 K. Expansion through a turbine brought the pressure down to 7.38 kN m^{-2}, with further passage through a condenser and a feed pump. The lower temperature was 313 K. The Carnot cycle operating between 598 K and 313 K would have an efficiency of 0.476, calculated from equation (6.38). In the case of the superheated Rankine cycle, the ideal thermodynamic efficiency comes out to be 0.337, with the difference largely due to the sacrifice involved in using constant pressure than constant temperature heat input.

Modern turbine systems use superheated steam with the upper working temperature something of the order of 800 K, when the Carnot efficiency approaches 0.6; on the basis of the above example, we might reason that the ideal Rankine efficiency would be around 0.5 under these circumstances. The actual achievement under real irreversible operation can be as high as 0.47.

Generators can convert up to 99% of the mechanical energy of the turbine into electricity. With the 88% heat-transfer efficiency mentioned earlier, the overall thermal efficiency of electric current generation can be 0.88 x 0.47 x 0.99 = 0.41 for a boiler burning fossil fuel. An average thermal efficiency for present power stations is probably around 0.33, so that some two-thirds of the fuel energy content is 'dumped' as waste heat.

Nuclear power plants generally operate at a lower maximum thermal efficiency, because present day reactors cannot be run as hot as can boilers using fossil fuel. An overall thermal efficiency of around 0.30 is probably typical for nuclear station, entailing the rejection of some 70% of the fuel energy as waste heat.

It is worth noting that the difference between the efficient fossil-fuel burning power stations and the nuclear power stations is more accentuated than the apparent difference between the wastages of 60% and 70% of the fuel energies. For plants of the same generating capacity, the nuclear version produces about 50% more waste heat than the fossil fuel plant. This arises from the fact that the nuclear station must consume 40/30-times more fuel energy to produce a specified amount of electricity; then it wastes 70/60-times more of this energy, producing a total of 40 x 70/30 x 60 = 1.56 of 'dumped' energy compared to unity for the fossil fuel plant.

It is not any law of thermodynamics which enforces the dumping of waste heat; the energy content of the 'wet' steam at the end of the adiabatic expansion, 3′–4′, in Figure 57, still represents more than 50% of the original fuel energy content over water itself at the starting temperature. This is low grade energy in the sense that it is difficult to find useful applications for water at around 310 K, as potential uses tend to have economic limitations. However the district heating scheme in use in Malmö, Sweden is of interest. This involves a less efficient turbine extraction of energy at the expansion stage with a consequent drop in the thermal efficiency of electric power generation from the potential 33% to near 19%. But hot water at near 370 K is then the by-product and about 67% of the fuel energy can be applied through insulated pipelines to heating buildings; as such this would represent an 86% utilization of the fuel energy. The length of pipeline need not be too critical; with suitable insulation temperature loss can be kept down to 1 K per 15 miles of pipe. Of course there is a large capital outlay in laying the piping system but this is justified by a two- to three-fold increase of fuel reserves.

Another suggestion of merit is for the establishment around power stations as operated at present of greenhouse assemblies for growing high value crops. In the light of our remarks on the high potential photosynthetic yields for higher latitudes in Section 4.5, one may be impressed by the ability of this scheme to provide conditions with high ambient temperature, boosted carbon dioxide levels [to overcome the absence of a CO_2-priming mechanism for C_3 plants (Section 4.4)] and abundant supplies of water (currently applied to cooling).

The static power station suffers from the limitations on thermodynamic efficiency imposed by the comparatively-low upper working temperature of the steam, while at the same time the generation of pollutants takes place in a combustion occurring at normal flame temperatures. Had the initial source of heat also a peak temperature not much above 800 K, then the emission of nitric oxides would have been vanishingly low. Moreover, because combustion in power stations provides a means of extracting the energy content of fuels with a high sulphur content without immediate undesirable side-effects, output of sulphur dioxide also tends to be high. We shall see in Section 6.3 the potential solutions offered in these respects by the fluidized bed combustor, where a much closer match between the initial combustion temperature and the upper working temperature can be achieved in principle.

In conclusion to this section, it may be noted that the application of electricity to general space heating represents a considerable wastage of fuel, hardly appropriate to the so-called 'energy crisis'. When 60 to 70% of the fuel energy content has been wasted necessarily in the conversion of heat energy into electricity in the first place, it is reprehensible for the end-use to be reconversion to heat. A reasonably efficient gas- or oil-fired central heating system can achieve double the efficiency of conversion of the fuel energy into space or water heat compared with application of electricity for the same purpose. Moreover, even in transmission from the site of generation to the site of usage, resistive losses for electricity in cables over more than a hundred miles become substantial, a further source of decreased efficiency. Between 10,000 and 15,000 MW of the present winter peak demand of electricity in Britain is applied to heating, something of the order of 30% of the total electrical power generated. The overall efficiency from the original coal to heat supplied to a living room with a twin-bar electric heater on a cold winter day is no more than 20%. Even when the transportation costs of solid or liquid fuels for domestic use are taken into account, the economics of electric space heating are thoroughly unfavourable.

6.3 The Fluidized-Bed Combustor

Looking to the future, it is to be hoped that more efficient application of the energy of a fuel to its various uses will be achieved, in conjunction with the abatement of emissions of pollutant species. In this section and the next, particular attention is focused upon two means of reaching these objectives, at least in part. Both the fluidized-bed combustor and the fuel cell are at present in the developmental stage and have yet to find widespread application. Several major problems in these areas remain to be solved but the promise is outstanding. We shall seek to gain some insight into the origins of the better potential performance characteristics of these new systems, compared with the conventional systems which have been considered in Sections 6.1 and 6.2. Such is the current research effort that substantial progress in bringing matters to successful fruition may well have been achieved by the time that this book appears in print.

Surfaces of ceramic materials, e.g. silica, will promote heterogeneous combustion of gaseous fuels and air at temperatures in the region of 1000 K. Finely-divided ceramic materials are more active in this respect and can support combustion at temperatures as low as 700 K. The actual combustion reactions occur in whole or in part upon the solid surface with adsorption prior to reaction, and the products are desorbed rapidly as gases.

This is then a different mode of effecting heat release from a fuel compared to the homogeneous gas-phase flames which have been discussed earlier. One marked difference is that the hot solid surface will have a high emissivity compared with a gas; black, or rather grey, body radiative losses will produce a significant term corresponding to radiative cooling in the heat-balance equation for such a heterogeneous combustion system [Kirchhoff's law (Sections 1.3 and 3.1)]. This contrasts with the relatively minor radiative cooling term in the heat balance equation for a homogeneous flame (Section 5.4). Hence for this reason alone the peak temperature reached in heterogeneous combustion systems would be expected to be considerably lower than that for a comparable homogeneous flame.

For an overall combustion reaction to be self-sustaining, the rates of all the elementary reactions involved must be sufficiently large at the temperature concerned. Since heterogeneous combustion can be self-sustaining at temperatures much lower than those for homogeneous combustion, it is necessary for the overall reaction on the solid surface to be much faster than the corresponding homogeneous overall reaction at the same temperature. However it is the general basis of catalysis that chemical reactions proceed much more rapidly in the presence of an active solid surface than in its absence. Hence we may expect the same general principle to be operative in combustion systems.

Fluidization finds extensive application in industrial processes, even if the underlying theoretical basis is not fully understood as yet. In a fluidized bed a mass of solid particles is made mobile by means of an upward flow of gas; the pressure drop through the bed balances its weight. The bed behaves like a liquid and the solid particles are thoroughly agitated, so producing a uniform temperature distribution and high heat-transfer coefficients between the walls or embedded pipes and the 'fluid'. In a gas fluidized bed there is a distinct minimum fluidization flowrate; the bed expands in volume to a limited extent as increasing gas flowrate induces fluidization before it becomes unstable when further increase in flowrate results in the formation of large bubbles of gas. The smaller bubbles corresponding to stable operation pass upwards through the bed in analogy with bubbles rising in a liquid. The movement of the solid particles under fluidizing conditions is rapid and apparantly random, but in reality this is only an effect engendered by the rising gas bubbles, which draw up a 'tail' of particles in their wake. The resultant upward flow of particles in one localized zone is balanced by a downward drop of particles in other parts of the bed.

In a fluidized bed of, say, sand particles supporting the combustion of a gaseous hydrocarbon fuel in air (premixed prior to entry) the motion of the particles is driven principally by the increase in volume flowrate of the gases within the bed rather than by the initial flowrates of fuel and air. For a bed operating at a temperature of 1200 K say, produced by combustion of a stoichiometric mixture of propane and air, according to the equation:

$$C_3H_8 + 5\,O_2 + 20\,N_2 \longrightarrow 3\,CO_2 + 4\,H_2O + 20\,N_2$$

The total volume flowrate increases by a factor of approximately 4 due to the temperature rise, and this is ampified by the slight increase in the total number of molecules in the products and also by the pressure gradient. A working demonstration of a fluidized-bed combustor operating on the basis of the above reaction may be established with the design shown in Figure 58.

In setting up the demonstration, a flow of propane from a lecture bottle is established, which when ignited burns as a diffusion flame at the top of the quartz chimney. The air flow is then started and upon gradual increase the flame eventually descends within the chimney to burn at the upper surface of the sand with pronounced green and blue chemiluminescent emission. This should be allowed to continue for a few minutes. If there is no sign of the surface grains becoming red and beginning to 'boil', then the flowrates of both gases should be gradually increased, preserving the correct equivalence ratio as far as possible. A useful indication on this score is the appearance of a diffusion flame at the top of the chimney. That indicates that the mixture is too rich in fuel; the air flowrate should therefore be increased until this flame just disappears. Eventually the blue—green flames at the top surface of the sand will vanish and the upper layers of the bed become agitated and red hot. With experience the fluidized combustion zone can be made to extend downwards for about 30 to 40 mm with stable performance thereafter. Under optimum conditions this demonstration can be made to proceed quietly, but more frequently there is a continuous popping and crackling due to imperfect operation; larger bubbles of the fuel—air mixture always tend to form to some extent and the noise comes from the resultant homogeneous explosions. In operation the fluidized zone is quite spectacular, with the sand taking on the appearance of molten lava.

In practical boiler operation, the fluidized bed would contain submerged pipes as the steam generation system. Conductive heat transfer to these can be made the dominant heat loss component, overwhelming the radiative losses for temperatures of 1300 K or less in general. This contrasts with the totally radiative heat-loss situation of the above demonstration.

The fluidized-bed combustor offers good heat contact between combusted gases and the solid grains of ceramic material because of the large solid surface area per unit volume and mass, and because there is no permanent channelling. On the heat transfer side there are two independent processes, namely gas to solid and from the whole bed to submerged surfaces of pipes.

In the case of heat transfer from the hot gases to the sand, the rate may be

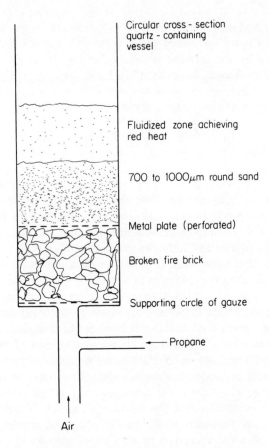

Figure 58. Bench demonstration fluidized-bed combustor design.

regarded as faster than that of the combustion reaction itself and limited accordingly. Hence we may consider that all of the gas-to-sand heat transfer is accomplished in a thinnish layer at the bottom of the fluidized part of the bed. In this circumstance, good heat transfer does not depend upon fluidization *per se*. The more critical factors are relatively low gas velocity upwards through the bed, the high heat capacity of the solid compared to the gases, and the large surface to volume ratio.

On the other hand, heat transfer from the bed to submerged pipes has a high rate as a direct consequence of fluidization. A dry-packed (i.e. non-fluidized) bed of grains is usually a good thermal insulator because there is only a limited surface contact between adjacent grains. In fact little heat could flow by conduction even if the solid of which the grains were composed had a high thermal conductivity, which silica and silicates have not. In the static packed bed, heat transfer by convection is also severely restricted because the motion of interstitial gas is restrained by the grains. Therefore if combustion could be effected within an assembly of immobile grains, then large thermal gradients would rapidly develop around submerged pipes but without effective heat transfer.

Consider a static packed bed with large thermal gradients established around submerged pipes, and what would happen if suddenly the positions of the grains were significantly changed. A hot grain from a zone remote from a pipe might arrive beside a pipe and would quickly transfer its excess heat content as compared to the pipe temperature. A cold grain from beside a pipe might arrive in a zone remote from a pipe and would then rapidly acquire the high temperature of other grains within the immediate vicinity. Thus we see that the instantaneous states produced by such mass transport have both very large localized temperature gradients with resultant high heat-transfer rates. As such these can be considered to be unsteady states.

However, if the grains are kept in continuous motion, the whole system is maintained in a permanently unsteady state, and the more rapid the grain transport, the steeper the localized temperature gradients and the greater the heat transfer rates. This is the secret of the high heat-transfer potential rates of fluidized beds, and continuous and rapid movement of relatively high heat capacity grains is the essential prerequisite. However, the effective grain residence time in a particular locality is the main determining factor, predominant over the heat capacity of the solid or the thermal conductivity of the gases.

In small scale operation, heat-transfer coefficients approaching $1000 \ W \ m^{-2} \ K^{-1}$ can be achieved to submerged pipes, compared with typical values below $100 \ W \ m^{-2} \ K^{-1}$ achieved in a gas convector boiler.

Therefore the fluidized-bed combustor concept presents the double advantage of a high heat intensity together with a relatively low peak temperature. As we have seen in the preceding sections, the formation of nitric oxide in combustion systems is sharply dependent upon the peak temperature. Consequently it is to be expected that fluidized-bed combustion units will have much reduced nitric-oxide emission indices compared with more conventional systems. The typical index for a conventional oil-fired furnace, using the normal homogeneous flame, is around 500 parts per million of the exhaust gases. However a fluidized-bed combustor operating on oil will cut this by a factor of up to 6. Typical nitric-oxide emission indices for coal-fired, fluidized-bed, combustors are in the range 60 to 180 parts per million, while for a fluidized-bed combustor using gaseous propane fuel the index may be as low as 40 parts per million. In fact it could be that the main source of nitric oxide emissions from fluidized beds is the imperfect operation represented by the explosions within larger gas bubbles, where much higher localized peak temperatures will be produced, and also the original nitrogen content of the fuel.

Of particular interest from the practical point of view is the fact that crushed coal when used as fuel generates ash grains which serve as the ceramic material of the fluidized bed. As a consequence, temperatures are kept below $1300 \ K$ in this type of fluidized-bed operation in order to prevent melting and fusion of the ash, but this condition is easily met in practice. The lower temperature limit is around 1050 K for the efficient burning of pulverized coal, otherwise high levels of carbon monoxide appear at exhaust and mean a loss of thermal efficiency. The optimum operating temperatures are in the region of 1100 to 1150 K, arising from the peak sulphur-absorbing ability of limestone additions, to be discussed in the next paragraph, occurring under these conditions. A further point is that, because the ash remains soft and non-erosive, a gas turbine can be used to extract additional energy from the hot combustion gases, which would not be possible in a conventional furnace where the ash is fused into hard aggregates. Moreover the temperatures achieved in the fluidized-bed combustion of coal are not high enough to volatilize alkali salts, again in contrast to conventional furnaces. Condensation of these salts on turbine blades would therefore be another factor in preventing the extraction of the additional energy in existing furnaces. It is considered at present that efficiencies approaching 50% for electricity generation could be achieved in such a mixed cycle, fluidized bed–gas turbine operation. With pulverized coal as fuel a continuous feed would be used, with new fuel fed in at the top and spent ash drawn off below. Experimental systems based on this principle are already in action (see, for example, Squires, 1972). Currently, it is projected that commercial operating boilers of 30 MW generating capacity will come into use in the next few years, using pulverized coal with particles up to 6 mm in diameter. It has also been demonstrated that the fluidized-bed combustor can burn coals with a high ash content (up to 75%), where use in a conventional system would cause severe clogging problems. This is of great potential value to a country like India, where much of the coal is of this type.

There is a proven capability for the almost complete elimination of sulphur emission, even for the dirtiest fuels in this connection. Attempts to retain sulphur at source in conventional furnaces by incorporation of limestone into the feed have failed generally to achieve much. However, if granulated limestone or better dolomite is added continuously to a fluidized-bed combustor, then the sulphur is near completely retained as sulphate. Dolomite is a true chemical species, $CaCO_3MgCO_3$, and in the so-called half-calcined form is represented as $[CaCO_3 + MgO]$, where there is an intimate intermingling of microscopic crystallites of MgO with the calcium compound. Because of the large surface-to-volume ratio presented in the fluidized bed and the close contact of the gas phase with the solid, the principal reaction:

$$[CaCO_3 + MgO] + SO_2 + \tfrac{1}{2} O_2 \longrightarrow [CaSO_4 + MgO] + CO_2$$

proceeds very favourably, with the possibility that MgO also reacts to some extent. Hence the sulphur content of the fuel can be retained with better than 80% efficiency and the inert calcium sulphate can be drawn off. Limestone performs best in unpressurized beds, where over 90% of the sulphur can be retained, but in this, only 60% of the limestone becomes sulphated because of pore size limitations and the extent of pore penetration into the particles.

Trace elements within the fuel are also retained by a fluidized-bed combustor, due to the involatility of the element or its compounds under the comparatively low temperature conditions. For example, vanadium is such a component of oil and can be retained with up to 98% efficiency within the solid phase of the bed.

There is also a point to be made about the relative sizes of fluidized-bed boiler systems compared to the conventional ones in current use. In the former the use of pressure above atmospheric improves the heat-transfer properties while at the same time the physical size of the combustion unit can be reduced for the same heat output. In fact the fluidized-bed boiler can be considerably smaller in principle, even with atmospheric pressure operation, than the cathedral-like boilers of present day power stations. With supercharging (i.e. operation above atmospheric pressure) the size of the fluidized-bed boiler required would be no larger than the average detached house. With the reduced need for cooling towers resultant upon the higher overall efficiency, the prospects indicate a significant improvement in the landscape around the power stations of the future, coupled with reduced thermal and chemical pollution.

With all of these outstanding advantages, it is surprising that no fluidized-bed power stations are even in the planning stage, far less operational. The basic problem hindering widespread usage is that of the controlled operation of the scaled-up fluidized bed. Large bubbles tend to form in large beds and the resultant homogeneous explosions within can eject the entire bed or damage the submerged pipes. However we can be hopeful that the problems will be overcome in the near future, when there will be a large place for fluidized-bed combustors in the generation of our energy.

6.4 Fuel Cells

The potential advantage of a fuel cell over a combustion engine lies in the fact that it converts the chemical energy of a fuel directly into electric power without the preliminary conversion into heat demanded by the latter. Thus the fuel cell avoids the theoretical limitation to its thermal efficiency imposed upon combustion or steam cycles by the laws of thermodynamics.

We may start our necessarily brief review of the principles of fuel cells by referring to the basic design of the simplest version, namely the hydrogen—oxygen fuel cell. The overall reaction achieved is simply:

$$H_2 + \tfrac{1}{2} O_2 \longrightarrow H_2O$$

the same as for the combustion of hydrogen. The difference is that the fuel cell may operate at ambient temperatures and delivers the free energy change of the reaction as electric current, with no high temperatures involved at any stage. Figure 59 shows the basic elements of the hydrogen—oxygen fuel cell.

At present this is the most highly developed fuel cell and has been applied in the American space programme. The electrolyte may be either strong alkali or strong acid. The electrodes are typically porous carbon with the catalyst, usually heavy metal oxide, nickel, platinum, or spinel material, embedded within the pores.

In point of fact the voltaic pile and its modern offspring, the 'dry' battery, are fuel cells by the strict definition that they convert chemical energy into electric

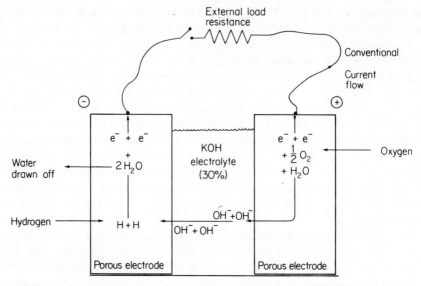

Figure 59. Schematic design of a hydrogen–oxygen fuel cell and the mechanism of electron transfers.

current directly. However the 'fuel' in these cases is an expensive metal, such as zinc or lead, which demands large refining energy derived from conventional power plants. Of course, on the basis of what was said in Section 5.2 on the production of the synthetic fuel hydrogen, it could be argued that the fuel cell of Figure 59 is not much different in this sense. It is, however, the ultimate aim to develop a fuel cell which will operate using the hydrocarbon fuels in common current usage. At the time of writing this is hardly a reality: only partial successes have been reported, and these were achieved in laboratories with highly refined fuels.

We shall concentrate our attentions in this limited examination of fuel cells principally upon the hydrogen–oxygen cell. In so doing we may hope to gain understanding of the basic general principles and of the problems which remain to be overcome.

The reaction between hydrogen and oxygen does not proceed spontaneously at ambient temperatures. A stoichiometric mixture of hydrogen and oxygen can be made up and, if left undisturbed, no water will be formed even over a time scale of years. However, if an ignition device (such as an electric spark) is applied, then explosive combustion is effected in a fraction of a second. This emphasizes the essential chemical kinetic block upon the homogeneous reaction. Thermodynamically the reaction to yield water is highly feasible at 298 K; the Gibbs free energy change ($\Delta G^{\ominus}$) is large and negative, being $-237.19 \text{ kJ mol}^{-1}$ for liquid water product. Under normal circumstances, however, there is no kinetic route for the process to be effected, but if a few grains of platinum black are allowed to come into contact with the hydrogen–oxygen mixture at 298 K, explosion can again ensue. The platinum black serves to catalyse the overall combustion reaction, i.e. it furnishes a heterogeneous route for satisfaction of the thermodynamic feasibility. It is this unblocking of the kinetic pathway by the presence of a catalytic solid surface which forms part of the theory of the hydrogen—oxygen fuel cell. The embedded catalysts render fuel and oxidant (separately) into adsorbed forms amenable to the electrochemical processes which follow.

The pores of the electrodes enable the creation of a three-phase equilibrium of gas, liquid, and solid. Capillary action determines whether a pore of a particular radius (r) is flooded by electrolyte or blown clear of liquid by the gas pressure applied from the side opposite to the electrolyte. Consider first the uniform pore shown in Figure 60. Hydrostatic equilibrium requires that the hydrostatic pressure of the electrolyte, P_2, and the capillary pressure, P_c, must balance the gas-side pressure, P_1. Now the capillary pressure depends upon the pore radius according to the relationship:

$$P_c = 2\sigma(\cos\theta)/r \tag{6.39}$$

Here σ is the surface tension of the electrolyte, and θ is the wetting angle and is close to zero for the typical electrolytes used in fuel cells. Hence the larger the pore, the smaller is the capillary pressure and the greater the likelihood that it will be blown clear of electrolyte i.e. when $P_1 > P_c + P_2$. For normal values of σ and the applied pressure $P_1 - P_2$, the critical pore radius is of the order of 5 to 10 μm in diameter. So, if the pore is too large, the capillary effect cannot resist the applied pressure and the gas simply blows through the empty pore. If a pore is smaller than the critical size, the capillary action floods it with electrolyte and blocks the passage of gas. Neither of these two extremes would prove effective in a fuel cell electrode. In practice however, one would not expect materials like carbon, or sintered or pressed materials, which form the structure of the porous electrode, to exhibit uniformity of pore radius. Rather we should expect irregular variation of cross-section even along one particular pore, together with a branching structure

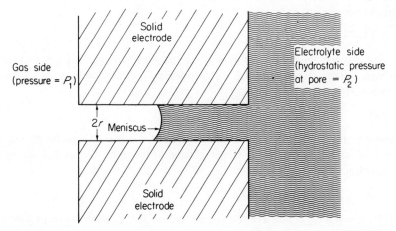

Figure 60. Pore diagram to represent the basis of operation of a gas diffusion electrode.

linking smaller pores to larger. Moreover in a working system the gas-side pressure, P_1, will fluctuate slightly about its mean value. As a result, parts of the internal porous structure are alternately flooded and blown clear. The active catalytic material is located along the pore walls and is therefore partly within these 'tidal' regions. When a pore is momentarily blown clear of electrolyte, a thin film of the liquid phase is retained, connecting with the main bulk of the electrolyte, and the catalyst there is in intimate contact with both gas and liquid phases. Further the electrodes may be waterproofed with a layer of paraffin wax, approximately a monolayer thick; this allows ions and individual water molecules to penetrate to the internal surfaces, but inhibits the flooding of the pores.

Within a partially flooded pore, the gas molecules are adsorbed onto the active catalyst prior to the electron-transfer stage whereby ions are generated. In the case of hydrogen, the adsorption stage produces chemisorbed atoms in contact with a film of electrolyte. This may be represented simply as:

$$H_2(g) \rightleftharpoons H_{ads} + H_{ads}$$

$$H_{ads} \rightleftharpoons H^+(aq) + e^-$$

where H_{ads} is the chemisorbed hydrogen atom and e^- is an electron. In alkaline electrolyte (such as 30% KOH) the hydrogen ions will react with hydroxyl ions almost instantaneously, so the overall reaction at this electrode (the anode – negative pole) may be written as:

$$H_2 + 2\,OH^- \longrightarrow 2\,H_2O + 2\,e^-$$

The oxygen electrode (the cathode – positive pole) has a considerably more complex mechanism. In alkaline electrolytes the theoretical reaction is written as:

$$O_2 + 2\,H_2O + 4\,e^- \longrightarrow 4\,OH^-$$

If this is combined with the anodic reaction for four electrons produced, it generates the overall reaction:

$$2\,H_2 + O_2 \longrightarrow 2\,H_2O$$

but the oxygen electrode reaction given above involves at least two distinct elementary steps, passing through a hydroperoxy ion intermediate, HO_2^-. The true primary reaction is:

$$O_2 + H_2O + 2\,e^- \longrightarrow HO_2^- + OH^-$$

producing only 2 electron equivalents per mol of O_2. It is only in the presence of a good catalyst for effecting the decomposition of the hydroperoxy ion according to the equation:

$$HO_2^- \longrightarrow OH^- + \tfrac{1}{2}O_2$$

that the combined effect at the cathode approaches equivalence with the overall, theoretical four-electron reaction above.

The two-step nature of the process at the oxygen electrode poses a major obstacle to the development of hydrocarbon fuel cells. Since carbon dioxide is the desired end-product, it is clear that such cells must use acid rather than alkaline electrolytes to prevent carbonate formation. But the HO_2^- ion, which may be decomposed reasonably easily in alkaline media, is much more stable in acid media. Consequently it is much more difficult to obtain the theoretical free-energy release in hydrocarbon fuel cells.

The Efficiency and Power Output of Fuel Cells

Two situations must be distinguished at the outset. If an external voltage is applied to the electrodes of a cell which just balances the electromotive force (e.m.f.) and results in zero current flow, the cell is said to be under open circuit operation and operates reversibly in the thermodynamic sense. Thermodynamic criteria alone apply under these circumstances; kinetic criteria are inapplicable since there is no net reaction rate. However, it is closed circuit operation, when the cell delivers current, and hence power, and necessarily operates irreversibly, which is of principal interest to us. There is net chemical reaction and the rates at which the various reactions proceed turn out to be limiting for the performance in large part.

As a first approach, the open circuit or reversible operation will be considered under the heading of the thermodynamics of fuel cells. Following that we shall deal with the actual operating characteristics under the heading of the kinetics of fuel cells.

The Thermodynamics of Fuel Cells

For any electrochemical cell operating reversibly the basic relationships are:

$$\Delta G = -\Delta W_{rev} = -z.E.F = \Delta H - T.\Delta S \tag{6.40}$$

Here ΔG is the Gibbs free energy change for a cell reaction where z electron equivalents of charge are passed per mol of reactant consumed. The value of z is 4, for example, for the theoretical hydrogen–oxygen cell reaction above, if ΔG is expressed with respect to 1 mol of O_2. On the other hand z would only have a value of 2 in this cell if the HO_2^- ion is stable and achieves a substantial concentration or activity, and ΔG is again expressed with respect to 1 mol of O_2. As already mentioned in Section 4.3(ii), F is the Faraday and is equal to 96,494 coulomb. ΔW_{rev} is the theoretical work done by the cell and is the maximum possible work yield because of the reversible operation. E is the e.m.f. corresponding to the voltage measured across the cell terminals with a high impedance voltmeter. In combination, E and F have units of joules per equivalent.

By the Second law of Thermodynamics for isothermal operation:

$$T.\Delta S = -q_{rev} \tag{6.41}$$

where q_{rev} is the quantity of heat evolved in the overall cell reaction for reversible operation. In general for fuel cell reactions which are practicable, $T\Delta S$ is small compared to ΔH, so that heat release or absorption is small under reversible operating conditions. Since the Gibbs free energy released by the reaction can only appear as electrical work or as heat released, this immediately provides us with the key to understanding the high efficiency of conversion of chemical energy of a fuel into electrical energy which is theoretically possible for a fuel cell. In equation (6.1) the thermodynamic efficiency was defined as:

$$\eta = \Sigma \Delta W / Q$$

where $\Sigma \Delta W$ was the net useful work done and Q was the heat provided. For a fuel cell under reversible operation, $\Sigma \Delta W = \Delta W_{rev}$ and $Q = -\Delta H$. Thus we obtain the expressions for the theoretical thermodynamic efficiency of a fuel cell:

$$\eta_i = -\frac{\Delta W_{rev}}{\Delta H} = \frac{\Delta G}{\Delta H} = 1 - \frac{T \cdot \Delta S}{\Delta H} \tag{6.42}$$

The subscript i applied to η indicates the ideality of reversible operation. For the reaction between hydrogen and oxygen to produce liquid water at 298 K, the relevant changes are $\Delta H^{\ominus} = -285.84 \text{ kJ mol}^{-1}$ (of water or hydrogen) and $T.\Delta S^{\ominus} = -48.65 \text{ kJ mol}^{-1}$. This emphasizes the wide difference in magnitude of these quantities in this typical case. Therefore the ideal thermodynamic efficiency for the reversible operation of the hydrogen–oxygen fuel cell is obtained as 0.83 or 83% upon insertion of these parameters into equation (6.42). In principle then, this fuel cell could achieve an efficiency some three-times larger than that for a combustion engine burning hydrogen as fuel.

We may also at this stage calculate the corresponding standard e.m.f. for the hydrogen–oxygen cell, using equation (6.40), noting that $z = 2$ for the theoretical reaction consuming 1 mol of hydrogen or forming 1 mol of liquid water.

$$E^{\ominus} = \frac{\Delta G^{\ominus}}{z \cdot F} = -\frac{(\Delta H^{\ominus} - T \cdot \Delta S^{\ominus})}{z \cdot F} = +\frac{237190}{2 \times 96494} = 1.229 \text{ V}$$

Here we have revealed another problem in the use of fuel cells, in that the delivered voltage of an individual cell is relatively low compared with the voltages required for propulsion, etc. Therefore a battery of fuel cells connected in series is required for such purposes. It is appropriate to point out in this connection that, under irreversible power-producing operation, the current delivered by an individual cell is proportional to the geometric area of the electrodes. Since there is an obvious limitation upon the physical size of an electrode from the practical point of view, the current is increased by parallel connection of cells. Hence the final working pack is envisaged as sets of series-connected individual cells connected in parallel across the end terminals to produce the high voltages and currents required.

Perhaps even more interesting than the hydrogen–oxygen fuel cell with its negative entropy change, largely determined by the decrease in the number of gas-phase molecules from reactants to products, are reactions where a positive entropy change is involved. For example, propylene has been used as the fuel substance in a cell operating at high temperature, where water vapour is the product in the equation:

$$C_3H_6(g) + 9/2\,O_2(g) \longrightarrow 3\,CO_2(g) + 3\,H_2O(g)$$

There is an increase in the number of gas phase molecules from five and a half on the left hand side to six on the right; as a result ΔS is a positive quantity. Reference back to equation (6.42) and remembering that ΔH is a negative quantity for this reaction shows that the theoretical efficiency can exceed unity: for this reaction at $T = 1000$ K, $\eta_i = 1.04$. Explanation comes from equation (6.41); a positive value of ΔS corresponds to a negative value of q_{rev}, i.e. heat is taken from the surroundings and converted to electrical work.

Even more spectacular hypothetical situations can be thought of, worthy of mention even if exceedingly impractical. If a reaction of solid carbon could be induced in a fuel cell according to the equation:

$$C(s) + \tfrac{1}{2}O_2\,(g) \longrightarrow CO(g)$$

this would involve a large fractional increase in the number of gas-phase molecules from one-half on the left to one on the right. Hence ΔS would be relatively large and positive, and calculation shows that the ideal thermodynamic efficiency at 1000 K is no less than 178%!

We have now anticipated the operation of fuel cells at temperatures well above ambient. Obviously water-based electrolytes cannot be used, and the electrolyte is usually a fused salt such as sodium carbonate or lithium carbonate, which melts at 800 K or above. The great potential advantage in high-temperature operation of a fuel cell is kinetic rather than thermodynamic in origin, but the raising of the temperature alters the theoretical efficiency on the basis of the T factor in equation (6.42). Since ΔH is always negative, the direction of change of η_i with temperature is determined entirely by the sign of ΔS. A positive ΔS, as in the case of the propylene cell reaction above, results in increasing efficiency with rising temperature. A near zero ΔS, corresponding to a reaction like:

$$CH_4 + 2\,O_2 \longrightarrow 2\,H_2O(g) + CO_2$$

with equal numbers of gas-phase reactants and products, gives rise to an ideal efficiency close to unity, invariant with temperature. Finally a negative ΔS, such as that for the hydrogen–oxygen reaction, produces a decreasing ideal efficiency with rising temperature. In most practical cases ΔS is either zero or negative, so that there is no thermodynamic advantage in high temperature operation and often a slight penalty.

Finally on the thermodynamic side, the effect of increased operating pressure needs to be considered. The differential Gibbs relationship:

$$dG = +V.dP - S.dT$$

becomes the incremental form for a reaction taking place at constant temperature:

$$\Delta G = + \Delta V \cdot dP \tag{6.43}$$

Thence

$$\left(\frac{\partial \Delta G}{\partial P}\right)_T = + \Delta V$$

and since $\Delta G = -z \cdot E \cdot F$:

$$\left(\frac{\partial E}{\partial P}\right)_T = \frac{\Delta V}{z \cdot F} \tag{6.44}$$

where ΔV is the change in volume from reactants to products. Making the ideal gas assumption, where Δn is the change in the number of moles from left to right in the conventional form of the chemical equation:

$$\Delta V = \Delta n \cdot R \cdot T/P$$

Upon integration of equation (6.44) between pressures P_1 and P_2, we obtain the e.m.f. difference:

$$E_{P_2} - E_{P_1} = -\frac{1}{z \cdot F} \int_{P_1}^{P_2} \frac{\Delta n \cdot R \cdot T}{P} \, dP$$

$$= -\frac{1}{z \cdot F} \cdot \Delta n \cdot R \cdot T \cdot \ln(P_2/P_1) \tag{6.45}$$

In the hydrogen–oxygen fuel cell operating at 298 K, the effect of increasing the pressure from 1 to 70 atm may now be calculated. If we consider the product to be liquid water, then $\Delta n = -3$ in combination with $z = 4$. Hence upon inserting the numbers into equation (6.45) we obtain:

$$E_{70} - E_1 = \frac{3 R \cdot T}{4 F} \cdot \ln 70 = 0.082 \text{ V}$$

Hence compared to the standard reversible e.m.f. of 1.229 V, the effect of increased pressure is relatively small. Even then it is not all gain in terms of efficiency since work has to be done on the gases in the compression stage. Further the containing casing needs to be stronger and inevitably heavier, making the overall power-to-weight ratio less favourable.

The Kinetics of Fuel Cells

The limitation to the current density delivered by a fuel cell on load is of kinetic origin. In turn this provides an upper limit to the power which may be drawn, since the power is the product of voltage and current. The parameter used to measure the rate of electrochemical reactions is the current density (i) or current delivered per unit area of electrode surface, presumed uniform. Thus if V is the voltage corresponding to a current density i from an electrode of geometrical area A, then the power delivered (P) is given by:

$$P = V.i.A \tag{6.46}$$

The overall electrochemical process occurring at the porous electrode of a fuel cell can be resolved into a series of distinct stages, any one of which can be rate determining in closed circuit operation.

(i) The gas-diffusion stage. In the instances of the hydrogen electrode discussed in the preceding section, the gas must diffuse from the gas side through the pores to the active sites. In the case of a sufficiently rapid consumption of the hydrogen, the partial pressure within the pores will become less than that in the bulk gas and the overall rate will be limited by the mass transport of gas through the pores.

A more complex situation arises when the reactant gas is consumed to yield a product gas. This creates a countercurrent diffusion problem, where the product gas must diffuse out against the flow of the incoming reactant gas. An example of this situation arises in the anode of a carbon monoxide fuel cell, where the product is carbon dioxide.

(ii) The chemisorption stage. The gas within the pores in contact with the active sites must pass to the chemisorbed form, which will usually involve atomization, as in the case of hydrogen, or fragmentation in the case of hydrocarbons. The Arrhenius activation energy for chemisorption of hydrogen onto a fuel cell catalyst such as platinum can be small enough to allow a relatively high rate at ambient temperatures. However the chemisorption of hydrocarbons frequently involves a much larger activation energy barrier, perhaps as high as 40 kJ mol^{-1}. Such imposes a severe limitation upon the rate of chemisorption which may be achieved at low temperatures so that this stage can become rate determining for the overall electrode process. Furthermore in a hydrocarbon fuel cell using aqueous electrolyte, the water can inhibit chemisorption of the fuel by blocking active sites. The advantages of using hydrocarbons in a high temperature fuel cell with fused salt electrolyte become more obvious; at temperatures of 800 K or more the activation energy barrier will be considerably less limiting for the rate of chemisorption while the non-aqueous electrolyte lessens the inhibition of active sites.

(iii) The electron-transfer stage. Once the fuel or oxidant species has been chemisorbed on a site which is in contact with both the gas and liquid phases, electron transfer from or to the liquid phase generates or destroys ions and opens the gateway to the electrochemical reactions. The contact between the liquid and the active surface must result in the creation of an electrical potential gradient in the immediate vicinity of the surface. This occurs by way of a flow of charge on the first establishment of contact until the potential difference across the boundary layer of the liquid phase compensates for the difference in chemical potentials of a species in one phase relative to the same species in the other phase. As a result there is a separation of charges across the interface, establishing a so-called electrical double layer. In effect this acts like a capacitor where the electrode forms one plate

and the layer of charges in the solution forms the other plate. From our point of view, the important element here is the creation of a resistance to transfer of electrons across the interface between the phases, clearly vital to the operation of a fuel cell.

In an electrochemical reaction occurring at an interface, the electrical potential at any point along the reaction coordinate would be expected to affect the free energy of any transient state involving an ion or an electron. Thus the total free energy of any intermediate state may be resolved into the sum of a chemical term and a term dependent upon the local electrical potential. At some point along the reaction coordinate a maximum free energy is attained, and this represents the effective barrier to reaction. It is very clear that the height of this barrier must depend upon the extent of the charge separation at the interface, termed *polarization*. Therefore we may conclude that the rate of overall electrochemical reactions, limited by the rate of the electron-transfer stage, will depend upon the degree of polarization and that this will appear as a limitation to the delivered current density.

We can predict the effect of polarization upon the terminal voltage simply by realizing that a cell delivering current is operating irreversibly in the thermodynamic sense. Also the more current is drawn, then the greater is the degree of irreversibility. Reference back to equation (6.40) shows the relationship between the maximum electrical work obtainable and the e.m.f. for open circuit operation. The second law of thermodynamics tells us that if a system is operated irreversibly, i.e. at finite rate, then the work obtainable from the system is less than that obtainable for reversible operation, i.e. infinitely small rate. Substituting the terminal voltage (V) in place of the reversible e.m.f., the relationship becomes:

$$W_{irrev} = z \cdot V \cdot F < W_{rev} = z \cdot E \cdot F \quad \text{i.e.} \quad V < E \tag{6.47}$$

Therefore we perceive that the effect of polarization, which may be regarded as a manifestation of irreversible operation of the cell, will be that lower terminal voltages will be available the greater the current density drawn from the cell.

Detailed analysis of polarization effects is beyond the scope of this book, but we may quote the final results with some advantage. The fuel cell is composed of two half cells, each of which possesses its own polarization characteristics. A parameter termed the *overpotential* is invoked, which is the difference between the operating potential $[V(\frac{1}{2})]$ and the reversible potential $[E(\frac{1}{2})]$ for the half cell:

$$\text{Overpotential} = V(\frac{1}{2}) - E(\frac{1}{2}) \tag{6.48}$$

The overpotential then simply expresses how far away from the equilibrium or reversible condition the half cell is. In most electrochemistry textbooks, overpotential is given the symbol η, but to avoid confusion with our use of this symbol for thermodynamic efficiency, we shall retain $V(\frac{1}{2}) - E(\frac{1}{2})$ to denote overpotential. The construction of equation (6.48) satisfies the sign convention for half-cell potentials. The anodic reaction proceeds as an oxidation and so has a negative conventional reduction potential. $V(\frac{1}{2})$ is then less negative than $E(\frac{1}{2})$ so that anodic overpotential is positive. Correspondingly cathodic overpotential is negative. It therefore becomes clear that the two half-cell potentials move together as increasing current densities are drawn. The anode becomes less negative and the cathode less positive. The difference between the two is naturally the terminal voltage of the whole cell.

Even under the heading of chemical polarization, the developed overpotential may originate from distinguishable steps within the general electron-transfer stage. The magnitude of the overpotential and its detailed variation as a function of current density are dependent upon the precise nature of the rate-determining elementary step in the process, but the general form of equation:

$$V(\frac{1}{2})-E(\frac{1}{2}) = a + b.\ln i \tag{6.49}$$

with a and b as empirical constants is often found to hold over considerable ranges of the current density (i) for practical power delivery situations. This will be the condition of main interest to us in our concern with energy technology.

When it is the electron-transfer elementary step itself which is rate determining, equation (6.49) is known as the Tafel equation. But when the rate-controlling step is another, such as the reaction between discharged ions on the surface, the form of equation (6.49) often still applies, but with different values of a and b. Perhaps it is best for us to regard equation (6.49) as an empirical form, useful for describing the results of chemical polarization, without enquiring into its theoretical origin within the detailed mechanism.

(iv) The liquid diffusion stage. When the ion gas been generated at the active site, it must then diffuse through the flooded part of the pore and thereafter from the electrode face into the bulk of the electrolyte. In the reverse case, the ion will move in from the bulk of the electrolyte and enter the pore. These are liquid diffusion phenomena and it can happen that the rate here is too slow to cope with the rate of ion production or discharge. Under such rate-determining conditions, a concentration gradient will develop between the interface and the bulk of the electrolyte. This gives rise to the phenomenon termed concentration polarization, to be clearly distinguished from the chemical polarization discussed above. In concentration-polarized half cells there is a distinct limiting current density, denoted i_L. In the circumstance where ions are moving in the electrode to be discharged at the active sites, this will evidently occur when the concentration of ions at the electrode surface falls to zero. In the reverse case of ion production, the ion concentration tends towards infinity in the vicinity of the surface.

It is conceptually easier to take for discussion the case of ions being discharged at the surface. There will be a layer of thickness δ extending from the contact surface into the bulk electrolyte across which the concentration of the ions in question increases from its value C_s at the surface to the bulk concentration C_b. To be exact we should consider activities of ions rather than concentrations. The rate of diffusion of ions to the electrode will be proportional to the activity gradient, which is taken as $(a_b - a_s)/\delta$, with a denoting activity in the location indicated by the subscripts. The current density corresponding to the diffusion flux is:

$$i = \frac{D \cdot z \cdot F}{\delta} (a_b - a_s) \tag{6.50}$$

where D is the diffusion coefficient. As the current density increases it is clear that a_s must tend to zero; thus i_L is produced by putting a_s equal to zero in equation (6.50) so that we obtain:

$$i_L = \frac{D \cdot z \cdot F}{\delta} \cdot a_b \qquad (6.51)$$

The corresponding half cell potential, $V(\frac{1}{2})$, is then developed on the basis that the rates of the processes actually operating across the electron transfer interface are fast enough for the application of localized equilibrium criteria, namely the Nernst equation. For zero current density the bulk activity, a_b, extends to the contact surface and the Nernst equation takes the form:

$$E(\tfrac{1}{2}) = E^{\ominus} + \frac{R \cdot T}{z \cdot F} \cdot \ln a_b \qquad (6.52)$$

where $E^{\ominus}$ is the standard e.m.f. for unity activity. When current is drawn and the activity of the ion falls to a_s at the contact surface, the analogous Nernst equation becomes:

$$V(\tfrac{1}{2}) = E^{\ominus} + \frac{R \cdot T}{z \cdot F} \cdot \ln a_s \qquad (6.53)$$

From these two equations the concentration polarization overpotential is given by:

$$V(\tfrac{1}{2}) - E(\tfrac{1}{2}) = \frac{R \cdot T}{z \cdot F} \cdot \ln(a_s/a_b) \qquad (6.54)$$

From equations (6.50) and (6.51) in combination we derive:

$$a_s/a_b = [1 - (i/i_L)]$$

and with substitution into equation (6.54), the final expression is:

$$V(\tfrac{1}{2}) - E(\tfrac{1}{2}) = \frac{R \cdot T}{z \cdot F} \cdot \ln[1 - (i/i_L)] \qquad (6.55)$$

This equation gives the relationship between the overpotential of a half cell and the delivered current density, when the rate-determining step in the overall electrolyte.

In equation (6.51) it is evident that the smaller is the thickness of the diffusion layer, δ, then the lower in magnitude will be the overpotential for a given current density, and hence the higher is the power which may be obtained. In general, for porous electrodes the diffusion zone of consequence is confined in large part to the meniscus which the electrolyte forms in each pore. This is one origin of the high power delivery potential of fuel cells because the meniscus value of δ is very small, often of the order of 10^{-7} m, compared with the normal diffusion layer thickness of 10^{-3} to 10^{-4} m for the planar electrodes of more conventional electrochemical cells. Therefore, in general, the limiting current density for a porous electrode is very much larger than that for a planar electrode.

The typical forms of plots of overpotential and half-cell potential versus current

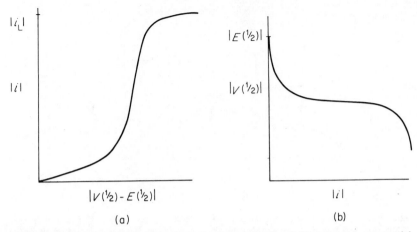

Figure 61. Generalized forms of plots of current density ($|i|$) versus overpotential ($|V(\frac{1}{2}) - E(\frac{1}{2})|$) (a) and potential difference ($|V(\frac{1}{2})|$) versus current density (b) for a half-cell. The modulus bars remove the conventional signs.

density are shown in Figures 61 (a) and (b). Under normal circumstances chemical polarization is the dominant source of overpotential for moderate current densities; it is only for high current values, when i begins to approach i_L, that concentration polarization becomes limiting. With chemical polarization dominant, the current density (i) grows exponentially with overpotential, according to the recast form of equation (6.49):

$$i = \exp[(V(\tfrac{1}{2}) - E(\tfrac{1}{2}) - a)/b]$$

Once concentration polarization becomes important at high current densities, the form tends to the recast equation (6.55):

$$i = i_L \langle 1 - \exp\{z \cdot F \cdot [V(\tfrac{1}{2}) - E(\tfrac{1}{2})]/R \cdot T\} \rangle$$

It is the transition from the first of these forms to the second which produces the inflexion in the curves plotted in Figure 61.

When two half cells are combined to produce the working cell, the polarization characteristics of each combine to produce the actual terminal voltage, V, as shown in Figure 62(a). This emphasizes the fact that the half-cell polarization characteristics are unrelated; in the figure the cathode half-cell potential is shown to be governed by concentration polarization, while the anode is only moderately chemically polarized at current densities just below that at which the terminal voltage becomes zero. Figure 62(b) shows the delivered power density, which is simply the product $V \times i$ from Figure 62(a), as a function density. From this we see that there is a definite maximum power density which may be obtained from a fuel cell, and evidently the operation of the fuel cell with current densities in excess of that corresponding to the maximum power density will be self-defeating.

In the hydrogen—oxygen fuel cell it is chemical polarization at the oxygen electrode (cathode) which causes the main voltage loss. The hydrogen electrode can

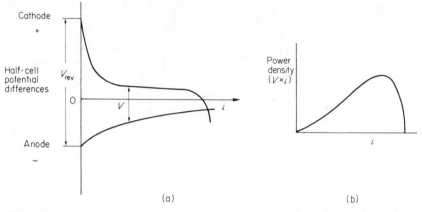

Figure 62. (a) Typical fuel-cell polarization characteristics, where V is the delivered voltage and i is the delivered current density. V_{rev} is the reversible operation potential difference ($i = 0$). (b) Corresponding typical power density ($V \times i$) variation as a function of current density (i).

be run with almost no polarization. Often the voltage loss due to polarization at the oxygen electrode is more than 25% for any reasonable current density in the aqueous-electrolyte, low-temperature, cell. However, since these losses must arise from energy barriers in the reaction coordinate, it will be anticipated that the scale of polarization will be considerably reduced in the high temperature versions of the hydrogen—oxygen fuel cell. Even then the irreversible nature of the operation when current is drawn results in decreasing thermodynamic efficiency with increasing current density. This is illustrated by some performance characteristics for a hydrogen—oxygen fuel cell listed in Table 6.3 below. This cell was run at 473 K and at a pressure of 28.5 atm; the electrolyte was 45% KOH.

The data in the table indicate that the current density for maximum power output per unit cell weight has not been reached at the highest current density approaching 20,000 A m^{-2}. The energy efficiency has dropped below 50% at this level so that the fuel cell is tending towards the same order of efficiency which may be attained in conventional engines, for example the 40% operating efficiency of a diesel engine (Section 6.1). It follows that one must be cautious in advancing the view, deriving from the high theoretical efficiencies for reversible operation with $i = 0$, that a fuel cell must achieve a much higher energy efficiency than other power sources. However it is clear that the hydrogen—oxygen fuel cell does provide a means for achieving uniquely-high efficiencies of utilization of the chemical energy of the fuel when only moderate current densities are drawn; this entails a considerable sacrifice in terms of the power output per unit weight.

The energy effectively wasted by an operating fuel cell appears as heat rejected to the surroundings and the electrical energy yield is less than $-\Delta G$. The heat evolved may be resolved into two terms. Firstly there is the relatively small reversible-operation heat term of equation (6.41), which may be either loss or gain depending upon the sign of ΔS for the reaction concerned. Secondly, on the basis

Table 6.3. Performance characteristics of a hydrogen–oxygen fuel cell under load ($T = 473$ K, $P = 28.5$ atm) (45% KOH electrolyte)

Current density(i)/A m^{-2}	0	269	1080	2150	4300	6450	10,760	16,140	19,550
Voltage on load (one cell)/V	1.188	1.055	0.995	0.950	0.880	0.820	0.725	0.630	0.585
Energy efficiency/%		81.5	82.2	79.2	73.6	68.9	61.0	53.0	49.2
Power output/kW m^{-2}		0.261	1.05	2.02	3.77	4.28	7.79	10.2	11.4
Power output per unit weight/W kg^{-1}		8.55	34.3	66.5	123.5	173	225	333	374

Data from A. M. Adams, F. T. Bacon, and R. G. H. Watson, 'The high pressure hydrogen–oxygen cell', in *Fuel Cells* (Ed. W. Mitchell Jr.), 1963, Chapter 4, p. 176. Reproduced by permission of Academic Press Inc.

of the First Law of Thermodynamics, any deficit between the chemical energy released and the electrical work yielded must appear in the form of heat evolved. Hence we may write the energy conservation equation for the total heat evolved (q):

$$q = -T.\Delta S + z.F.(E - V) \tag{6.56}$$

where E is the open circuit e.m.f. It is obvious that, depending upon the sign of ΔS and the relative magnitude of the two terms, there can be net heat absorption or evolution by the cell on load. As the terminal voltage, V, depends strongly upon the current density as shown in Figure 62(a), the direction of heat flow can sometimes reverse with increasing current density. In the practical fuel cell, however, working at useful levels, we can be certain that the second term in equation (6.56) is much larger than the first; heat is therefore invariably evolved as the energy conversion efficiency is well below its theoretical value. The energy loss is expended in driving the reactions over the polarization energy barriers. It is also worth noting that any electrolyte has a finite conductance and thence a resistance to the passage of current. With increasing current densities there is an increase in resistive heating of the electrolyte, resulting in a voltage loss, in what is termed Ohmic polarization.

In the case of hydrogen fuel there are relatively few complications to the electrode operation in a fuel cell. However if the hydrogen is produced electrolytically using electricity from conventional power stations, then the fraction of the energy content of the original fuel substance stored in the hydrogen can be no more than 0.3 at present. Hence the true overall efficiency of the operating hydrogen—oxygen fuel cell could be no more than 20% at best. Therefore it is necessary to look at the potential for the direct use of fossil fuels such as natural gas or oil, in fuel cells when the overall efficiency could be several times larger.

In considering, for example, hydrocarbon electrodes in fuel cells, it must be acknowledged that there are additional sources of inefficiency in the use of the more complex fuel substances and not all of the fuel may be capable of being used electrochemically at maximum efficiency. Potential sources of inefficiency may be listed:

(i) The occurrence of side reactions yielding fewer electron equivalents per mol of fuel than the theoretical reaction.
(ii) Non-electrochemical decomposition of the fuel substance. For example, for the reasons discussed earlier in connection with the activation barriers to the chemisorption of hydrocarbons, it is envisaged that hydrocarbon fuel cells will be run at high temperatures. However, when a hydrocarbon comes into contact with the typical catalysts such as platinum or nickel, other non-electrochemical processes like cracking, isomerization, and dehydrogenation are induced in addition to the desired electrochemical consumption.
(iii) Direct chemical reaction between the fuel and oxidant due to inefficient separation through the electrolyte.

For an electrode of geometric area A, the electrochemical rate is expressed by

$i.A/z.F$. If it is presumed that non-electrochemical reactions are the main source of inefficiency and that these are kinetically first order in terms of the reactant concentration at the electrode, C_s, then a so-called Faradaic efficiency, η_F, may be defined:

$$\eta_F = \frac{i/z \cdot F}{(i/z \cdot F) + k_D \cdot C_s} \tag{6.57}$$

Here k_D is the rate constant per unit area of electrode surface for the non-electrochemical processes. It is evident from the form of equation (6.57) that increased current density, i, will actually lead to an improvement in the value of η_F; the non-electrochemical rates would not be expected to vary with the current density. This Faradaic efficiency will then be multiplied with the other efficiency parameters to produce the overall efficiency.

Steady progress is being made in the development of hydrocarbon fuel cells, but it will be many years before they become a practicable proposition for general usage. The successful electrodes developed so far are almost exclusively based upon platinum and there is no doubt that essentially complete oxidation can be achieved in the presence of acid electrolytes. However, a general phenomenon with such electrodes is the much greater tendency to polarize severely compared to the hydrogen electrode. Some illustrative performance characteristics for a propane—oxygen cell operating at 473 K are compared with those for an ambient temperature hydrogen oxygen cell using the same platinum based electrodes in Figure 63. With the propane cell, polarization has reduced the terminal voltage by half at a current density of around 500 A m^{-2} whereas the corresponding current density for the hydrogen cell is around 6000 A m^{-2}. It has been found that the best electrolytes for direct hydrocarbon fuel cells are strong acids of high boiling point with anions that are chemically inert, of which sulphuric acid and phosphoric acid in the cells shown in Figure 63 are examples. Useful gains in power outputs have been achieved by using hydrofluric acid as the electrolyte, and with a propane cell it has been possible to move towards power outputs of the order of 1 kW m^{-2} of electrode surface. These electrolytes are generally so corrosive as to make platinum virtually the only established electrocatalyst which can be used.

This last mentioned fact raises a further problem of the availability and to some extent the cost of platinum. Estimates have been made that the minimum platinum requirement to obtain 1 kW of power from a hydrocarbon fuel cell is around 0.3 kg, with a cost at the time of writing of around £900. Even assuming that the cost factor can be overcome, it is doubtful if there is enough platinum in the world's reserves to meet the demand for general usage. The estimated average amount of platinum in the crustal rocks is of the order of 0.005 parts per million; from Table 1.4 we see that this amounts to $(5 \times 10^{-9}) \times 2.4 \times 10^{22}$ kg, i.e. 1.2×10^{14} kg. On the above basis this could give rise to 4×10^{14} kW of power if applied *in toto* to hydrocarbon fuel cells, compared with a world wide annual demand of the order of 10^{13} W at present. Of course, it is highly doubtful if more than an exceedingly small fraction of that platinum reserve is recoverable. This is the same limitation as is encountered when considering the use of platinum in exhaust-gas catalytic

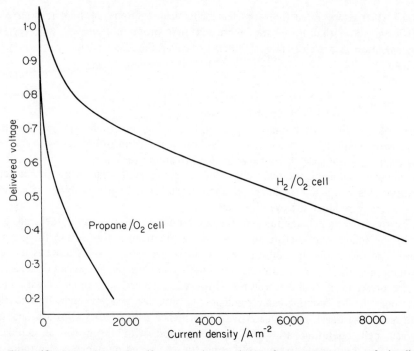

Figure 63. Comparison of performance characteristics of a propane–oxygen fuel cell operating at 473 K and those of a hydrogen–oxygen fuel cell operating at ambient temperature. Both cells used the same types of General Electric platinum electrodes with concentrated phosphoric acid and 5 N-sulphuric acid electrolytes respectively. Based upon data given in H. A. Liebhafsky and E. J. Cairns, *Fuel Cells and Fuel Batteries*, John Wiley & Sons Inc., New York, 1968, pp. 293, 492. Reproduced with permission of John Wiley & Sons Inc.

convertors suggested as fittings for internal combustion engines to limit pollutant emissions. If all vehicles were required to be so fitted, again it seems dubious that the minable platinum could meet the demand. Therefore it appears that development of the hydrocarbon fuel cell will be further hindered by the need to seek out alternative catalysts to the noble metals.

Alcohols have proved to be somewhat more amenable as fuels in this connection, but a general problem has been that the oxidation of the alcohol in fuel cells with aqueous electrolytes often does not go beyond the corresponding acid anion, carbonate for methanol and acetate for ethanol. This will accumulate in the electrolyte and is not easily removed. Nickel-based electrodes can be used in alcohol cells with alkaline electrolytes, and yield essentially the same potentials as cells which use hydrogen as fuel. In the case of ethanol, for example, this can be interpreted on the basis of an oxidation mechanism proceeding at the anode as:

$$C_2H_5OH + OH^- \longrightarrow CH_3COO^- + 4\,H_{ads}$$
$$4\,H_{ads} + 4\,OH^- \longrightarrow 4\,H_2O + 4\,e^-$$

The electrochemical reaction itself is then identical to that represented earlier in this section for consumption of two hydrogen molecules in the hydrogen–oxygen cell.

However, catalysts have been developed which allow the achievement of virtually complete oxidation of methanol in a methanol–air cell. Use of acid electrolytes, such as sulphuric acid, has ensured that carbon dioxide gas rather than carbonate is the end-product. The reactions may be summarized as:

Anode: $\quad CH_3OH + H_2O \quad \longrightarrow \quad CO_2 \quad + 6 H^+ + 6 e^-$

Cathode: $\quad 6 H^+ \quad + 6 e^- \quad + 1.5 O_2 \quad \longrightarrow \quad 3 H_2O$

Overall $\quad CH_3OH + 1.5 O_2 \quad \longrightarrow \quad CO_2 \quad + 2 H_2O$

The performance of such methanol fuel cells have reached encouraging levels. For example, (Tarmy and Ciprios, 1965), a terminal cell voltage of 0.4 V has been obtained at a current density of 1000 A m^{-2} with an operating temperature of 333 K. At a current density of 500 A m^{-2}, the average single cell in a 16-cell battery produced 0.41 V at the terminal. The typical individual polarization losses under these conditions were 0.31 V for the methanol electrode and 0.42 V for the air electrode, with a 0.06 V Ohmic loss. Moreover it has been demonstrated that the operating lifetime of such cells exceeds 3000 h with only small performance debits.

Fuel cells may use a wider variety of fuels than conventional combustive power systems. For instance the use of nitrogen–hydrogen fuels may be contemplated whereas the potential production levels for nitrogen oxides rule these out completely for normal combustion.

The potential advantage of using hydrazine (N_2H_4) in a fuel cell is the inert nature of the products of the reaction:

$$N_2H_4 + O_2 \quad \longrightarrow \quad N_2 + 2 H_2O$$

The appearance of N_2 as the product in place of the CO_2 product with hydrocarbons offers the prospect of relief if the build-up of the atmospheric CO_2 level and the resultant 'greenhouse' heating is considered as a long term problem (Section 1.3).

The theoretical open circuit e.m.f. [calculated from ΔG for the above reaction from equation (6.40)] is 1.56 V at 298 K, but in practice only 1.16 V can be obtained, close to the hydrogen-oxygen open circuit e.m.f. The situation here is similar to that in the alkaline-electrolyte alcohol cell discussed above; the hydrazine acts effectively as a source of adsorbed hydrogen atoms rather than as an electrochemically-active fuel in its own right.

The advantages offered by hydrazine as a practical fuel are that it is as easy to handle and store as gasoline, since it is always used in the form of hydrazine hydrate, $N_2H_4.H_2O$, containing 64% of N_2H_4 on a weight basis, but it is considerably less inflammable than gasoline. The prime disadvantage is that hydrazine vapour is toxic, to a greater extent even than carbon monoxide.

Moreover it is a synthetic fuel, produced industrially by partial oxidation of ammonia, and is currently too expensive for general application.

Ammonia itself may then look somewhat better from the points of view of relatively low toxicity and cost. Again the potential overall cell reaction is attractive for the same reasons as hydrazine:

$$4\,NH_3 + 3\,O_2 \longrightarrow 2\,N_2 + 6\,H_2O$$

However investigations have shown that the ammonia electrode suffers from extensive polarization even at modest current densities. The polarization has been ascribed to the involvement in the mechanism of nitrogen atoms strongly bound to the catalytic surfaces. These effectively surface nitrides lead on to N_2 very sluggishly and thus impose a very low limiting current density.

This brief survey of the critical features concerned in fuel cell development should give a basic understanding of the problems to be overcome. For a more comprehensive approach the reader may pursue some of the relevant sources cited in the reading list below.

6.5 Concluding Remarks

At this point we come to the end of our survey of present energy technology. We have seen that combustion systems represent our current prime source of energy, largely on the grounds of the availability of suitable fuels and well-tried service to our needs. However in the current terms of the so-called 'energy crisis' and faced with finite reserves of fossil fuels, a search for more efficient means of energy generation with concurrent reduction of pollutant emission indices is demanded. Fluidized-bed combustion and fuel cells provide hope for future success in these objectives. On the face of it, nuclear reactors appear as the obvious major source of power in the future but it is clear that the present high level of heat 'dumping' is far from satisfactory from the point of view of thermal pollution. At the same time, the problems associated with the safe disposal of radioactive wastes and the dangers of radioactive leakage must be borne in mind.

Bibliography

1. *'The Internal Combustion Engine in Theory and Practice'* Volume 1, 2nd edition, Chapters 1–5, by C.F. Taylor, M.I.T. Press, Cambridge, Mass., 1966.
2. *'Chemistry in the Environment'*, A collection of readings from *Scientific American*, Section III, W.H. Freeman and Company, San Fransisco, 1974.
3. *'Thermodynamic Cycles and Processes'*, by R. Hoyle and P.H. Clarke, Longmans, London, 1973.
4. *'Engineering Thermodynamics'*, by D.H. Bacon (particularly Chapter 4), Butterworth, London, 1972.
5. 'Fluidisation', by J.B. Lewis and B.A. Partridge, *Nature*, **216**, 124–127 (1967).
6. 'New Systems for Clean Power', by A.M. Squires, *International Journal of Sulphur Chemistry*, Part B, Volume 7, 85–98 (1972).

7. *'An Introduction to Electrochemical Science'*, by J.O'M. Bockris, N. Boncociat, and F. Gutmann, Wykeham Publications (London) Ltd., 1974.
8. *'Fuel Cells'*, by A.B. Hart and G.J. Womack, Chapman and Hall, London, 1967.
9. *'Fuel Cells'*, Ed. W. Mitchell, Academic Press, New York, 1963.
10. *'Fuel Cells and Fuel Batteries − A Guide to their Research and Development'*, by H.A. Liebhafsky and E.J. Cairns, John Wiley & Sons, New York, 1968.

Chapter 7

The Major Cycles of the Atmosphere

Our atmosphere and its component gases take part in a dynamic balance in the main, with exchange of species between air and sea and land. For example the famous White Cliffs of Dover are composed of chalk, substantially calcium carbonate, and are therefore the result of the fixation of carbon dioxide from the atmosphere via solution in the sea, incorporation into the skeletons of marine organisms and finally consolidation of the deposits on the sea bed into rock. There can be no doubt that similar processes are going on in the seas of today, creating the rocks of the future. Another example of the continuous cycling between earth and atmosphere is the action of denitrifying bacteria and nitrifying symbionts, which achieve a substantial balance. However around the turn of the century, in the absence of any knowledge of the organisms which fix nitrogen from air to soil, many eminent scientists viewed the recognized denitrification processes as auguring disaster for agricultural productivity, which is so dependent upon nitrogen in the soil. Later came the realization that the denitrifying bacteria constituted but one element in the cycle of nitrogen between earth and atmosphere. In fact the denitrification prevents the establishment of a closer approach to full chemical equilibrium, where most of the atmospheric nitrogen would be rendered to ions (mainly nitrite) in solution in the oceans or withdrawn into sedimentary deposits.

Our principal interests lie in those parts of the atmospheric cycles where man's production or use of energy gives rise to possible anthropogenic perturbations of the natural cycles. Our aim will be to assess the significance of the anthropogenic terms, whether they are comparatively trivial and can therefore be ignored on a global basis, or whether the scale of anthropogenic injection into the natural cycle is large enough to give cause for concern. For example one area where anthropogenic perturbation could be becoming significant is the industrial fixation of nitrogen (mainly by the Haber synthesis of ammonia) in the production of artifical fertilizers. The concern is whether the rates of the natural denitrification processes can keep pace with the increasing total rate of fixation, which on the industrial side has been doubling every 6 years or so recently. Moreover, the major source of naturally-fixed nitrogen is likely to be legume crops such as peas, beans, and alfalfa with their microbial symbionts within the root nodules. Recommendation for increasing the world's food supply usually emphasize legumes without considering the associated increase in the rate of nitrogen fixation within the soils.

One consequence which could follow is a decrease in the pH of the sea due to increased nitrate content, with release of carbon dioxide by perturbation of the carbonate equilibria and the inhibition of carbonate skeleton formation by marine organisms.

7.1 The Water—Hydrogen Cycle

The amount of precipitable water in the atmosphere is controlled basically by the vapour pressure function relating to the ambient temperatures within localized volumes of air. However, on considering the huge natural turnover of water involved in rainfall, etc., we can be fairly certain that the amount of water released into the atmosphere by man's combustion of fossil fuels is negligible. The oceans, rivers, glaciers, swamps, soils, etc., and the atmosphere contain an estimated 1.4×10^{21} kg (Table 1.4) of water, which corresponds to around 1.5×10^9 km^3 of liquid water in its various forms. Of this only about 10 parts in a million are, on average, in the vapour form at a given time while oceans and seas contain all bar some 3% of the total water; about two-thirds of the remainder is frozen in glaciers and icecaps, mainly at the poles. The average residence time for water in the evaporation—precipitation cycle is approximately 10 days. The overwhelming mass of water moves in its cycle chemically unchanged, so that a separate hydrogen cycle must be distinguished and our interest devolves more to the minor components of the overall water cycle in which the chemical bonds in the molecule are broken.

The photodissociation of water vapour by ultraviolet radiation in the upper atmosphere is one such instance and this has significance in determining the atomic and molecular hydrogen-mixing ratios there. In turn these govern the escape rate of hydrogen to space from our planet. The development of this aspect will be reserved for Chapter 9, but we may remark at this stage that the water vapour content of the stratosphere and above is unrelated to the general water cycle. Water is generated chemically in the upper atmosphere, largely through the agency of methane production at the Earth's surface and its subsequent diffusion upwards through the tropopause. The lapse rate of the troposphere, producing the low temperature of 217 K at the tropopause, completely inhibits any significant transfer of water vapour as such from troposphere to stratosphere. Thus the stratosphere is very dry compared to the lower troposphere and the mixing ratio of water vapour there lies in the range 10^{-5} to 10^{-6}, well below the average value of 10^{-3} close to the ground.

When photosynthesis takes place, water is incorporated into the cellulose of the plant, as we have seen in Chapter 4, with the bonds being broken. Moreover, as was pointed out in Chapter 4, the oxygen evolved comes uniquely from water rather than from carbon dioxide. Hence the processes involved might be viewed as part of the hydrogen turnover cycle. However, the amount of water actually bound chemically into the plant structure is a comparatively small fraction of the amount of water which transpires through the leaves. At present, evidence is accumulating which favours the viewpoint that the rate of growth of plants is strongly dependent upon the rate of water transit through the plant. Moreover the solar energy used to effect transpiration can be as much as 40-times greater than the solar energy fixed by photosynthesis.

For a typical crop plant, the weight of water which will have passed through it via transpiration during the growing season may be as much as two orders of magnitude larger than the fresh-harvested weight. Furthermore, up to three-quarters of this weight can be water in transit so that the dry weight of the crop can be as little as 0.25% of the total weight of water which has been transferred through it from the ground to the atmosphere. The general formula for carbohydrate is $(CH_2O)_n$ so that the ratio of the weight of water fixed to the dry plant weight is as 18 to 30. Hence eventually it turns out that only some 0.15% of the total amount of water which has entered the plant structure actually provides the hydrogen content of the structure.

Why is it that the rate of transpiration governs the rate of growth of the plant? One element of the answer must lie in the fact that the best condition for the movement of carbon dioxide from the atmosphere into the plant leaf is when the stomatal apertures on the leaf surface are fully opened. This state which allows easy inflow of carbon dioxide however, also enhances the outflow of water vapour. At the end of its transit through the plant, the change of state of water from liquid to vapour is effected by, and the rate is limited by, the availability of solar energy. It seems, therefore, that the rate at which carbon dioxide can be brought in to take part in photosynthesis is more limiting than the rate at which water can be raised from the roots. In Section 4.4 it was said that the evidence on this score was the enhancing of growth rates observed in common plants (C_3 photosynthetic mechanism) when the surrounding air was artificially enriched with CO_2. The excess water, over that required for photosynthesis, is used rather like a cooling system in the necessary transpiration concurrent with the intake of carbon dioxide. It follows that the smaller the degree of stomatal opening, the lower is the rate of intake of carbon dioxide, the lower is the rate of transpiration, and the lower is the rate of photosynthesis for C_3 plants, limited by the carbon-dioxide supply rate.

In Section 4.5 we referred to potential photosynthesis rates, governed entirely by the integrated incident solar irradiance during the growing season of the plant. We must now impose the rider that any hope of achieving this demands the supply of large amounts of water to the roots. Since the usual agency of provision will be rainfall associated with periods of cloud cover, and thence reduced incident irradiance, we can begin to appreciate that the potential photosynthetic yield is unlikely to be produced on a large scale. Only in isolated instances, such as irrigation schemes or the chance happening of a large ground water supply close to the surface, can the crop yield have any chance of approaching the limit.

The energy supply for transpiration comes from the photons initially absorbed by the pigment molecules in the plant structure, with energy transfer to chlorophyll as detailed in Section 4.2. As is pointed out in Section 4.1 and 4.3(i), only one photon out of a fairly large number abosrbed by a canopy leaf is effective in photosynthesis. It is the energy of the remainder which powers the vaporization of water and therefore it is perhaps not so surprising that the bulk of the water uptake is involved in transpiration rather than photosynthesis on this basis of the partition of the absorbed energy.

The fraction of Earth's water contained within the plant and animal populations is exceedingly small: a rough estimate suggests that it is less by one order of magnitude than the water vapour content of the entire troposphere; this places it at something of the order of one millionth part of the total water content of the planet.

The effect of a limited supply of transpiration water upon crop yield may be illustrated by the situation which occurred in Britain in 1975. This was one of the summers of highest integrated solar irradiance with very few periods of clouded skies and consequently little rainfall. However, the average crop yields fell substantially (of the order of 30%) compared with those obtained in the damper and less sunny summers of preceding years, demonstrating clearly that the integrated incident irradiance is only one factor in the complex process of plant growth. Let us attempt a rough calculation of the minimum water requirement for a crop subjected to clear sky conditions for a month in Britain. From Figure 12 the total daily radiant energy incident on a horizontal surface at the top of the atmosphere, averaged over the summer growing period, is around 40 MJ m^{-2} for a latitude of 53.5° N. In Section 2.2 we suggested that, for cloudless skies, some 70% of this irradiance, i.e. 28 MJ m^{-2}, would be the 24-hour average incident at the surface. We may reduce this by 15% to 24 MJ m^{-2}, on the basis of the green meadow albedo of Table 2.3, to obtain the irradiance absorbed by the crop. We shall ignore any reemitted radiation and convective heat transfer terms and consider that all of this absorbed energy is used in transpiration, neglecting the small fraction effective in photosynthesis. The latent heat of evaporation of water at 288 K is 2.464 kJ g^{-1}. In 30 days, therefore, the weight of water which could be evaporated from an area of 1 square metre is $30 \times 2.4 \times 10^4/2.464 = 2.9 \times 10^5$ g. The equivalent volume of liquid water is therefore approximately 0.3 m^3 so that the required rainfall is 0.3 m or close to 12 inches in normal weather parlance. We can therefore appreciate the enormous strain put on the growth of the crop by a prolonged period of summer drought.

The decay of organic materials under anaerobic conditions gives rise to hydrogen gas itself as the most important primary product, together with carbon dioxide of course. The evolved hydrogen is then used by several groups of microorganisms, which generate the eventual product gases such as methane and hydrogen sulphide, the forms in which most of the hydrogen from photosynthesized materials is released to the atmosphere. Nitrous oxide and nitrogen are other products of these anaerobic consumptions. Figure 64 represents the typical major decay processes occurring in a swamp or pond environment.

Methane is the major product of the hydrogen evolution process. As already mentioned, the diffusion of this gas into the stratosphere provides the main source of water there, so that we might consider this as another element, albeit minor, in the water cycle.

This biological generation of molecular hydrogen may be regarded as a

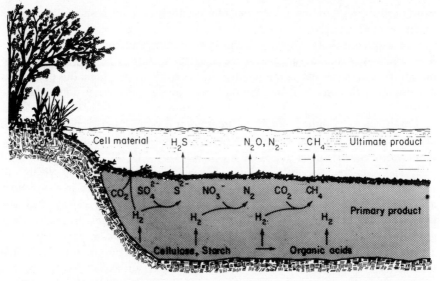

Figure 64. The ecology of hydrogen production and consumption in a fresh-water environment. The gas is produced by anaerobic decomposition of organic materials in deeper mud layers: the ultimate products which are emitted to the atmosphere depend upon the availability of the various hydrogen-acceptor species shown. From H. G. Schlegel, 'Production, modification and consumption of atmospheric trace gases by micro-organisms', *Tellus,* **XXVI,** 12 (1974). Reproduced by permission of the Swedish Geophysical Society.

by-product of the anaerobic existence of the cells of microorganisms. As in photosynthesis, the metabolism of the cells has the production of ATP as the prime objective; in so doing, the required energy is extracted from intramolecular disproportionation reactions whereby carbon dioxide and highly reduced organic compounds (e.g. alcohols and fatty acids) are produced. The anaerobes then accomplish the disposal of reducing power in the evolution of gaseous hydrogen. Therefore the absence of oxygen to take part in the more usual oxidative release of energy in surface-living organisms and plants does not bar such microorganisms from gaining their vital power through alternative pathways. For example, when glucose reacts with oxygen, the maximum available energy yield is 2870 kJ per mol of glucose, but in microorganisms living under anaerobic conditions the reaction of glucose with the nitrate ion can yield 2280 kJ mol^{-1} when the nitrogen reappears as nitrous oxide, and some 2380 kJ mol^{-1} if the nitrogen end-product is molecular nitrogen itself. Hence life can be just as viable under anaerobic conditions when the microorganism has the necessary energy-deriving mechanism in its metabolic cycle.

The ability to produce hydrogen is found in many microbial species, and is catalysed by the secretion of the hydrogenase enzymes. Amongst such species are *Clostridia* and various phototrophic bacteria and anaerobically adapted algae, the last two exhibiting photoproduction of hydrogen gas. Such ability has evoked much interest recently in connection with the possibility of direct production of hydrogen from solar radiation. If the hydrogen so produced could escape without being consumed simultaneously or subsequently by other microorganisms associated with the same anaerobic environment, hydrogen could well have become a major constituent of our atmosphere rather than a variable trace component.

The hydrogen-producing bacteria are very widely distributed in the soils and it is the upper layers which seem to act as the sink for conversion of the hydrogen rising from the layers below. Methane-generating bacteria inhabit the anaerobic zones of marshes, ponds, etc., in a more-or-less symbiotic relationship with the hydrogen producing bacteria. It is such bacteria, reducing carbon dioxide to methane, which are cultivated in sewege fermentation tanks. It may be that before long man may begin to tap this additional source of energy on a large scale.

The presence of sulphate in the anaerobic environment gives rise to the production of hydrogen sulphide by bacterial action. In the atmosphere the hydrogen sulphide is oxidized to sulphur dioxide on a fairly short time scale so that this may be regarded as a natural source of the latter gas. Present estimates for the lifetime of hydrogen sulphide in the troposphere average at around 1 day.

Estimates of the total source strength of methane production through biological mechanisms have ranged as high as 2×10^{12} kg year^{-1}, corresponding to 1.3×10^{14} mol year^{-1}. On comparison with the estimated annual turnover of 8×10^{13} kg of organic carbon by photosynthesis (Section 1.3) this implies that well over 1% of the total energy stored by photosynthesis in the biosphere can be expanded in the production of methane. If something of the order of 10% of this could be collected, then the available energy in combustion would be around 2×10^{16} kJ. Reference back to Table 5.3 shows that the 1977 worldwide consumption of natural gas corresponds to 4.3×10^{16} kJ, almost equal to the above estimate. Alternatively, looking at Table 5.4, the estimated reserves of natural gas locked below the Earth's surface are equivalent to 2100×10^{16} kJ, amounting to between 100- and 1000-years' accumulation of the annual methane production from the biosphere.

The group of gases produced principally by the biosphere, CH_4, H_2, N_2O, and also CO, have mixing ratios in the troposphere and residence times within a narrow range, quite distinct from all other constituents, as is illustrated by the data in Table 7.1. The *residence time* is defined as the reciprocal of the fraction of a gas which is removed from the atmosphere in unit time.

The main point to be made on the basis of these data is that there must be efficient sinks for the last four gases, i.e. sites where the gases are destroyed at rates matching the production rates. Otherwise these gases would have mixing ratios well above those observed, when the relatively high production rates are taken into account. These may be derived on the basis of the equality to the removal rate, which is expressed as the product of the reciprocal of the residence time and the total amount of the gas in the atmosphere. For example, for methane, one-fifth of the total methane, corresponding to 0.22 p.p.m. is removed each year. Table 1.4 tells us that the total mass of the atmosphere is 5×10^{18} kg, so that 0.22×10^{-6} p.p.v. corresponds to approximately 12×10^{11} kg. This is a somewhat lower annual turnover rate than that quoted above, but the extent of uncertainty is such as to make either reasonable. The corresponding annual turnover rates for the

Table 7.1. Volume mixing ratios and estimated residence times of atmospheric gases

Gas	Average mixing ratio in troposphere/p.p.m.	Residence time (primarily tropospheric, excluding exchange with ocean)
Argon	9300	Accumulated during Earth's lifetime
Neon	18	Accumulated during Earth's lifetime
Krypton	1.1	Accumulated during Earth's lifetime
Xenon	0.09	Accumulated during Earth's lifetime
Nitrogen	7.8×10^5	Around 10^6 years
Oxygen	2.1×10^5	Around 10^4 years
Carbon dioxide	330 (1977)	Complicated by solution in oceans
Methane	1.1	Around 5 years
Hydrogen	0.4	Perhaps 5 years (?)
Nitrous oxide	0.25	Around 10 years
Carbon monoxide	0.1	Between 0.1 and 1 year (see text)

Data are taken mainly from C. Junge, 'The cycle of atmospheric gases – natural and man made', *Quarterly Journal of the Royal Meteorological Society*, 98, 711–729 (1972), with updated estimates of the mixing ratio of carbon dioxide and the residence time of carbon monoxide. Reproduced by permission of the Royal Meteorological Society.

other gases, according to the data of Table 7.1, are 4×10^{11} kg of H_2, 1.3×10^{11} kg of N_2O and 5×10^{11} kg of CO. We may conjecture the locations of the sinks for hydrogen and methane, which are our principal interest in this section.

Measurements of the hydrogen mixing ratios at a series of altitudes extending into the stratosphere show that it remains almost constant even up to 50 km. This suggests that hydrogen is not consumed with any high efficiency within the stratosphere. This situation may be contrasted with that for carbon monoxide in Figure 21; the stratosphere is evidently an efficient sink for this gas and the sharp concentration gradient across the tropopause implies net upward transport. Further, measurements of the dissolved hydrogen content of the surface waters of the ocean have shown substantial solution equilibrium with the air above. Hence it is unlikely that the sea constitutes a sink for hydrogen. This leaves the land surface as the only possible sink medium and some support is given to this postulation by the knowledge that the soil contains aerobic bacteria which oxidize hydrogen. These are termed autotrophs, which are defined as organisms which use carbon dioxide as their sole source of carbon and these perform a metabolism chemical conversion represented by the overall equation:

$$2\,H_2 + CO_2 \longrightarrow (CH_2O) + H_2O$$

where (CH_2O) is once again the carbohydrate building block.

With methane the sinks are more obvious. Methane is used as a growth substrate by many soil bacteria under aerobic conditions and over one hundred species of methane-using bacteria have been identified after isolation. In point of fact, many of these species show a high specificity for methane and are incapable of accepting longer chain hydrocarbons. This is not to say that there are no other bacterial species which can exist on the higher hydrocarbons. There is no doubt that methane is also consumed by reactions in the stratosphere into which it diffuses. Above the tropopause there is a decrease in the methane-mixing ratio, marking the onset of consumption reaction which increase in rate with altitude. More will be said of these reactions in Chapter 9.

Thus we complete our present view of the water-hydrogen cycle. It must be borne in mind that the overwhelming proportion of the cycle is purely physical in nature but that the minor chemical adjuncts involved in photosynthesis and biological decay have substantial importance within the theme of this book.

7.2 The Carbon Cycle

In our preceding discussion of the water—hydrogen cycle, we have necessarily impinged upon part of the carbon cycle. It is evident that photosynthesis must be a major element of this section and has provided the organic fuels upon which most of our energy technology is based. However we have yet to consider the magnitude of the anthropogenic intrusion into the natural cycle. Man is at present conducting a vast and largely uncontrolled experiment whereby the ability of the carbon cycle to cope with the sudden mobilization (comparatively speaking) of the huge quantity of carbon fixed or stored within the Earth's crust is being tested.

We have seen that the carbon fixed from the atmosphere by photosynthesis is sooner or later restored to the atmosphere by the decay of organic matter. An approximate time scale for the second process can be obtained from measurements of the ratio of radioactive carbon-14 to normal carbon-12 in the organic matter of the soils. After the initial fixation of the carbon with the ratio of isotopes found in new growth, which reflects the ratio in the carbon dioxide of the atmosphere, the radioisotope decays with a half-life of 5760 years. Thus the decrease in the $^{14}C/^{12}C$ ratio provides one of the standard dating procedures; it has yielded rates for the oxidation of the organic matter in the soil with time scales ranging from decades in tropical soils to centuries in more temperate locations. The radionuclide ^{14}C is produced naturally in the atmosphere by cosmic radiation but the level was almost doubled by the atmospheric nuclear-weapon testing of the 1960's.

Carbon Dioxide

The longer term variations in the carbon dioxide level of our atmosphere are of principal interest to us. In this connection we must gain perspective from the fact that only a few tenths of one percent of the total amount of carbon in the crust of the Earth (total of the order of 10^{19} kg) is in rapid circulation in the biosphere. The huge bulk of the crustal carbon lies immobilized in the form of inorganic carbonate deposits (chalk, limestone, dolomite, soda) and in the organic fossilized relics of prehistoric photosynthesis (coal, oil, etc.) The time scales for the laying down of these deposits extend over hundreds of millions of years (Chapter 1) so that current formation rates can be regarded as negligible components of the carbon cycle for present purposes.

Returning to the generation of the radionuclide ^{14}C through nuclear testing, it was observed that the $^{14}C/^{12}C$ mixing ratio in the atmospheric carbon dioxide decreased subsequent to the ending of the tests far more rapidly than was compatible with photosynthetic turnover. The explanation was that the hydrosphere, mainly the oceans, contains about 60-times as much carbon dioxide dissolved as in the atmosphere and of course this preserved much more the normal $^{14}C/^{12}C$ ratio; the dynamic exchange of carbon dioxide across the ocean—atmosphere interface then led to the apparent removal of ^{14}C from the carbon dioxide of the air. Clearly this allows an estimate to be made of the residence time for carbon dioxide in the atmosphere, which comes out at between 5 and 10 years as opposed to nearer 20 years if vegetation scavenging were the only turnover mechanism. The ocean is by no means a homogeneous system from the point of view of carbon dioxide exchange with the atmosphere. It is only the so-called mixed layer with a mean depth of 70 m which is actively involved. The deeper layers exchange rather slowly with the mixed layer and can be virtually ignored in this connection. A further complication arises from the oceanic biomass, primarily concentrated in the mixed layer, composed very largely of the microorganisms termed phytoplankton. These are the start of the oceanic food chain in that they achieve the initial fixation of carbon within their short lifetime of the order of months at most. Eventually the dead microorganisms sink and decompose to produce dissolved organic matter; this is then oxidized to carbon dioxide and largely retained in the mixed layer. Thus the dead organic matter originating from phytoplankton creates a virtually self-contained cycle effective in the carbon dioxide balance of the mixed layer of the oceans.

In his article in *Scientific American* (cited at the end of this chapter), Bolin has produced a clear picture of the carbon circulation pattern in the biosphere as a whole. This is reproduced as Figure 65, with the estimates of the reservoirs and annual rates of exchange expressed in mass units of 10^{12} kg. This figure emphasizes the dominant role of the oceans in determining the carbon dioxide mixing ratio of the troposphere and, as it turns out, the stratosphere and the mesosphere which have the same mixing ratio.

However, the carbon dioxide content of the atmosphere is increasing by a significant amount each year. Measurements over the decade from 1959 to 1969 showed a rise in the average level from 314 p.p.m. to 321 p.p.m., corresponding to an average rise of 0.8 p.p.m. per annum. There is also a seasonal variation with a maximum in the northern hemisphere spring and a minimum in the autumn, caused by the photosynthetic fixation of carbon dioxide during the growing season on the larger land surface area of the northern hemisphere $(1 \times 10^{14} \text{ m}^2)$ compared with that of the southern hemisphere $(0.5 \times 10^{14} \text{ m}^2)$. The resultant variation about the mean is around 3 p.p.m. on either side. The total amount of carbon uptake by land plants has been estimated for $10°$ latitude belts and these data are listed in Table 7.2. This table emphasizes also the importance of the equatorial region from $10°$ N to $10°$ S in the photosynthetic budget, with around 35% of the annual carbon dioxide fixation or oxygen evolution taking place there. In fact the Amazon jungles are often referred to as the Earth's 'oxygen factory' and the steady deforestation there may therefore give rise to some long term concern.

Let us consider whether the rising carbon dioxide content of the atmosphere corresponds to the anthropogenic output from fossil fuel combustion. The total mass of carbon dioxide in the atmosphere on the basis of a mixing ratio of 330 p.p.m. is approximately 2×10^{15} kg out of a total mass of 5×10^{18} kg (Table 1.4). For 1960, man's total energy usage derived from fossil fuels was some 1.2×10^{17} kJ (from Table 5.3, increased by 30% for Communist world contribution). Accordingly, using the arguments of Section 5.1, we may estimate the

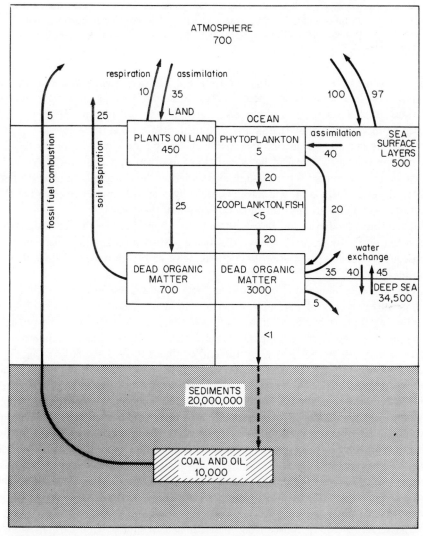

Figure 65. Carbon circulation in the biosphere, with annual turnover rates and reserves estimated in billions (10^9) of metric tons (10^3 kg). From B. Bolin, 'The carbon cycle', *Scientific American*, September 1970, p. 130. Copyright © 1970 by Scientific American, Inc. All rights reserved. Reproduced with permission.

Table 7.2. Total carbon dioxide uptake by land plants/10^{12} kg year^{-1}

Latitude ($°$)	0-10	10-20	20-30	30-40	40-50
Northern hemisphere	23.5	14.3	11.7	9.2	10.3
Southern hemisphere	26.4	18.0	8.8	4.0	0.7

Latitude ($°$)	50-60	60-70	70-80	80-90	Overall
Northern hemisphere	9.2	4.8	0	0	83.0
Southern hemisphere	0	0	0	0	57.9

From C. E. Junge and C. Czeplak, 'Some aspects of the seasonal variation of carbon dioxide and ozone', *Tellus*, **20**, 422–434 (1968). Reproduced by permission of the Swedish Geophysical Society.

anthropogenic CO_2 output for 1960 at 8×10^{12} kg. Hence the rise in the atmospheric level should have been near to $8/2000 = 0.4\%$ or an average annual increase of 1.3 p.p.m. if nothing else entered into the picture. However, the observed increase was only 0.7 p.p.m., or roughly half, so that it seems that close to 50% of the anthropogenic carbon dioxide release passes into solution in the sea within the annual time scale. The rate of emission of anthropogenic carbon dioxide is increasing by some 5% per annum; extrapolation then suggests that the present (1977) level of 330 p.p.m. will have increased by some 50 p.p.m. by the end of the century, taking into account the buffering action of the oceans.

Therefore we have concluded that for carbon dioxide the anthropogenic contribution to the natural cycle is a significant perturbation and that there are two consequences of increased atmospheric loading of carbon dioxide which we should consider.

The first is an enhanced 'greenhouse effect', discussed earlier in Section 1.3 and due to the trapping of part of the Earth's thermal reradiation spectrum by the infrared absorption bands of carbon dioxide. Various estimations of the possible increase in average surface temperature have been made; the general conclusion is that the rise will be below 1 K by the end of the century. It must be recognized that, although this appears insignificant in itself, the resultant recession of the polar ice-caps and glaciers could produce a rise in sea-level with disastrous consequences for low-lying parts of the land surface, for example, London, where there is current concern over the possibility of abnormally high tides produced by reinforcement of equinoxial tides by strong easterly winds. However, the actual temperature effect to be expected is by no means easy to predict. Besides increasing the carbon dioxide level of the atmosphere, the burning of fossil fuels may well increase the dust content of the atmosphere: the consequent decrease in solar irradiance at the surface could well balance the enhanced greenhouse effect or even lead to a fall in average temperature.

The second possible consequence of an increased atmospheric carbon dioxide burden concerns the possibility of disturbing the delicate biological balance of the mixed upper layer of the oceans. It has been postulated that continuous influx of carbon dioxide into the ocean waters could be leading towards a situation where the mixed layer may become undersaturated with respect to calcium carbonate within the next few decades. It is considered that under these conditions the carbonate concentration could decrease in favour of bicarbonate formation in seawater of lowered pH, i.e. greater acidity. Marine organisms living in the mixed layer would then be unable to evolve their calcium carbonate skeletons, resulting in a biological disaster.

The carbonate–bicarbonate–carbon dioxide atmosphere–ocean system depends mainly upon three equilibria represented as follows:

$$HCO_3^- \rightleftharpoons CO_3^{2-} + H^+ \tag{7.1}$$

$$H_2CO_3 \rightleftharpoons HCO_3^- + H^+ \tag{7.2}$$

$$H_2CO_3 \rightleftharpoons CO_2(g) + H_2O(l) \tag{7.3}$$

Combination of equations (7.1) and (7.2) produces the alternative equation:

$$H_2CO_3 + CO_3^{2-} \rightleftharpoons 2 HCO_3^- \tag{7.4}$$

We may also define the total carbon dioxide content of the mixed layer of the ocean, $[CO_2]_T$, as:

$$[CO_2]_T = ([H_2CO_3] + [HCO_3^-] + [CO_3^{2-}])$$

Solubility data for seawater at ambient temperature generate a relationship between H_2CO_3 (perhaps better represented as $CO_2 \cdot H_2O$ to emphasize the weakness of binding) concentration in the mixed layer and the mixing ratio of carbon dioxide in the atmosphere (p.p.m. CO_2) as:

$$[H_2CO_3]/\text{p.p.m. } CO_2 = 2.89 \times 10^{-8} \text{ mol dm}^{-3} \tag{7.5}$$

Insertion of the present 330 p.p.m. value gives $[H_2CO_3] = 9.5 \times 10^{-6}$ mol dm^{-3}.

The 1975 value of $[CO_2]_T$ is 2.7×10^{-3} mol dm^{-3}. A combination of the equilibrium constant data for equations (7.1) and (7.2) produces the equilibrium constant for equation (7.4):

$$[HCO_3^-]^2/[CO_3^{2-}][H_2CO_3] = 1230 \tag{7.6}$$

Solution of this equation on the basis of the 1975 value of $[CO_2]_T$ yields $[HCO_3^-] = 2.3 \times 10^{-3}$ mol dm^{-3} and $[CO_3^{2-}] = 4.3 \times 10^{-4}$ mol dm^{-3}. The corresponding value for the pH of the 1975 mixed layer is 8.22($= -\log_{10}[H^+]$).

Consider the situation then in prospect for the end of the century, taking the mixing ratio of CO_2 to have reached 380 p.p.m. in the atmosphere at that time. A pessimistic estimate for $[CO_2]_T$ (obtained simply by summing the CO_2 anthropogenic emissions projected for the years up to 2000 AD) could be as high as 3.0×10^{-3} mol dm^{-3}, making the reasonable assumption that transfer of carbon dioxide to deeper oceanic layers will be insignificant. The calculated values are $[HCO_3^-] = 2.5 \times 10^{-3}$ mol dm^{-3} and $[CO_3^{2-}] = 4.6 \times 10^{-4}$ mol dm^{-3} and the predicted pH for seawater is 8.207.

Thus we see that the dissolution of the anthropogenic carbon dioxide in the oceanic mixed layer will have little effect upon the carbonate equilibria therein and no significant alteration in pH will occur within the foreseeable future. It can be seen that the increase of around 15% in the atmospheric carbon dioxide mixing ratio in prospect by the end of this century will be paralleled by a near corresponding increase in the bicarbonate content of the mixed layer; little change in carbonate concentration or in pH is expected, therefore making the biological disaster unlikely on this score at least. The paper by Whitfield cited at the end of this chapter gives a more detailed discussion and originally indicated the fallacy in the arguments of the 'prophets of doom' on this aspect.

Carbon Monoxide

We now turn our attentions to more minor components of the carbon cycle, where it might be expected that the anthropogenic source could be important. Until recently, it was considered that the anthropogenic source strength of carbon monoxide was overwhelming, amounting to around 3×10^{11} kg per annum and with gasoline combustion accounting for almost 60% of this. The average global mixing ratio is 0.1 p.p.m. so that the total mass in the atmosphere is about 6×10^{11} kg and matches a 1- to 2-year residence time in the troposphere. Strong evidence that most of the carbon monoxide is anthropogenic in origin is the fact that the northern hemisphere has an average level of 0.14 p.p.m. with a peak as a function of latitude at around $50° N$ of some 0.20 p.p.m., compared with the average level of 0.06 p.p.m. for the southern hemisphere: most of the major industrial nations lie in the northern hemisphere. However it might appear odd that a species with a residence time of the order of a year does not become better dispersed than this, the first contrary evidence. Carbon monoxide is known to be produced naturally by animals, plants, and bacteria: one outstanding instance is the siphonophore known as the Portuguese man-of-war, where up to 13% of carbon monoxide has been found in the float bladder. In another instance, this gas has been found at levels of up to 5% in the float bladder of the brown algal species, *Nereocystis luetkeana*. Also, comprehensive measurements in the oceans have revealed that the surface water is supersaturated with carbon monoxide with respect to the air in content by factors exceeding 20 on the basis of solution equilibrium for the gas. On deeper sampling, it was found that the carbon monoxide content of the oceans is concentrated in the upper layers and decreases sharply with depth. For example the dissolved concentration at a depth of 50 m can be six-times less than that at the surface, while below 100 m the concentration is over 35-times less. A further interesting aspect is that there is a distinct daily variation in the carbon monoxide surface concentration, passing through a maximum around noon. Moreover the amplitude of the variation depends upon the solar irradiance; on cloudy days the surface concentration is several times less than on sunny days. All of this points clearly to production of carbon monoxide associated with the photosynthetic activities of microorganisms like phytoplankton. In laboratory experiments on the irradiation of natural seawater to which has been added dissolved organic matter produced by phytoplankton, carbon monoxide was produced, thereby demonstrating the point. However, on overall estimation, it has to be conceded that the oceanic source can only yield something of the order of 3% at best of the anthropogenic source strength. No other direct natural sources of consequence are known at present.

Despite the absence of major natural sources of carbon monoxide as such, the analysis of air trapped in the Greenland ice dated as far back as 1800 shows unexpectedly high levels of carbon monoxide and provides strong qualitative evidence for a much larger natural source than had been supposed previously.

Moreover, there is a major sink for carbon monoxide at the surface of the soil, capable of consuming much more of this gas than would be required to balance the anthropogenic output. Despite a continual increase in the rate of anthropogenic emission of carbon monoxide over the years, the background level measured at locations far removed from industrial areas has remained substantially constant. It is therefore clear that this must be maintained by a dynamic balance, with efficient mechanisms available for removal. As we have seen the oceans are net producers of carbon monoxide, so that we must look elsewhere for sink operation. Within the troposphere there is no gas phase mechanism capable of achieving the required consumption rate. Reference back to Figure 21, where the sharp decrease in carbon monoxide mixing ratio at the tropopause is obvious, shows that there is a significant consumption mechanism in the stratosphere; the rate of transfer from the troposphere to the stratosphere is too slow to enter into the present argument, based on a time scale of around 1 year. So it is to the land surface that we must look for our major sink. Key experiments were performed at the Stanford Research Institute at Irvine in California (Inman, Ingersoll, and Levy, 1971). In one instance, a potting soil mixture was found to be capable of depleting carbon monoxide in a test atmosphere from a level of 120 p.p.m. to near zero within 3 h, a spectacular performance indeed. However, if the soil was steam sterilized or had antibiotics or 10% w/w saline solution added, then its ability to remove carbon monoxide was destroyed, but such sterilized soil, inoculated with non-sterilized soil, acquired activity which increased gradually with time eventually reaching that of the unsterilized soil. It was also shown that maximum activity occurred with the temperature was maintained at 303 K, with uptake rates falling sharply for changes of 5 K on either side. Different soils showed varying activities but the highest activities generally resided in those soils containing higher organic material levels and having higher acidities. It therefore becomes clear that the phenomenon is due to microbiological activity rather than to any physical mechanism. Based upon an extensive study of the *in situ* rates of carbon monoxide uptake for soils at a wide variety of locations in the U.S.A., Ingersoll (1972) estimated that the total global capacity of soils was around 1.4×10^{12} kg per annum, an order of magnitude larger than the anthropogenic output. On this account, taking the average global mixing ratio in contact with soils to be 0.1 p.p.m., the average residence time comes down to around 4 months and the flux of carbon monoxide into the atmosphere must be up to 10-times larger than the anthropogenic flux.

Both aerobic and anaerobic carbon monoxide-consuming soil bacteria are known. The successful isolation of aerobic bacteria, which oxidize carbon monoxide, to carbon dioxide, has been achieved, initially by Russian workers, and some of the strains are rather similar to the hydrogen consuming bacteria (Section 7.1). All of them demand the presence of oxygen as such so that the site of natural activity for carbon monoxide consumption must be at or near the soil surface. Of the anaerobic bacteria, one species, *Methanosarcina barkeri*, achieves the conversion represented as:

$$4 CO + 2 H_2O \longrightarrow 3 CO_2 + CH_4$$

This probably proceeds by way of a primary step involving the generation of hydrogen. There is also some evidence that carbon monoxide can be utilized by photosynthetic bacteria; it has been observed that radioactively-labelled ^{14}CO can be incorporated by various green algae and even by barley leaves.

Having established with some confidence the nature of the principal sinks for carbon monoxide, we must enquire into the nature of the major natural source required to match the removal rate. It seems certain that this must arise from the primary production of methane and its oxidation within the troposphere. In the preceding section we have discussed the microbiological production of methane and the source strength was of the order of 10^{12} kg per annum. It is now considered that a large fraction of this gas is oxidized within the troposphere, possibly through the agency of the initial attack of hydroxyl radicals, and any realistic overall mechanism leads inevitably to carbon monoxide in large part. It must be borne in mind that the rate of the oxidation is slow enough for another significant fraction of the methane flux to penetrate to the stratosphere, where it serves as a principal source of water vapour. Accordingly the tropospheric residence time for methane of around 5 years is not unreasonable, but it could be as low as 2 years within the present levels of uncertainty of the natural source strength.

Through the arguments above, we have arrived at the conclusion that the anthropogenic source strength of carbon monoxide is probably relatively small compared to the natural source strength. The evidence is fairly strong for a residence time in the troposphere measured in months rather than years, which fits in with the differing background levels of the two hemispheres, but it is not yet clear as to how a gas like methane, with a long enough residence time to be thoroughly mixed throughout the troposphere, can in its oxidation generate the asymmetric distribution of carbon monoxide between the hemispheres. Perhaps it is some function of the proposed coreactant, hydroxyl radical, concentrations but little is known on the background levels of that.

Hydrocarbon Species

In conclusion to our survey of the carbon cycle, it is worth noting that hydrocarbon species are also generated by biological activity. For example, the phytoplanktonic activity in illuminated seawater, discussed above in connection with the oceanic production of carbon monoxide, gives rise to hydrocarbon species also. In one instance, a bacteria-free culture of the ultradiatom *Chaetoceros galvestonensis* produced increasing concentrations of carbon monoxide, and C_2 to C_4 alkanes and alkenes, as a function of illumination time. Plant species also generate significant amounts of volatile hydrocarbons and the major released species from trees are ethylene, C_{10} monoterpenes and C_5 isoprenes. Present thought suggests that the natural emission rates of hydrocarbons (excluding methane) could exceed by an order of magnitude the total anthropogenic source strength, but of course rather different species are involved in each category. Many of the hydrocarbon species are quite reactive in air; ethylene, the most abundant in this group, has a rather short residence time as is evidenced by a background tropospheric mixing ratio of less than 10^{-3} p.p.m. There is also evidence that soils can take up such species as ethylene and acetylene fairly rapidly through microbiological activity.

As an aside, it should be noted that microbiological activity is not considered to have played much part in the formation mechanism of the bulk of the world's petroleum. The alkanes above C_4 therein are thought to have arisen from the thermal cracking of organic materials buried within sedimentary deposits at depths greater than 1 km where temperatures exceed 325 K. Strong support for this contention comes from the enhanced concentrations of the higher alkanes detected within sediments in proximity to underground igneous activity.

7.3 The Oxygen Cycle

The free oxygen of our present atmosphere has been generated mainly by biological processes in the prehistory of the Earth. Photosynthetically-produced oxygen enters the atmospheric reservoir, effectively generated from the dissociation of water molecules (Chapter 4), and the cycle is closed by respiration of plants and organisms. From Table 7.2 we find that an estimated 1.4×10^{14} kg of carbon dioxide is taken up by land plants each year and this may be approximately doubled on the basis of the numbers given in Figure 65 to take into account the phytoplanktonic photosynthesis in the mixed layer of the oceans. From the photosynthetic equation:

$$CO_2 + H_2O + Light \longrightarrow (CH_2O) + O_2$$

we see that for every 44 g of CO_2 fixed, 32 g of O_2 is released. Hence we deduce an annual release rate for O_2 of some 2×10^{14} kg. The total oxygen content of the atmosphere is near to 10^{18} kg (Tables 1.4 and 1.5), so that the annual production represents some 2×10^{-4} or 0.02% of the whole. Thus the residence time of oxygen in the atmosphere is around 5000 years, since the removal rate must balance production.

The three major repositories for oxygen not incorporated into living matter are therefore water, carbon dioxide, and oxygen itself as far as the biosphere is concerned. We have considered the cycles of the first two in the preceding sections and in so doing we have covered much of the oxygen cycle. Moreover, in Section 1.3, the evolution of oxygen in the Earth's atmosphere was discussed. Oxygen is not found either in the volcanic outgassings of the interior of our planet, nor is it found in the vesicles of volcanic rocks. Therefore we can guess that oxygen cannot be a primary component of the atmosphere in the sense that carbon dioxide and water vapour are.

We may ask at this stage a critical question — how may oxygen accumulate in the atmosphere to such a considerable extent as compared to other secondary components, if it is the intermediate in a cyclic process of photosynthesis—respiration—decay of carbohydrates? The organic material produced is decomposed and oxidised back to carbon dioxide and water, withdrawing oxygen in the process, and should forbid any large accumulation in the atmosphere. However, this postulation forgets about the large amounts of carbon buried in the Earth's crust, devoid of associated oxygen, for example, coal, oil, shale. When estimates have been made of the total masses of oxygen and carbon which originated as outgassings and are contained in sedimentary rocks and deposits, the hydrosphere, the biosphere, and the atmosphere, the proportions turn out to be close to the 32:12 in CO_2 itself. This would be a logical state of affairs if the primary injection of the element oxygen to the mobile system came from outgassed carbon dioxide over the lifetime of the planet. Table 7.3 gives estimates for the distribution of oxygen and carbon in the mobile system and in each category the positive or negative excess of oxygen over carbon compared to carbon dioxide.

The relatively small excess of carbon over oxygen shown in Table 7.3 is very likely to be within the error limits of the estimates when the difficulty of making them is considered.

If we look back to Table 1.3 we find that oxygen is in huge excess over carbon within the Earth as a whole. Of the total mass of 1.7×10^{24} kg of oxygen in our planet, only around 6×10^{19} kg is in the mobile system. However, the vast bulk of

Table 7.3. Distribution of mobile oxygen and carbon on the earth and the excess of oxygen to carbon with reference to carbon dioxide (masses in 10^{17} kg)

Component	Oxygen	Carbon	Excess O/C
Carbonate rocks	500	120	+180
Organic matter in sedimentary rocks, including fuels	20	120	−300
Sedimentary sulphates	30	0	+ 30
Iron oxides	20	0	+ 20
Sulphate in seawater	30	0	+ 30
Atmosphere, hydrosphere, and biosphere (excluding water as such)	11	1	+ 10
Others (e.g. nitrate deposits)	1	0	+ 1

Total excess of oxygen to carbon compared with CO_2 − 29

Data are derived from a diagram in P. Cloud and A. Gibor, 'The oxygen cycle', *Chemistry in the Environment*, a collection of readings from *Scientific American*, 1973, Chapter 4, p. 38, W. H. Freeman, San Francisco.

the oxygen is locked irretrievably away as in silicates, which are immune to separation of silicon and oxygen for all purposes.

It therefore becomes clear that the free oxygen content of our present atmosphere arises effectively from the dissociation of carbon dioxide with much of the carbon remaining buried. At the same time the formation of carbonate rocks amounts to a burying of oxygen, but Table 7.3 emphasizes the excess of buried carbon over buried oxygen in this sense. We may also appreciate the 'fringe' nature of the oxygen vital to life, amounting to only about 2% of the oxygen in the mobile system.

This section then carries a warning for the long term future of man's existence on Earth. Projections have been made of the possible consumption patterns of fossil fuels, generally going through a maximum at around the year 2200. Up to the present day we have mobilized but a small fraction (around 2%) of the original fossil fuel reserves. As this fraction increases, the carbon dioxide mixing ratio in the atmosphere will rise (projected to a value of the order of 1500 p.p.m. over the next few centuries) and, even ignoring the consequences of that in itself, the eventual result is the precipitation of carbonate in the oceans. Table 7.3 throws light on the effective conversion of withdrawn carbon to withdrawn oxygen so resulting. The two largest excess quantities will thus move positively and it can be envisaged that the oxygen content of the atmosphere could be significantly fixed as carbonate well before the fuel reserves approach exhaustion on an overall basis. This oversimplified extrapolation is only presented to draw attention to the spurious concept of total fuel reserves often talked about. There can be little doubt that the real degree of consumption of fossil fuels will be even lower on the basis that a mixing ratio of 1500 p.p.m. of carbon dioxide in the atmosphere would almost certainly produce an intolerable environment.

As mentioned in Table 7.3, water has been excluded from the assessment of mobile oxygen, largely because it is a primary component of volcanic gases also. Over the period of existence of the secondary atmosphere of the Earth, water can be regarded as having remained as such to a large extent. Only the escape of hydrogen from the upper atmosphere to space can have produced free oxygen from water, and this can be taken to be insignificant compared to the content of the oceans. It may appear anomalous that, in the discussion of photosynthesis in Chapter 4, it was stated that the oxygen liberated was actually derived from water rather than from carbon dioxide; yet the carbon/oxygen system controls the accumulation of oxygen in the atmosphere. However, it must be remembered that in the formation of coal for example, the overall process of degradation of carbohydrate may be represented as:

$$(CH_2O) \xrightarrow[\text{compression}]{\text{heat}} C(s) + H_2O \uparrow$$

Thus water is restored to the oceans eventually with the oxygen atom originally derived from carbon dioxide.

Finally in this section, we must justify the view that the carbonate rocks are part of the mobile system, even if it involves a very long time scale. The limestones and the like, formed beneath the sea in geological epochs long ago, have been thrust up to the surface by earth movements, a spectacular instance being the limestone near to the top of Mount Everest. Acidic rainwater will then dissolve the surface slowly away, restoring carbon dioxide to the atmosphere from which it was originally derived.

7.4 The Nitrogen Cycle

Nitrogen gas composes 78.084% of our atmosphere and from Table 1.4 we can deduce that the mass of the nitrogen in the atmosphere is close to 4×10^{18} kg. Rough estimates suggest that this represents approximately 18% of the total mass of nitrogen atoms within the Earth system. The remainder is largely locked away within the rocks and deposits of the crust and, in general, this is immobile in terms of the nitrogen cycle to which we address ourselves at present.

Within the mobile system, only a very small fraction of the total nitrogen is in forms other than N_2 and the transfer rates are rather small. This is borne out by the estimated residence time of N_2 in the atmosphere of around 10^6 years. But the fixation of nitrogen is absolutely vital to life and our supply of food is perhaps more determined by the rate of nitrogen fixation than by any other factor on a general basis. Amino acids, the building blocks of proteins, incorporate the fixed nitrogen into the higher forms of life.

Our interest in this cycle is concentrated upon those components which achieve importance within the general areas of energy and atmospheric chemical dynamics. We are already aware that combustion achieves a degree of nitrogen fixation in the emission of nitric oxide. Also we have seen in Section 7.1 that bacteria within soils can synthesize nitrous oxide, and it will become apparent in Chapter 9 that this is of significance in the chemistry of the stratosphere into which it is transported. A large amount of nitrogen is intentionally fixed each year by man in the production of synthetic fertilizers; in 1973 some 9×10^8 kg of ammonia were fixed through the Haber—Bosch process on a world-wide basis, involving the expenditure of around 2×10^{16} J of energy, mainly derived from fossil fuels. It is particularly interesting to note the massive expenditure of energy occasioned by a complex interaction of technical and economic factors in practice required to drive the *exothermic* Haber reaction:

$$N_2 + 3 H_2 \xrightarrow{\text{catalyst}} 2 NH_3 \quad \Delta H^{\ominus}_{298} = -92.048 \text{ kJ per mol } N_2.$$

In fact recent analyses (Pimentel and coworkers, 1973) of the energetics of the production of maize in the U.S.A. have suggested that the energy input to the crop as nitrogen fertilizer can amount to almost 12% of the final energy yield. Moreover, if all the energy input terms (e.g. gasoline, machinery, drying, electricity, etc.) are added together, then the final harvested maize energy content is only about three-times larger. This marks a worrying trend in current agricultural practice in that more and more fossil fuel energy is applied year by year to produce a given crop yield. In this section we shall attempt to assess the relative magnitude of the various components of the nitrogen cycle.

The microorganisms which fix nitrogen into soils naturally are of two broad categories, viz. independent and symbiotic. The latter type are microorganisms which depend directly upon the higher plant with which they are associated for their supply of vital energy. Amongst these are the symbionts located within the roots nodules of the legume crops, e.g. alfalfa, peas, beans, which are, for the planet as a whole, probably the greatest source of naturally-fixed nitrogen. The annual intensity of nitrogen fixation by such crops can reach 3.5×10^4 kg km^{-2}, about two orders of magnitude larger than the corresponding intensity for the independent organisms. In comparison the annual addition of fixed nitrogens rained out of the atmosphere on the average amounts to no more than 500 kg km^{-2}.

Molecular nitrogen can be considered to be highly inert from the chemical point of view on both thermodynamic and kinetic grounds. The $N \equiv N$ bond has a dissociation energy of 941.8 kJ mol^{-1} and is one of the strongest bonds. As such it is then difficult to find reactions involving N_2 as reactant which are not heavily endothermic and which therefore do not demand a large investment of energy, but even when an exothermic reaction is found, such as the Haber reaction above, there is usually a kinetic limitation associated with the difficulty of activating N_2 to take part in the reaction. As a result, the Haber reaction is immeasurably slow at ambient temperatures, even in the presence of catalysts. Temperatures of the order of 700 K are required to overcome the kinetic limitation in the Haber–Bosch synthesis of ammonia.

How then do the microorganisms achieve the activation of N_2 at ambient temperatures?

The assimilation of nitrogen is catalysed in living organisms by the binary enzyme complex known as nitrogenase. This has been shown to consist of two separable protein structures. Component I has molybdenum and iron atoms bound within it and has a molecular weight of the order of 220,000. It is capable of existing in an oxidized form, a half-reduced form, and a fully reduced form. Component II has a lower molecular weight, around 60,000, and has several iron and sulphur atoms per molecule. The active site of nitrogenase is by no means substrate specific; the enzyme has been shown to be capable not only of reducing molecular nitrogen to ammonia but also nitrous oxide to water and molecular nitrogen, and even acetylene to ethylene, and hydrogen ions to gaseous hydrogen.

In the nitrogenase fixation process the nitrogen is reduced first to ammonium ions. The energy demands of such an energy-consuming process are met by the same energy carrying species, Adenosine Triphosphate (ATP), as was encountered in the same general role in our discussion of photosynthesis (Chapter 4). In the case of the symbiotic microorganisms, the host plant donates the ATP as its part of the coexistence arrangement. In the independent microorganisms, such as blue-green algae and photosynthesizing bacteria, the ATP is the result of photosynthesis within the microorganism itself. Magnesium ions appear to be an essential cofactor so that we may represent the primary fixation process as:

$$N_2 + 8\,H^+ + 6\,e^- \xrightarrow{\;\;Mg^{2+}\;\;} 2\,NH_4^+$$

$$n\,\text{ATP} \qquad\qquad n\,\text{ADP} + n\,P_i$$

The number, n, of ATP molecules required is uncertain, perhaps being in the range of 6 to 15, but experiments have suggested that it can vary with conditions. Substances effecting the required electron transfers must also be a part of the reaction system. It is worth mentioning that in Australia there were regions where legume crops could not be cultivated, for no obvious reason. These tracts of land were found eventually to be highly deficient in molybdenum: it was therefore hardly surprising, in view of the molybdenum content of component I, that fertility developed almost magically upon minor applications of a molybdenum dressing to the land.

Thus microorganic fixation of nitrogen is only possible because energy carried in the form of ATP molecules can be applied to the activation of the N_2 molecule. As this is highly expensive in terms of energy, it is not surprising that nature has evolved a regulatory system whereby nitrogenase synthesis is inhibited in the presence of adequate external concentrations of ammonium or nitrate ions. Under these conditions a switch is made to assimilation of these less energy expensive ions to satisfy the nitrogen requirements of the microorganism.

The symbiotic microorganisms can denote up to 90% of their fixed nitrogen to the host plants, the second element of the coexistence bargain. On the other hand, the independent organisms excrete up to 80% of their assimilated nitrogen in various forms such as ammonia, hydroxylamine derivatives, and amino acids. In turn these can be assimilated by higher plants in the vicinity.

Once ammonia or ammonium ions have been generated, they can be incorporated into a plant, which may become food for an animal. Thus the nitrogen works its way up the biological chain, with some fraction being excreted at each stage. A particular instance may be used to exemplify this stage of the nitrogen cycle.

A study has been performed of the nitrogen conversion in a grazed alfalfa pasture near Canberra, Australia, during March 1974 (Denmead, Simpson, and Freney, 1974). The field concerned had an area of around 4×10^4 m^2 and during the measurement period was evenly grazed by 200 sheep. Measurements of the upward flux of ammonia showed an average daily loss of nitrogen in this form of 26 kg m^{-2}. Of the annual fixation of nitrogen, it was estimated that about 5000 kg km^{-2} remained in the soil and the equivalent of about 2000 kg km^{-2} remained in the animals. Thence the estimated annual loss as ammonia was estimated to be of the order of 10^4 kg km^{-2}, which figure expressed on a daily basis almost matched the measured ammonia flux. The sheep were returning an estimated 100 kg km^{-2} per day to the soil as excreta, probably the major source of the ammonia flux. This study establishes the importance of gaseous losses in the nitrogen economy of pastures, suggesting that these are important contributors to the cycle of atmospheric nitrogen.

Within soils there is also a bacterial chain which effects energy abstractions from the conversions of ammonia or ammonium ions to nitrogen or nitrous oxide. Such bacteria are termed *chemoautotrophs* and they obtain their vital energy from inorganic compounds. The first member of the chain, the genus *Nitrosomonas*, relies upon the c.272 kJ mol^{-1} made available to the conversion of ammonia to nitrite ion in the presence of oxygen:

$$2\,NH_3 + 3\,O_2 \longrightarrow 2\,HNO_2 + 2\,H_2O$$

The *Nitrobacter* group of microorganisms rely upon the extraction of energy from the oxidation of nitrite ions to nitrate ions. This conversion furnishes some 72 kJ mol^{-1} and provides for *Nitrobacter*.

In soils there are a large number of facultative bacteria for the denitrification which follows under anaerobic conditions. Amongst these species are *Pseudomonas denitrificans* and *Micrococcus denitrificans*. These possess the ability to use the nitrate ion to effect the oxidation of organic matter and so to produce the energetic needs of their existence. The energy yields available from this type of metabolism have already been mentioned in Section 7.1. It is evident that the application of synthetic fertilizers such as nitrates must lead to increased rates of the denitrification processes and hence increased emissions of nitrous oxide from the surface to the atmosphere.

Similar microorganic action in the surface waters of the oceans also give rise to denitrification products such as nitrous oxide. This postulation is supported by the slight supersaturation of the ocean surface layers with respect to the nitrous oxide content of the air, just as is the case for carbon monoxide (Section 7.2).

In the troposphere nitrous oxide is a rather inert gas from the point of view of chemical reactivity but, as shown in Table 7.1, the average tropospheric mixing ratio is only 0.25 p.p.m., implying the existence of a fairly efficient sink for nitrous oxide at the Earth's surface. The oceans, by virtue of the slight supersaturation of their surface layers, are ruled out as sinks for nitrous oxide. Further microbiological activity at the land surface appears to be the major sink since laboratory studies have shown that many denitrifying species are able to grow with nitrous oxide as the sole electron acceptor for oxidation. In fact some workers have postulated that as much as 93% of the nitrous oxide produced by soils is reabsorbed there. The remainder is likely to be transported into the stratosphere where it is mainly destroyed by ultraviolet photodissociation:

$$N_2O + h\nu \longrightarrow N_2 + O$$

an aspect which will be considered in Chapters 8 and 9. The data in Table 7.1 suggest an annual evolution of nitrous oxide of the order of 10^{11} kg.

The residence time of ammonia in the atmosphere on the other hand is very short, only a matter of around 10 days. It is no coincidence that this is close to the average residence time for water in the atmosphere (Section 7.1); the high solubility of ammonia in water (62.9 kg/100 kg at 293 K) probably means that the most important removal mechanism is solution in rain water. Reaction with sulphur dioxide, nitric acid vapour, and other gases may also be a significant removal term. It therefore becomes apparent that the emission of ammonia probably contributes little to the turnover of atmospheric nitrogen in the chemical sense, since much of it is returned to the soil or ocean for further conversion by microorganisms. In view of the short atmospheric residence time, it is not surprising that tropospheric background levels are highly variable and this point will be emphasized in Section 8.6, when tropospheric aerosols are considered.

There is almost no information on the production rates of nitric oxide and nitrogen dioxide by natural agencies. Background mixing ratios are very low, perhaps as little as 0.001 p.p.m., but to date there have been insufficient measurements for a reliable figure to be quoted. It is possible that these oxides are produced by microorganic activity in soils and there appears to be a strong case for generation by the gas-phase oxidation of ammonia; estimates of the latter term have ranged up to 7×10^{10} kg of N in the form of NO and NO_2 per year. The anthropogenic source strength of the two oxides (represented collectively as NO_x) has been estimated at around 2×10^{10} kg of N per annum. Amidst the general uncertainty of the natural source strength of NO_x, it does appear likely that the anthropogenic source contributes less than 20% of the total at present.

In the atmosphere, nitric oxide will become oxidized to nitrogen dioxide, by reaction with oxygen if nothing else. The primary removal mechanism for nitrogen dioxide is precipitation, usually in the form of nitric acid or other nitrates. In this event the residence time would be of the order of 10 days or less, just as argued for ammonia, and on the basis of a speculative average mixing ratio of 0.001 p.p.m., the removal rate would be of the order of 4×10^{11} kg per annum. This, then, supports the contention that the natural sources of NO_x dominate the anthropogenic contribution.

In conclusion to this survey of the atmospheric nitrogen cycle, it appears clear that for the atmosphere as a whole, natural sources are dominant for the production of both the oxides of nitrogen and ammonia. The problems which arise from the anthropogenic sources of NO_x are due to highly localized source strengths and inhibited dispersal. Rainwater has a significant concentration of both ammonium ions and nitrate ions. The annual rate of deposition of this 'natural' fertilizer has been estimated to be some 2×10^{11} kg, containing 7×10^{10} kg of fixed nitrogen. This is some two orders of magnitude larger than the annual yield of fixed nitrogen from the Haber–Bosch process worldwide. The synthetic fertilizer, however, is applied to localized agricultural areas whereas the natural downward flux is disseminated over oceans, mountains, and polar icecaps and so is to a large extent 'wasted' from the agricultural point of view.

7.5 The Sulphur Cycle

The element sulphur gives rise to major problems for man, which have already reached the scale of national concern. Sulphur is present in most coals and crude oils, and the combustive generation of energy leads invariably to the emission of sulphur dioxide at exhaust. In the atmosphere, sulphur dioxide eventually becomes oxidized to the sulphate ion, largely in the form of sulphuric acid. The consequent acidity in rainfall has produced unfortunate results such as the poisoning of soils and the leaching of the faces of buildings. In the light of these and other circumstances it is essential to examine the global sulphur cycle in order to assess the significance of the anthropogenic term, which can be controlled in principle.

We can make a rough estimate of the anthropogenic source strength from the projected figures for the 1977 world energy demand given in Table 5.3. These envisage consumption of oil equivalent to 1.3×10^{17} kJ for the non-Communist world and this may be increased to 1.7×10^{17} kJ for an approximate global figure. On conversion to mass on the basis of the heat of combustion of alkanes given in Table 5.1, this corresponds to 3.5×10^{12} kg of oil. A similar procedure for the solid fuel data gives a mass of 1.7×10^{12} kg. A reasonable average for the sulphur content of these fuels would be 2% so that the estimate for the anthropogenic emission of sulphur dioxide for 1977 comes out at 2×10^{11} kg, probably accurate to within a factor of 2. Other sulphur sources of an anthropogenic nature, such as ore smelting, contribute less than one-quarter to the total anthropogenic source strength.

Sulphate ions are present in seawater at a concentration of approximately 1 kg of sulphate in 3.7×10^{-10} km^{-3}, which would appear to be the logical sink of atmospheric sulphur. The total volume of seawater is 1.5×10^9 km^3, so that assuming complete mixing of sulphate throughout the oceans, the sulphate content is some 1.4×10^{18} kg in terms of sulphur. Since the anthropogenic release of sulphur is of the order of 10^{11} kg per annum, it can be seen that this will be literally a drop in the ocean.

Volcanoes come to mind as perhaps the most obvious natural sources of sulphur emission to the atmosphere, but all estimates of the annual injection rate from these are of negligible proportions in comparison with the anthropogenic release rate, perhaps around 5%.

The primary natural source of sulphur release to the atmosphere is the biogenic activity which generates hydrogen sulphide. This gas is the primary product of sulphur metabolism in the soil by microorganisms and arises from essentially two sources. It is a product of the decomposition of the amino acids containing thiol groups on a minor scale, but by far the major source of hydrogen sulphide is the process of sulphate respiration by bacteria; wherever organic material decays under anaerobic conditions with sulphate available, copious evolution of the gas takes place. Only two strictly anaerobic genera of bacteria, *Desulfovibrio* and *Desulfotomaculum*, account for the bulk of the evolution. Their metabolic processes depend upon the use of organic acids as hydrogen donors for the subsequent reduction of the sulphate. Common locations for this bacterial activity are estuarine muds and coastal marshlands, where the sulphate is derived from seawater itself; the blackening of the muds is largely due to the precipitation of iron sulphide following the generation of the hydrogen sulphide. As described in Section 7.1, we expect primary generation of hydrogen gas from buried organic matter. In the presence of this hydrogen the sulphur-using bacteria can respire according to the overall equation:

$$4\,H_2 + 2\,H^+ + SO_4^{2-} \longrightarrow H_2S + 4\,H_2O$$

It is to be noted that even resting and non-growing bacterial cells continue to reduce sulphate, with the result that large amounts of hydrogen sulphide are in fact produced by quite small numbers of the bacteria. It is evident also that any marshland, lake, or pond bottom, or even the soil itself, will gain sufficient sulphate from drainage water to give rise to hydrogen sulphide evolution.

It has been demonstrated (Lovelock, Maggs, and Rasmussen, 1972) that another volatile compound of sulphur, dimethyl sulphide, $(CH_3)_2 S$, is released into the atmosphere by both living and decaying vegetation, from the soil, and by marine algae such as *Laminaria*. Living intact leaves of trees such as oak and pine were found to give off measurable amounts of dimethyl sulphide but the emission rate was found to be up to two orders of magnitude larger for decaying leaves. Moreover, as a result of marine algal activity, there are measurable concentrations of dissolved dimethyl sulphide in samples of seawater taken off the south coast of England.

It is not yet clear as to the proportions of hydrogen sulphide and dimethyl sulphide which are evolved on an annual basis; there have been suggestions that they could be approximately equal but it is likely that hydrogen sulphide evolution is more prevalent. Most estimates of the annual natural emission of hydrogen sulphide alone suggest a value of the order of 100×10^9 kg of sulphur with a slightly larger fraction being evolved by the oceans than by the land. This figure could be approximately doubled to take into account dimethyl sulphide production, unrecognized at the time.

We conclude, therefore, that the major production of sulphur compounds by natural processes results in sulphide gases in the first place, but background mixing ratios of hydrogen sulphide are extremely small, usually below 10^{-3} p.p.m., and are highly variable. It is therefore immediately clear that the residence time of hydrogen sulphide in the troposphere must be very short, perhaps as little as 1 day, and the implication is that the emitted hydrogen sulphide is fairly rapidly oxidized to sulphur dioxide. The effective coreactant in this could be ozone, inducing the reaction:

$$H_2 S + O_3 \longrightarrow SO_2 + H_2 O$$

However presently available kinetic data for this reaction in the gas phase seem unable to match the required rate of conversion, but there is considerable evidence becoming available that heterogeneous oxidation of hydrogen sulphide on atmospheric particulate materials may be the answer to this problem.

It is probable that dimethyl sulphide is also subject to fairly rapid oxidation in the troposphere, although little is known of the mechanistic pathway which this would take. By analogy with the mechanism of oxidation at temperatures above ambient in laboratory studies, the intermediate species could be dimethyl sulphoxide and methane sulphonic acid. However the sulphur must end up as sulphur dioxide for the most part.

The important conclusion from our point of view, which comes from comparing the effective emission rates of sulphur dioxide to the atmosphere, is that the anthropogenic term is a major one, certainly approaching a magnitude comparable to the natural term on a global basis. This is then the first instance which we have uncovered in this chapter where there is potentially significant perturbation of a natural cycle on a global scale by man's activities, principally the combustion of fossil fuels.

Sulphur dioxide is very soluble in water, with a solubility of about 10 kg per 100 kg of water at ambient temperatures. Consequently it seems likely that a major portion of the tropospheric content of this gas is removed by solution and precipitation scavenging. However, sulphur dioxide can be photooxidized rapidly (particularly in polluted environments where much of the anthropogenic source is located), and a significant fraction of the sulphur rained out may be present as sulphate aerosol prior to scavenging; under typical background tropospheric conditions something like 60% of the sulphur content can be in the form of sulphate with equal parts of the remainder present as hydrogen sulphide and sulphur dioxide. Vegetation is capable of absorbing sulphur dioxide as such from the atmosphere. All species of plants can be damaged by exposure to substantial levels of the gas and in fact some species like lichens are sufficiently sensitive as to serve as monitors of the average levels in contact: more will be said of the use of lichens in this connection in Section 8.5. The plants take up the gas through the stomata in the leaves; as was remarked in Section 7.1, the stomatal opening is vital to the transpiration of the plant and to its intake of carbon dioxide for photosynthesis. Accordingly it is not unexpected that the maximum rates of sulphur dioxide uptake are detected under maximal stomatal opening conditions such as high humidity or high incident solar irradiance. The absorbed sulphur may be incorporated into the plant structure, as for example in the protein content of wheat which contains peptide units derived from the thiol-group amino acid methionine. Eventually such sulphur finds its way to the soil when the plant dies, the leaves fall, or when it is eaten and excreted by animals. It is also known that sulphur dioxide can be absorbed directly by soils and of course it can go directly into solution in lakes, ponds, and marshy environments. Taken together, various estimates of the annual rate of withdrawal of sulphur dioxide by these land-based systems must come close to matching the annual rate of emission from land-based sources since the residence time for the troposphere is rather short, on the average about 4 days but as short as 1 hour in certain environments, particularly urban or industrial ones. Estimates of the net transfer rate from the atmosphere over land to the atmosphere over the oceans generally lie in the range of 10 to 20% of the total gaseous sulphur injection rate by land. The final land-based sink for the sulphur cycle is the solution of sulphate in rivers and run-off to the oceans.

Hence the land sulphur budget involves two substantial input terms: precipitation of sulphur dioxide or sulphate dissolved in rain to the ground (possibly with a minor addition by dry-deposition of sulphate aerosols), and the combined uptake of sulphur dioxide by plants and soils. These are balanced mainly by two large terms deriving from biogenic evolution of gaseous sulphides and river run-off of sulphate. Figure 66 summarizes the land-based cycle; the numbers inserted are feasible annual rates in units of 10^9 kg of sulphur and are only intended to convey an order of magnitude impression for the year 1977.

The sulphur budget over the oceans involves one additional large term representing the direct solution of sulphur dioxide in the water. Not only does

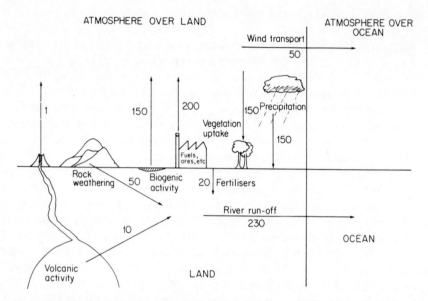

Figure 66. The sulphur budget for the land–atmosphere system. The annual turnover rates are indicated in units of 10^9 kg S and are the author's estimates for 1977.

sulphur dioxide have the relatively high solubility in water mentioned earlier but the slight alkalinity of seawater (present pH = 8.22) should aid the solution of the anhydride sulphur dioxide.

Let us consider first the exchange of a sparingly soluble gas such as oxygen from the atmosphere into solution across the water–air interface. The greatly reduced rate of diffusion of dissolved gas molecules compared to those in the gas phase above the water creates effective liquid phase resistance to the transfer across the interface. In other words, when a small element of liquid at the interface has dissolved so many gas molecules, it becomes momentarily saturated and cannot accept any more until a few of those within the element have diffused downwards and thus out of the element. In the interim, if a further gas phase molecule does enter the element, then one can imagine that another must be ejected back into the gas phase to alleviate the momentary supersaturation.

In the case of a gas, like sulphur dioxide, which changes chemically by forming ions upon entry to the liquid phase, transport out of the element is enhanced because the normal dissolved gas concentration gradient becomes supplemented by the concentration gradient of ionic species, in this case the sulphite, SO_3^{2-}, ion. Under these circumstances it could be the case that the gas molecules enter solution as soon as they come into contact with the liquid surface.

Hence, it is the rate of diffusion of a chemically reactive gas from the bulk gas phase to the interface which may be rate determining for the overall absorption into the bulk liquid phase. Recent work (Liss, 1971) has put this aspect on a more quantitative footing. The exchange rate of a gas between gas and liquid phases is expressed in terms of the 'resistance' of the interface to gas transfer; the resistance, R, is defined as the ratio of the concentration difference of the gas molecules between the phases to the flux of the gas. This leads to an equation:

$$R_T = R_L + R_G/K_h \tag{7.7}$$

where R_T is the total resistance on a liquid phase basis and R_L and R_G are the individual resistances of the liquid and gas phases respectively. K_h is the Henry's Law constant, defined as:

$$K_h = \frac{\text{Equilibrium concentration in the gas phase}}{\text{Equilibrium concentration of unionized dissolved gas in the liquid phase}}$$

When the gas behaves as does sulphur dioxide and rapidly hydrates and ionizes, the liquid phase individual resistance is lowered from the value appropriate to an inert gas like oxygen. This is expressed by a factor denoted by α, defined simply as the ratio of R_L for the chemically unreactive situation over the effective R_L for the reactive situation. Values of α for sulphur dioxide have been calculated but the procedure is too complex for detailed derivation here. As might be expected, α, depends critically upon the ratio of total to ionic forms of the gas in solution, and also upon the hydration rate constant for the gas molecules and the molecular diffusivity of the dissolved gas. In turn these factors lead to a strong dependence of R_L and α upon the pH of the water, with resistance increasing, as expected, with decreasing pH. Table 7.4 sets out a selection of the results calculated for sulphur dioxide, paying particular attention to the ratio R_G/R_L.

Since the pH of the sea is 8.22, it can be concluded from the data of Table 7.4 that transfer of sulphur dioxide from the air to the sea will be entirely controlled by the rate of diffusion to the interface from the gas phase, in view of the high resistance therein compared to that in the liquid phase.

It is interesting to compare the behaviour of sulphur dioxide in this respect to that of other trace gases. In the case of carbon dioxide, a much less reactive gas, although the R_L value increases with increasing pH, it is always greater than R_G. Therefore exchange of carbon dioxide to the oceans is primarily controlled by the

Table 7.4. Parameters for the transfer of sulphur dioxide from the gas phase to water as a function of pH at ambient temperature

pH	α	$R_L/\text{h m}^{-1}$	$(R_G/K_h)/\text{h m}^{-1}$	R_G/R_L
2	2.7	3.7	0.88	0.24
4	169	0.059	0.88	14.92
6	2884	0.0035	0.88	251.4
8	2967	0.0034	0.88	258.8

From P. S. Liss, 'Exchange of SO_2 between the atmosphere and natural waters', *Nature*, **233**, 327–329 (1971). Reproduced by permission of Macmillan (Journals) Ltd.

resistance of the liquid phase. Consequently the rate of absorption from the atmosphere to the oceans will be much lower than that for the sulphur dioxide.

In fact, it appears that sulphur dioxide will behave more like water vapour in regard to transfer to the oceans; the evaporation or condensation of any pure liquid is controlled by resistance in the gas phase. Two important deductions follow: the vertical concentration gradients of sulphur dioxide and water vapour close to the ocean surface should be similar. Moreover the residence time of a sulphur dioxide molecule in the atmosphere should be of the same order (and probably less in view of photooxidation and vegetation scavenging, etc.) as that for water vapour, which is around 10 days. In fact 4 days is generally accepted as the average residence time for sulphur dioxide. Hence we may conclude that solution of the gas into the oceans will be an important term in the marine sulphur budget. This will be enhanced by the constant wave turbulence at the interface.

As has been mentioned earlier in this section, there is biogenic activity in the mixed layer of the oceans based upon sulphate respiration. Moreover it is known that sulphate-reducing bacteria populate parts of the ocean floor. However, the hydrogen sulphide which these latter generate must pass up through the layers of seawater containing dissolved oxygen. The residence time of dissolved hydrogen sulphide in these layers would appear to be sufficiently long for the bulk of the gas to be oxidized back to sulphate. Moreover the alkalinity of seawater will not favour release of the acidic hydrogen sulphide to the atmosphere. In accord with these arguments, no hydrogen sulphide has been detected at the ocean surface by currently available analytical techniques, but such preclusions do not apply to dimethyl sulphide, which has been detected in samples of seawater. Although there is little information on the global distribution of dimethyl sulphide generating capacity in the oceans, it does seem likely that this is the principal form of sulphur evolution from the oceans to the atmosphere, if only by elimination. In the absence of such a term for dimethyl sulphide, attempts to produce a mass balance for sulphur on a global basis have usually produced a deficit against balance in the atmosphere.

Figure 67 is the equivalent for the oceanic system of Figure 66 for the land. Again it must be emphasized that the values which are inserted are only feasible rather than backed by strong evidence. The uncertainties are even larger in this instance in view of the lack of knowledge of the significance of dimethyl sulphide in comparison with the other terms. However, in view of the slowness of transfer from the mixed layer to deeper layers, it does not seem unreasonable to suggest that the mixed layer releases sulphur in one form or another to the atmosphere at approximately the rate of transfer in the opposite direction.

In this figure, heed is paid to the 3.5×10^8 km^2 surface area of the oceans as opposed to the 1.5×10^8 km^2 surface area of land, but it may be considered that the precipitation rates in the two cases are of the same order for the following reasons. Most of the anthropogenic generation of sulphur dioxide occurs in the northern hemisphere where land and sea surface areas are similar (1.0×10^8 km^2 and 1.5×10^8 km^2 respectively). Moreover in view of the rather short residence time of sulphur dioxide in the atmosphere, the middle areas of the great oceans will

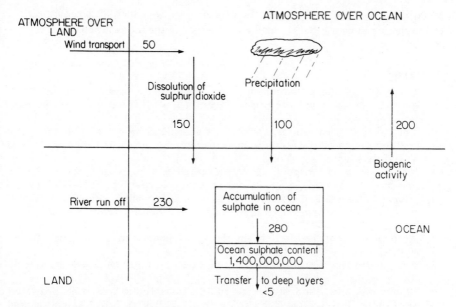

Figure 67. The sulphur budget for the ocean—atmosphere system. The annual turnover rates are indicated in units of 10^9 kg S and are the author's estimates for 1977.

not receive much anthropogenic contribution. Further, precipitation over the oceans will be less per unit area than over land because of the inductive effects of mountain ranges.

In Figure 67 the almost-closed cycle of the transfer of sulphate to the atmosphere in the form of sea-spray aerosol has been ignored; it is likely that the bulk of such sulphate returns to the oceans without contact with land. In any event, this sulphate remains chemically unaltered in the course of its transit.

Little is said here on the photooxidation of sulphur dioxide within the troposphere. This is a topic which is reserved for detailed treatment in the next chapter, as is the formation of dry sulphate aerosols.

7.6 Concluding Remarks

The aim of this chapter has been to provide an idea of the scale of anthropogenic pollution problems in the troposphere on a global basis in comparison with the natural sources and cycles, which were proceeding long before the present technological age. We have seen that there is no case where the anthropogenic contribution overwhelms the natural source strength of a species. For the hydrogen, carbon, oxygen, and nitrogen cycles, man's current levels of activity do not give rise to release rates which approach the order of magnitude of terms in the natural cycles. Only in the case of the sulphur cycle are anthropogenic terms encountered which are of a similar order of magnitude to the natural cycle terms.

We see, therefore, that atmospheric pollution, such as by oxides of nitrogen, is essentially a localized problem for the present. It is with such localized aspects of tropospheric chemistry that we shall be concerned in the following chapter.

Bibliography

1. 'Chemistry in the Environment'. A collection of readings from *Scientific American* (September, 1970 issue), Section I, W. H. Freeman and Company, San Francisco, 1973.
2. 'Emissions, Concentrations, and Fate of Gaseous Atmospheric Pollutants', by E. Robinson and R. C. Robbins, in *Air Pollution Control*, Part II ed, W. Strauss Wiley-Interscience, New York, 1972, pp. 1–93.
3. 'The Cycle of Atmospheric Gases – Natural and Man-made', by C. Junge, *Quarterly Journal of the Royal Meteorological Society* 98, 711–729, (1972).
4. 'Global Emissions and Natural Processes for Removal of Gaseous Pollutants', by K. H. Rasmussen, M. Taheri, and R. L. Kabel, *Water, Air, and Soil Pollution,* 4, 33–65, (1975).
5. 'Accumulation of Fossil CO_2 in the Atmosphere and in the Sea', by M. Whitfield, *Nature,* 247, 523–525 (1974).
6. 'Production, Modification and Consumption of Atmospheric Trace Gases by Micro-organisms', by H. G. Schlegel, *Tellus,* XXVI, 11–20 (1974).
7. 'Production of Carbon Monoxide and Gaseous Hydrocarbons in Seawater: Relation to Dissolved Organic Carbon', by D. F. Wilson, J. W. Swinnerton, and R. A. Lamontagne, *Science,* 168, 1577–1579 (1970).
8. 'Fixation of Atmospheric Nitrogen by Micro-organisms', by D. Kleiner, *Angewandte Chemie (International Edition),* 14, 80–86 (1975).
9. 'Atmospheric Ammonia', by J. C. McConnell, *Journal of Geophysical Research,* 78, 7812–7821 (1973).
10. 'A Study of the Sulphur Budget over Northern Europe', by H. Rodhe, *Tellus,* XXIV, 128–138 (1972).

Chapter 8

Photochemistry of the Polluted Troposphere

In the preceding chapter we concluded that there was little in the way of a global pollution problem created by anthropogenic emissions of typical pollutants like nitrogen oxides. Nevertheless we recognise that severe problems are generated on a localized basis by anthropogenic activity, as for example in the classical case of the Los Angeles basin, where there can be no doubt that it is principally the exhaust emissions of vehicles which are photochemically transformed into the obnoxious smogs which have proved to be the main stimulus to stringent emission control legislation in California. However, it must now be conceded that smog has become a worldwide hazard of urban environments, with episodes being a frequent occurrence in such diverse places as Tokyo, Ankara in Turkey, and Sao Paulo in Brazil. Accordingly in this chapter we attempt to come to an understanding of the conditions and chemistry which give rise to localized photochemical phenomena in the troposphere.

8.1 The Need for Photochemical Activation

It is light from the Sun which supplies the energy required to drive the chemistry of the troposphere. In the absence of such photochemical stimulation, always presuming that some heat source maintained the present ambient temperatures, there would be little if any chemistry since the activation energy barriers in reactions of stable molecules, are in general, large enough to make thermal activation at ambient temperatures vanishingly slow. Let us try to put this in perspective at the outset. Table 8.1 shows values of what is called the *Arrhenius factor*, $\exp(-E/R \cdot T)$, calculated for $T = 288$ K as a function of the *Arrhenius activation energy*, E.

We see that the Arrhenius factor falls dramatically as E increases. Now if we are to consider a bimolecular reaction of the general type:

$$X + Y \longrightarrow \text{Product molecules}$$

then the rate constant, k, for this reaction generally will vary with temperature according to the Arrhenius form:

$$k = A \cdot \exp(-E/RT)$$

Table 8.1. Values of the Arrhenius factor, $\exp(-E/RT)$, as a function of the Arrhenius activation energy, E, for $T = 288$ K

$E/\text{kJ.mol}^{-1}$	1.0	5.0	10.0	50.0	100.0	200.0
$\exp(-E/RT)$	0.66	0.12	0.015	8.5×10^{-10}	7.3×10^{-19}	5.3×10^{-37}

Here A is known as the Frequency or Pre-exponential factor. In the simplest collisional theory of bimolecular reactions, A is identified with a notional collision frequency of the reactant molecules, while the Arrhenius factor is the probability that the two colliding molecules possess between them sufficient kinetic energy to overcome the energy barrier, E, to reaction. For reactants which are typical stable molecules in the atmosphere (usually at most triatomic), the expected value of A is of the order of 10^9 dm^3 mol^{-1} s^{-1} and we shall presume exactly this value in the argument to follow.

Let us now consider some typical bimolecular reactions between common molecular species in the atmosphere. Table 8.2 sets out four of these with the activation energies which have been estimated for them. Molecules like ozone are not considered in this connection since they are generated photochemically. Comparison with Table 8.1 shows that the Arrhenius factors for such reactions are vanishingly small for ambient temperatures.

Consider the reaction of nitrogen and oxygen, the most abundant atmospheric constituents. The concentrations of these species are 78.084% and 20.948% of [M] respectively, where [M] denotes the total concentration irrespective of species. For a temperature of 288 K and a standard atmosphere pressure, the value of [M] is 4.23×10^{-2} mol dm^{-3}, so that [N$_2$] is 3.30×10^{-2} mol dm^{-3} and [O$_2$] is 0.886×10^{-2} mol dm^{-3}. The instantaneous rate of the bimolecular reaction concerned is then expressed:

$$\text{Rate} = -d[\text{N}_2]/dt = -d[\text{O}_2]/dt = k[\text{N}_2][\text{O}_2] \tag{8.1}$$

For suitably low extents of conversion (say 1%) we may assume that [N$_2$] and [O$_2$] remain constant and replace the differential form $-d[\text{N}_2]/dt$ by the incremental form $-\Delta[\text{N}_2]/\Delta t$, where $-\Delta[\text{N}_2]$ now represents the decrease in the concentration of nitrogen molecules within the finite time interval Δt. We might, in

Table 8.2. Activation energies (E) for some bimolecular reactions between atmospheric molecular species

Reaction		$E/\text{kJ mol}^{-1}$
CO $+ \text{O}_2$	$\longrightarrow$ CO$_2$ $+$ O	251
CO $+ \text{NO}_2$	$\longrightarrow$ CO$_2$ $+$ NO	115
SO$_2$ $+$ NO$_2$	$\longrightarrow$ SO$_3$ $+$ NO	106
N$_2$ $+$ O$_2$	$\longrightarrow$ N$_2$O $+$ O	538

Data collected from a variety of cources.

order to be dealing with realistic chemistry, impose arbitrarily a limiting $\Delta t = 100$ years $= 3.2 \times 10^9$ s as the maximum time scale for 1% reaction. What then is the minimum value of k in equation (8.1) which will produce this?

We rearrange equation (8.1) to produce:

$$-\Delta[N_2]/[N_2] = 0.01 = k \cdot [O_2] \cdot \Delta t = 2.85 \times 10^7 \cdot k$$

Thus the minimum value of k is 3.5×10^{-10} dm^3 mol^{-1} s^{-1}. On the basis of the typical frequency factor value of 10^9 dm^3 mol^{-1} s^{-1}, this value of k demands an Arrhenius factor of the order of 10^{-19}, which would correspond to an activation energy of close to 100 kJ mol^{-1} on the basis of Table 8.1. The actual activation energy of over 500 kJ mol^{-1} shown in Table 8.2 then precludes reaction between N_2 and O_2 from having any chemical significance. Although two of the other reactions in Table 8.2 might appear to have activation energies of the right order of magnitude, the corresponding rates will be made very small by the very low concentrations of the reactant species in the atmosphere.

It is necessary therefore to postulate that the chemistry of the atmosphere is photochemically induced. The fundamental reason underlying the unleashing of the highly complex chemistry which follows the photodissociation of just a few atmospheric components, is that the Arrhenius activation energies for the many exothermic reactions involving atomic or radical reactants are much lower than those shown in Table 8.2. Table 8.3 lists a few illustrative examples, some of them important in tropospheric chemistry, with averaged values of the rate constants at 288 K [$k(288)$] and the Arrhenius activation energy. Only ground electronic state species are considered for the present.

With a rate constant of 10^7 dm^3 mol^{-1} s^{-1} for a reaction, we may usually expect significant conversions in a time scale of hours at most. For example, consider the reaction between nitric oxide and ozone represented by the equation:

$$NO + O_3 \longrightarrow NO_2 + O_2 \qquad\qquad (14)$$

numbered to conform to the list in the Appendix, which has approximately this value for its rate constant at ambient temperatures. Within a polluted tropospheric locality, typical mixing ratios of these reactant species are of the order of 10 parts per hundred million (p.p.h.m.), i.e. the concentrations are 10^{-7} [M]. Thus we have

Table 8.3. Typical rate parameters for bimolecular reactions with an atomic or free radical reactant

Reaction	$k(288)/$dm^3 mol^{-1}.s^{-1}	$E/$kJ mol^{-1}
O + NO$_2$ $\longrightarrow$ NO + O$_2$	5×10^9	< 1
H + O$_3$ $\longrightarrow$ OH + O$_2$	2×10^{10}	< 1
OH + NH$_3$ $\longrightarrow$ H$_2$O + NH$_2$	8×10^7	6.7
O + H$_2$S $\longrightarrow$ OH + HS	1×10^7	6.3
NO + HO$_2$ $\longrightarrow$ NO$_2$ + OH	1×10^8	< 1

Data collected from a variety of sources.

[NO] = [O_3] = 4 x 10^{-9} mol dm^{-3} for our illustrative purposes. We may calculate the incremental time, Δt, for 1% conversion of each reactant, taking the concentrations to be constant. Thus we obtain the equation:

$$-\Delta[NO]/[NO] = 0.01 = k_{14} \cdot [O_3] \Delta t = 0.04 \cdot \Delta t$$

Hence Δt is 0.25 s in this instance and we may therefore expect reaction (14) to be of significance in urban environments.

However, it is not possible to generalize upon the order of magnitude of the rate constant required for a reaction to achieve significance in the overall reaction scheme for a polluted part of the troposphere. The elementary reaction rate depends upon the local concentrations of reactants, which will be variable from one set of conditions to another.

Now that the need for photochemical generation of reactive species has been justified, the next section develops the relevant aspects of photochemistry required for an understanding of atmospheric chemistry.

8.2 Basic Photochemical Considerations

It will be presumed that the reader is familiar with the elementary quantum theory of atoms and molecules.

A molecule may possess four different forms of energy. The first of these is kinetic energy possessed by virtue of translational motion and related to its instantaneous velocity, c. For thermal equilibrium, the condition which will apply to species in the atmosphere, the distribution of velocities, and hence energy of translation, will follow the Maxwell—Boltzmann function, given in Chapter 1 as equation (1.4). The quanta involved in translational energy, i.e. the energy separation of the discrete energy levels, are so small that this form of energy can be regarded as effectively continuous.

Progressing through the energy modes in order of increasing quantum size, the second form is rotational energy. A molecule possesses rotational energy by virtue of its spinning about an axis which passes through its centre of mass and about which the molecule has a finite moment of inertia. Hence a diatomic molecule will have two effective rotational modes about axes perpendicular to the line of centres; there will not be a rotational mode about the line of centres since atoms are considered as point masses and there is therefore no moment of inertia about this axis. Rotational quanta are generally equivalent to the range of energy of 10 to 100 J mol^{-1}.

Vibrational quanta are two to three orders of magnitude larger than rotational quanta and lie in the range 5 to 20 kJ mol^{-1} in general. Molecules possess vibrational energy through the oscillatory motions of the component atoms with respect to one another. On the basis of equation (1.3), the radiation which has photons of equivalent size to vibrational quanta lies in the wavelength range 24,000 nm to 6000 nm, which is in the near infrared region of the electromagnetic spectrum. Since the absorption of a photon depends upon a match between the energies of the photon and the separation of adjacent energy levels in the molecule, we expect the absorption and emission of near infrared radiation to involve vibrational excitation and de-excitation. As such this gives rise to no chemistry and is therefore not of interest for the moment; but we may note that such vibrational energy changes are the origin of physical phenomena like the carbon dioxide greenhouse effect (Section 1.3).

Electronic energy level separations are larger again and in general lie in the range of equivalent energies of 100 to 1000 kJ mol^{-1}. The matching photons lie in the wavelength range 1200 nm to 120 nm, which extends across the visible region of the spectrum, through the ultraviolet and into the vacuum ultraviolet region below 200 nm. Moreover the strengths of the bonds in atmospheric molecules cover the same energy range, extending from the lowest dissociation energy of ozone of 103.3 kJ mol^{-1} up to a lowest dissociation energy of 1069.4 kJ mol^{-1} for carbon monoxide. Therefore it is visible and ultraviolet radiation absorptions that would be expected to initiate chemical reactions in the atmosphere.

A basic law of photochemistry states simply that only radiation which is absorbed can produce chemical change in a system. This is only to be expected since it is the energy so derived which can be used to break chemical bonds and thus induce the primary photochemical step. The resultant generation of reactive species can then, in turn, lead to secondary chemical reactions which effect the remainder (if any) of the overall chemical reaction. For the moment we shall

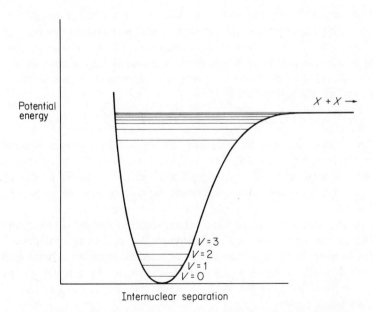

Figure 68. The typical form of the potential energy curve for the ground electronic state of a diatomic molecule. X_2. Vibrational energy levels are indicated as horizontal straight lines with the vibrational quantum number, V, indicated. At the dissociation limit the products are assumed to be the ground state atoms, X + X.

concentrate our attention upon the primary photochemical step and examine what may be involved therein.

Figure 68 represents the typical potential energy curve for the ground electronic state of a diatomic molecule, say X_2, or for one isolated bond of a polyatomic molecule.

The vibrational energy levels associated with this ground state molecule are shown as horizontal lines. The energy separations between these are virtually constant for low values of the vibrational quantum number, v, but, because of the finite dissociation energy of this anharmonic oscillator, there is convergence of the levels towards the continuum at the energy required to break the X—X bond and thus produce the free atoms X +X. In the classical view, the potential energy at any point in the course of a vibration follows the curve, and the horizontal line for the value of v concerned simply represents the conservation of total energy, potential plus kinetic. It is to this classical viewpoint that potential energy curves are directed rather than to the rather different quantum mechanical position. Nevertheless, the essential consequences of the latter treatment are superimposed in that only certain energies are allowed and the density of the energy levels increases as the dissociation limit is approached.

Term Symbols and Radiative Selection Rules

The ground state of any molecule will have a definite arrangement of its electrons within its molecular orbitals. Such orbitals will have differing energies and the ground state of the molecule is characterized by the maximum permitted packing of electrons into the lowest energy orbitals; the restrictions are the Pauli principle and Hund's rules. Thus for example the ground state configuration of the nitrogen molecule is represented as:

$$(\sigma_g 1s)^2 (\sigma_u 1s)^2 (\sigma_g 2s)^2 (\sigma_u 2s)^2 (\pi_u 2p)^4 (\sigma_g 2p)^2$$

where the $(\sigma_g 1s)$, etc., identify molecular orbitals and the superscripts denote the number of electrons in each. An orbital is filled when it contains two electrons with oppositely directed spins. The energy of the orbitals increases to the right as written, i.e. $(\sigma_g 2p)$ is the orbital of highest energy occupied by electrons in the ground state.

Now in this state there are no unpaired electrons, so that the net electron spin is zero. This information is conveyed by giving a value of zero to a parameter Σ, the quantum number for the spin component directed along the internuclear axis. It is conventional to talk in terms of $(2\Sigma + 1)$, which is then 1 for this state of nitrogen, and to refer to states as singlet, doublet, triplet, etc. Hence nitrogen has a singlet ground electronic state.

Let us now consider the configuration which corresponds to the first electronically excited state of the nitrogen molecule. In principle this excitation involves the promotion of one electron from the highest energy occupied orbital [i.e. $(\sigma_g 2p)$] up to the vacant orbital of lowest energy (i.e. $(\pi_g 2p)$). Thus the new configuration is:

$$(\sigma_g 1s)^2 (\sigma_u 1s)^2 (\sigma_g 2s)^2 (\sigma_u 2s)^2 (\pi_u 2p)^4 (\sigma_g 2p)^1 (\pi_g 2p)^1$$

It is immediately obvious that there can be two unpaired electrons in the highest orbitals. In fact it is Hund's rules which tell us that the lowest energy situation arises when these two spins are aligned in the same direction. So in this first excited state of the nitrogen molecule there is net spin; each electron contributes a conventional ½ to the value of Σ, so that $\Sigma = 1$ and $(2\Sigma + 1) = 3$ and this is a triplet state.

Correspondingly in other molecular situations, for example, nitric oxide, there may be just one unpaired electron producing $\Sigma = \frac{1}{2}$, $(2\Sigma + 1) = 2$ and hence a doublet state.

Now it is immediately clear that it would be exceedingly cumbersome to have to keep writing out configurations like those above in order to identify electronic states. It is hardly surprising, therefore, that a shorthand way of identifying the state and of conveying essential information on their nature has been devised. In this notation $(2\Sigma + 1)$ is one component and another parameter, generally denoted as Λ is also given, expressed specifically as an alphabetical code rather than a numerical value. The result is known as a *term symbol* and is written conventionally as:

$$^{2\Sigma+1}\Lambda \tag{8.2}$$

For our purposes it is unnecessary to know the derivation of Λ but only the code which is as follows:

(i) $\Lambda = 0$ gives rise to states designated Σ (not to be confused with the spin quantum number). Thus a configuration producing $\Lambda = 0$ and a net spin of zero (no unpaired electrons) is written as $^1\Sigma$. Increasing values of net spin with $\Lambda = 0$ gives rise to states denoted as $^2\Sigma$, $^3\Sigma$ and so on.

(ii) $\Lambda = 1$ gives rise to states designated Π. For no unpaired electrons in the molecule the term symbol is $^1\Pi$, with $^2\Pi$, $^3\Pi$ and so on denoting increasing numbers of unpaired electrons.

(iii) $\Lambda = 2$ gives rise to states designated Δ and we shall have a series of term symbols $^1\Delta$, $^2\Delta$ and so on as before.

We shall not encounter states having Λ greater than 2, although of course such states do occur under other circumstances.

Spectroscopists append other components to the term symbols to convey further information. These appear as subscripts, g or u, and superscripts, + and −, the latter attaching only to Σ states, both written after Λ. Thus the full term symbol for the ground state of the oxygen molecule (which has two unpaired electrons) is $X^3\Sigma_g^-$, with the X conventionally denoting that this is the ground state. We need not concern ourselves with how the g and − are derived nor with the information which they carry. The essential point for us is that term symbols which differ in any respect (e.g. $^3\Pi_g$, $^3\Pi_u$) represent different electronic configurations within the molecule and hence different states with different electronic energies.

Hence O_2 $(^1\Delta_g)$ is (as it turns out) the electronically excited state of oxygen with the lowest excitation energy above the ground state $O_2(^3\Sigma_g^-)$. We are merely required to recognize that $O_2(^1\Delta_g)$ is a different species from the ground state and as such has its own chemical reactivity towards atoms and molecules. This might be anticipated solely on the grounds that this electronically excited state has no unpaired electrons compared with the two in the ground state.

The first electronically excited state of oxygen should be written as $O_2(a^1\Delta_g)$ for completeness, but the preceding letter is often missed out. As mentioned above, X denotes the ground electronic state. The rule for allocating letters to higher states are that A, B, C, etc., are given to the states in ascending order of energy which have the same spin multiplicity (i.e. $2\Sigma + 1$) as the ground state. Hence in oxygen the ascending order of triplet states is $X^3\Sigma_g^-$, $A^3\Sigma_u^+$ and $B^3\Sigma_u^-$. States of different multiplicities than the ground state are preceded by small letters a, b, c, etc., in ascending order of energy. Thus for oxygen the ascending order of singlet states is $a^1\Delta_g$, $b^1\Sigma_g^+$. However, due to historical accidents and misinterpretations of the multiplicities of states, the rules are not always adhered to. For example, in nitrogen the first two triplet states were initially thought to be singlets, hence they continue to be known as $A^3\Sigma_u^+$ and $B^3\Pi_g$ even though the ground state is $X^1\Sigma_g^+$. The singlet states of nitrogen therefore carry small letters, e.g. $a^1\Pi_g$. Quite frequently also, an unrecognized state lying at an energy level between two established states has been discovered after allocation and widespread usage of the letters; this accounts for the appearance of dashed letters, as for example in the nitrogen state designated $a'^1\Sigma_u^-$. We shall miss out the letters except where it is necessary to make a clear distinction between two states.

The term-symbol parameters are useful in another respect. They allow us to predict the ease or difficulty with which a photon with energy matching the required electronic excitation energy (with vibrational and rotational excitation superimposed) can be absorbed. Conversely, they allow us to predict the ease or difficulty of the reverse process, the radiative decay of an electronically excited molecule to a lower energy state. For a molecule M therefore we are commenting upon the rate of the processes:

$$M + h\nu \longrightarrow M^*$$
$$M^* \longrightarrow M + h\nu$$

where M is the ground state and M* the electronically excited state. In this we are involved with what are called *selection rules*.

The origin of the selection rules for radiative transitions between electronically excited states and the ground state lies in fundamental quantum theory and need not concern us here, but the final results are of great importance. The conventional terminology refers to transitions as being 'allowed' or 'forbidden', but neither is absolute and perhaps 'easy' and 'difficult' might convey the meaning better. As far as radiative transitions in the atmosphere are concerned the diatomic molecular selection rules are as follows:

(i) Spin Multiplicity, $(2\Sigma + 1)$. Radiative transitions between states of the same

spin multiplicity are easy in the absence of contravention of other selection rules. Radiative transitions between states of different multiplicities are difficult.

The operation of this rule may be illustrated by reference to two radiative decay situations of consequence in the atmosphere.

Firstly we may consider $O_2(a^1\Delta_g)$, the first excited state of molecular oxygen, which is produced in the atmosphere by the photodissociation of ozone amongst other formation routes. The only decay route is to the ground state, $O_2(^3\Sigma_g^-)$ and it is obvious from the different spin multiplicities that this transition will be difficult. The radiative rate constant for the process:

$$O_2(^1\Delta_g) \longrightarrow O_2(^3\Sigma_g^-) + h\nu$$

would therefore be expected to be small. Conversely the radiative lifetime, the reciprocal of the radiative rate constant, would be expected to be long; in this case it is just over 1 hour. Thus where $O_2(^1\Delta_g)$ is produced in the atmosphere, it is anticipated that it will be a highly persistent species in the absence of other decay modes.

On the other hand, nitric oxide is excited in the upper atmosphere by the absorption of ultraviolet radiation from the ground state, $X^2\Pi$, to the first doublet excited state, $A^2\Sigma^+$. Since there is no change in spin multiplicity involved (and none of the following selection rules is contravened either), the radiative decay process represented as:

$$NO(A^2\Sigma^+) \longrightarrow NO(X^2\Pi) + h\nu$$

should be easy. This is reflected by the short radiative lifetime of $NO(A^2\Sigma^+)$ of around 2×10^{-7} s.

The rule applies equally to absorption and emission situations. As a result, it is hardly surprising that direct absorption of photons by $O_2(X^3\Sigma_g^-)$ is not an effective source of $O_2(a^1\Delta_g)$ in the atmosphere at large. In other words, ground state molecular oxygen does not have a significant absorption coefficient in the corresponding region of the spectrum near 1200 nm.

(ii) The Symbol Λ. The selection rule involving Λ is that changes of Λ in a transition of more than one in terms of its numerical value produce difficult transitions. Thus there is no preclusion on this count to zero changes, i.e. $\Sigma \leftrightarrow \Sigma$, $\Pi \leftrightarrow \Pi$, $\Delta \leftrightarrow \Delta$ nor to changes of unity, i.e. $\Sigma \leftrightarrow \Pi$, $\Pi \leftrightarrow \Delta$, etc., but $\Sigma \leftrightarrow \Delta$ will be difficult. This is then a further breach of the selection rules which makes the $O_2(^1\Delta_g) \leftrightarrow O_2(^3\Sigma_g^-)$ radiative transition difficult.

(iii) The Appended Symbols, g, u, +, −. The subsidiary parts of the term symbol refer to the symmetry properties of states, and preclusions can operate from these to radiative transition probabilities. Only Σ states carry the + and − superscripts and only difficult transition is $\Sigma^+ \leftrightarrow \Sigma^-$.

The g and u subscripts only appear in the term symbols of homonuclear diatomics and the difficult transitions are those going from like to like, i.e. g↔g and u↔u. Once again it is predicted that the $O_2(^1\Delta_g)$↔$O_2(^3\Sigma_g^-)$ transition will be difficult on this count.

However, in general only the spin rule (i) will be of importance in our discussion of atmospheric chemistry.

Atomic Term Symbols

So far we have only considered the electronic states of diatomic molecules. A similar set of term symbols are used to denote the electronic states of atoms.

The spin number, now denoted S, takes exactly the same values as Σ for the same number of unpaired electrons and, as might be expected, it is the spin multiplicity, $(2S + 1)$, which appears as a preceeding superscript in the term symbol. A parameter L corresponds to Λ, but the code of letters used are no longer Greek, i.e. $L = 0$ is termed an S state, $L = 1$ a P state, $L = 2$ a D state. The term symbol is now written as:

$$^{2S+1}L \tag{8.3}$$

comparable with the molecular term symbol of (8.2).

The ground electronic state of the oxygen atom is written in terms of electrons in atomic orbitals thus:

$$(1s)^2 \ (2s)^2 \ (2p)^4$$

As there are three degenerate 2p orbitals, we must consider the electron spins therein. Hund's rules tell us, as before, that parallel spins will produce the state of lowest energy. Thus the ground state of the oxygen atom has two of the 2p electrons paired off in one orbital and two unpaired electrons with parallel spins in the other two. Thus the state is a triplet and the term symbol is in fact ^3P.

The first electronically excited state of the oxygen atom would have in principle the two unpaired electrons of the ground state with opposite spins, producing a singlet state. In fact more detailed quantum mechanical treatment shows that there are two such states, designated by term symbols ^1D and ^1S, with the former the lower in energy of the two. Higher energy states will not concern us.

The selection rules for atomic transitions of atmospheric interest are listed:

(i) Spin Quantum Number, S. The spin multiplicity, $2S + 1$, must remain unchanged for an easy transition.

(ii) The Parameter L. Changes in L of one or zero correspond to easy transitions, with the exception of S↔S. Changes in L greater than one, as for a D↔S transition, make the radiative transition difficult.

(iii) The Laporte Rule. The Laporte rule precludes transitions between states which are derived from the same electronic configuration. Thus the transition between the $O(^1D)$ and $O(^1S)$ states mentioned above is made difficult on this ground in addition to the breach of rule (ii). A clearer case occurs with atomic nitrogen. The ground state is $N(^4S)$, with three electrons having parallel spins in the

three 2p orbitals. The reversal of the spin of one of these electrons generates $N(^2P)$ and $N(^2D)$ states based on the same electron configuration. Therefore the transition $N(^2P)\leftrightarrow N(^2D)$ is made difficult as a radiative process solely on the basis of the Laporte rule.

Term Symbols of Polyatomic Molecules

Polyatomic term symbols are different again but the spin multiplicity appears again as the left superscript. It turns out that the only selection rule which will be of importance to our context in assessing the ease or difficulty of transitions in polyatomic molecules is the spin conservation rule, where the multiplicity must not change for an easy transition. Hence, just as for atoms and diatomic molecules, singlet-to-singlet, triplet-to-triplet transitions in polyatomic molecules are easy, while a triplet-to-singlet transition would be difficult. So we may recognize immediately that the radiative decay of $SO_2(^3B_1)$ will not proceed readily to $SO_2(^1A_1)$, even if the overall term symbols are unfamiliar. Just as before, we use such term symbols as $^3B_1$ as a label denoting a particular triplet state and all that we require to know for the present purposes is its excitation energy with respect to the ground state, designated by another term symbol. Only a small number of polyatomic term symbols will be encountered in this book so that the reader will become familiar rapidly with those appertaining to a particular molecule and to the energy levels referred to by them.

The Primary Fates of Electronically Excited Molecules

(i) Fluorescence. From the above discussion of the selection rules, it is evident that an atom or molecule absorbing a photon of visible or ultraviolet radiation will be excited to a state of the same multiplicity as the ground state. Subsequently there is a substantial probability that the excited state will decay back to the ground state, reemitting the photon as it does so, within a short time scale, often of the order of 10^{-8} s. Such a process would be termed *fluorescence* and has no photochemical consequence. Nitric oxide fluoresces in the upper atmosphere as discussed above.

(ii) Physical Quenching. After the initial excitation, a second possibility is that the excited state may lose its energy to other molecules in the environment. While the molecule is in the excited state, it will suffer many collisions with other unexcited molecules, unless the pressure is exceedingly low. The collisional deactivation of an excited molecule XY^* may be written as:

$$XY^* + M \longrightarrow XY + M$$

where the fate of the energy is degradation to thermal or, less commonly, rotational or vibrational energy of the general species M. Physical quenching evidently has no chemical consequences. One example of interest is that in the atmosphere $O_2(^1\Delta_g)$ has a major removal pathway operating through physical quenching by ground state oxygen molecules, which determines the actual lifetime of this species rather than the radiative decay considered above.

(iii) Collisional Crossing. When an excited species collides with a general molecule M, there can be a transient perturbation of the electric and magnetic fields within the excited molecule. On occasions this can relax the selection rule for non-radiative transitions that one electronically excited state can only cross effectively into another at approximately the same energy if the two states have the same spin multiplicity. Thus the perturbation induced by collision can produce such as a significant extent of singlet to triplet conversion, competing with physical quenching and fluorescence decay of the singlet state. This phenomenon is referred to as collisional crossing. An illustrative example is found in sulphur dioxide, where absorption of a near ultraviolet photon effects excitation from the ground state $SO_2(^1A_1)$ to the first excited singlet state $SO_2(^1B_1)$. Collisional crossing then produces a significant population rate into the triplet state $SO_2(^3B_1)$, the lowest vibrational level of which lies at a slightly lower energy than that of $SO_2(^1B_1)$, under typical tropospheric pressure conditions.

(iv) Energy Transfer. If, during the lifetime of an electronically excited state, collision is made with another suitable species, then the original excitation energy can be donated at least in part to electronically exciting the acceptor. Such energy transfer processes account for the major part of the production of $O_2(^1\Delta_g)$ in polluted regions of the troposphere. Nitrogen dioxide can be one of the donor species, with the process represented:

$$NO_2(^2A_1) + h\nu \longrightarrow NO_2(^2B_1)$$
$$NO_2(^2B_1) + O_2(X^3\Sigma_g^-) \longrightarrow NO_2(^2A_1) + O_2(^1\Delta_g)$$

Irradiations of NO_2/O_2 mixtures in laboratory studies, using monochromatic radiation with wavelengths slightly longer than the nitrogen dioxide photodissociation threshold at around 400 nm, have demonstrated the generation of $O_2(^1\Delta_g)$ with efficiencies approaching 5% on the basis of the photon absorption rate. It has also been suggested that organic molecules such as benzene may also act as donors in this connection.

(v) Direct Chemical Reaction. An electronically excited species possesses an excess of energy compared to the ground state and this can result in a much greater chemical reactivity for the former. In the course of its existence the electronically excited species may react chemically with a collision partner. One established example of this is found with nitric oxide. When $NO(A^2\Sigma^+)$ is excited in NO–CO_2 mixtures with a suitable ultraviolet radiation source, the reaction:

$$NO(A^2\Sigma^+) + CO_2 \longrightarrow NO_2 + CO$$

is induced, involving the exchange of an oxygen atom.

(vi) Photodissociation. By far the most important primary photochemical act

in the atmosphere is photodissociation, where the energy of an absorbed photon causes a molecule to dissociate, usually producing atoms and/or radicals. For a particular absorbing molecule the precise nature of the species produced can vary with the wavelength of the radiation absorbed. Perhaps the best illustration of this from our present point of view is the photodissociation of ozone. The photodissociation processes are believed to be as follows:

$$O_3 + h\nu \ (1180 \text{ nm} > \lambda > 450 \text{ nm}) \longrightarrow O(^3P) + O_2(^3\Sigma_g^-)$$

$$O_3 + h\nu \ (350 \text{ nm} > \lambda > 313 \text{ nm}) \longrightarrow O(^3P) + O_2(^1\Delta_g)$$

$$O_3 + h\nu \ (\lambda < 313 \text{ nm}) \longrightarrow O(^1D) + O_2(^1\Delta_g)$$

To commence our discussion of the photodissociation process, let us consider a diatomic molecule. In Figure 68, the typical potential curve for the ground state of such a molecule was represented. The initial act on absorbing a photon must be for the molecule to transfer to an analogous potential curve of an electronically excited state, a process corresponding to the excitation of one electron from one molecular orbital to a higher one.

Besides the selection rules governing the ease of the transition, there is another important aspect to be considered within the *Franck–Condon principle*. This states that the transition of the electron from the lower to the higher energy orbital is accomplished in a time so short compared with the vibrational period of the molecule that this may be considered to take place at constant internuclear separation. On a potential diagram such as Figure 68, this means that the path of the transition to a higher energy potential curve is represented as a vertical line. The Franck–Condon principle goes on to predict the intensity of the resultant spectral absorption bands; bands are produced rather than lines at moderate resolution because of concurrent changes in rotational and vibrational energies. The statement is that the intensity of any band corresponding to a particular change in the vibrational quantum number (assuming that rotational structure is unresolved) from the lower to the upper state is proportional to the product of the quantum mechanical probabilities of the starting and finishing positions on the two curves. We shall not investigate the nature of the quantum mechanical probability and we shall only say that it is expected that a series of absorption bands, corresponding to a series of vibrational energy changes of varying probability superimposed upon the basic electronic energy change, will appear with varying intensities of absorption.

We expect the upper potential curve to have a convergence of vibrational energy levels towards its dissociation limit, just as represented in Figure 68 for the ground state. Moreover, on the basis of the Boltzmann distribution function which governs the equilibrium distribution of ground state molecules across the vibrational levels, it can be shown that the initial vibrational state for atmospheric absorption processes is overwhelmingly the lowest, with vibrational quantum number (v) equal to zero. The Boltzmann population distribution function for the number of molecules in level $v = 0$ (N_0) compared to the number (N_1) in the first vibrationally excited level is:

$$N_1/N_0 = \exp[-(E_1 - E_0)/R \cdot T] \tag{8.4}$$

Here E_0 and E_1 are the equivalent energies on the same scale of the two vibrational levels, which are non-degenerate. It has been suggested earlier that the typical value of $(E_1 - E_0)$ will be around 10 kJ mol^{-1}. Thus for ambient temperature taken as 288 K, equation (8.4) yields:

$$N_1/N_0 = 0.015$$

Therefore we can effectively ignore the populations of vibrational levels other than the lowest in the ground state and consider this as the starting point for the absorption of photons.

The precise nature of the straightforward photodissociation process depends upon the positions of the potential energy curves with respect to one another. Let us consider for the moment the situation represented in Figure 69.

The first point to be made by Figure 69 is that the dissociation of the excited state does not give rise to ground state atoms; this is the typical case. For a

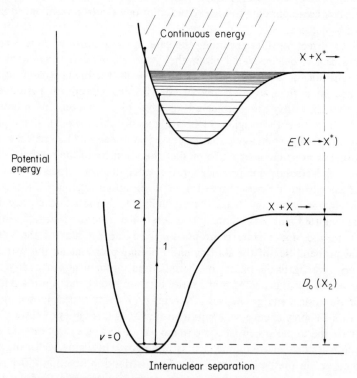

Figure 69. Typical forms of ground state and electronically excited state potential energy curves. Horizontal lines indicate vibrational energy levels, vertical lines (1 and 2) are transition pathways for electronic excitation. The products at the dissociation limit of the ground state of the X_2 molecule are assumed to be ground state atoms, $X + X$, and $D_0(X_2)$ is the dissociation energy from the lowest vibrational level ($v = 0$). The electronically excited state is considered to dissociate into a ground state atom, X, and an electronically excited atom, X*, with excitation energy with respect to X of $E(X \rightarrow X^*)$.

molecule X_2, at least one of the X atoms so produced is electronically excited, represented as X*. In the typical case the ground state of X_2 will dissociate to ground state atoms, X. Thus $E(X \rightarrow X^*)$ as represented in Figure 69 is the electronic excitation energy of X to X*.

On the basis of the Franck–Condon principle, electronic transitions between the two potential curves are to be represented as vertical lines proceeding from points along the lowest vibrational level of the ground electronic state; those numbered 1 and 2 are typical. The line 1 proceeds upwards and terminates close to its intersection with the upper potential curve, at a fairly high vibrational level. Hence the transition corresponding to line 1 proceeds between quantized levels; other transitions corresponding to lines quite close to line 1 would be expected to reach other quantized levels of the upper state. However there will only be a definite number of such lines governed by the definite number of vibrational levels. Each will have a distinct separation from its neighbours which decreases as the transition reaches vibrational levels near to the dissociation limit. Accordingly it is expected that the separations between the resultant spectral absorption bands as functions of wavelength, say, will decrease as the excitation produces molecules with energies closer and closer to the dissociation limit of the upper state, i.e. the bands converge to merge into a continuum when the transition just reaches the dissociation limit. The transition represented by line 2 in Figure 69 reaches a point above the dissociation limit of the upper curve. There is no quantization of energy in this region with the result that any photon above the continuous threshold can be absorbed in principle. This continuous absorption, which inevitably photo-dissociates X_2 into X + X*, is known as the photodissociation continuum.

Figure 70 shows the typical convergence spectrum of a diatomic molecule, as exemplified by the Schumann–Runge system of oxygen below wavelength 200 nm. This transition involves the excitation of O_2 from its ground state, $(^3\Sigma_g^-)$ to the excited state designated $(^3\Sigma_u^-)$; above the dissociation limit of this latter state the products are the ground state atom $O(^3P)$ and the first excited state atom $O(^1D)$. In Figure 70 we can see the convergence of the band structure to the continuum threshold located at around 175 nm wavelength. Below this wavelength continuous absorption is observed.

It is pertinent to investigate the energy requirements of such a direct photodissociation process. The dissociation energy of the ground state oxygen molecule into ground state atoms is $D_0^\ominus = 493.6$ kJ mol^{-1}, while the excitation energy of $O(^1D)$ with respect to $O(^3P)$ is 190.0 kJ mol^{-1}. According to Figure 69 we should combine these terms to produce the dissociation threshold of the excited state relative to the lowest vibrational level of the ground state; the resultant sum is 683.6 kJ mol^{-1}. From the recast form of equation (1.3) and incorporating the Avogadro number to take account of the molar energy involved, we predict the wavelength at which the photon energy is exactly equivalent to 683.6 kJ mol^{-1}:

$$\lambda = 6.023 \times 10^{23} \cdot h \cdot c/683600 = 175.1 \text{ nm}$$

where c is the velocity of light. The prediction thus agrees with Figure 70.

From the point of view of photochemistry in the troposphere, since radiation

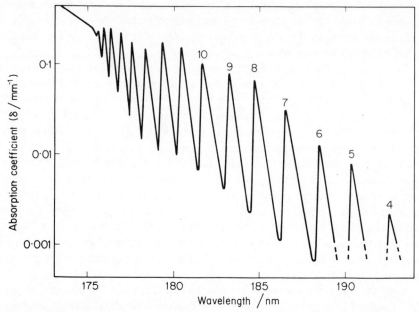

Figure 70. Decadic absorption coefficients of molecular oxygen as functions of wavelength in the Schumann–Runge system. The numbers refer to the vibrational level of the upper state. From K. Watanabe, E. C. Y. Inn, and M. Zelikoff, *Journal of Chemical Physics,* **21,** (June), 1028 (1953). Reproduced by permission of the American Institute of Physics.

with wavelengths below 290 nm does not penetrate the tropopause (Table 2.4), we can be certain that the Schumann–Runge absorption is of no consequence. The general argument leads us to an important conclusion. Because direct photo-dissociation usually involves production of an electronically excited produce, hence imposing an additional energy requirement over and above the dissociation energy of the ground state itself, such will almost invariably require untraviolet radiation with wavelength below 290 nm. Therefore for diatomic molecules, and often for polyatomic molecules, direct photodissociation is certainly not an effective source of atoms and radicals in the troposphere.

A photon of wavelength 290 nm corresponds to an equivalent energy of around 400 kJ mol^{-1}. Therefore in the troposphere only molecules with bond strengths of less than 400 kJ mol^{-1} may be dissociated by the absorbed energy from a single photon, however the process is accomplished. Multiple photon absorption is effectively forbidden by fundamental quantum mechanics under atmospheric conditions. It follows that our next objective is to discover what processes can give rise to indirect photodissociation in the troposphere.

The molecules upon which our attention is focused in connection with indirect photodissociation turn out to be triatomic or larger. Whereas a diatomic molecule has only one vibration and hence one set of vibrational energy changes superimposable upon an electronic energy change, more complex molecules will have several vibrational modes, all of which may have independent energy changes to superimpose upon the electronic energy change. Accordingly it is to be anticipated that polyatomic molecule absorption spectra will show a much higher density of vibrational band structure. The net result is often an apparently continuous absorption spectrum even though the photons induce transitions between quantized levels of the molecule.

The comparatively small molecule of nitrogen dioxide, with only three vibrators, displays an unresolvably continuous visible absorption spectrum with only minor evidence of banded structure. In principle, it is only for wavelengths less than 400 nm that there can be photolytic dissociation as a result of the bond strength of the molecule being 300.6 kJ mol^{-1}, about equivalent to the photon energy for 400 nm. The states involved in the nitrogen dioxide absorption process are the ground state, designated by the term symbol $^2A_1$, and, in the part of present interest, the electronically excited state $^2B_1$. The latter gives rise to product states $NO(X^2\Pi)$ and the excited atom $O(^1D)$ at its dissociation limit. Hence the direct photodissociation of the nitrogen dioxide molecule requires the sum of the above ground state dissociation energy with the excitation energy of $O(^1D)$ with respect to $O(^3P)$, viz. $300.6 + 190.0 = 490.6$ kJ mol^{-1}, corresponding to the photon energy of radiation of wavelength 244 nm. Evidently this process will not occur in the troposphere, but in fact it is well established that nitrogen dioxide does dissociate under the influence of the radiation available in the troposphere and at a relatively rapid rate. Inevitably the products of this process can only be $NO(X^2\Pi)$ and $O(^3P)$ in view of the range of photon energies concerned.

It is a phenomenon known as *predissociation* which allows nitrogen dioxide to be photolytically dissociated within the troposphere. The source of the predissociation phenomenon is generally the existence of a third potential energy curve or surface, other than those of the ground state or the directly excited electronic state, which intersects the excited state curve in the potential energy diagram. A typical situation is represented in Figure 71 for a diatomic molecule.

The important feature in Figure 71 from our point of view is that the third (or predissociating) state dissociates directly to ground state products and that it makes an intersection with the potential curve of the directly excited state close to the first dissociation limit. The basis of predissociation is then that the initial absorption of a photon generates a molecule in the excited state close to the point of intersection and a subsequent process, e.g. collision-induced crossing or non-radiative transition, effects transfer onto the potential curve of the predissociating state. Under the conditions of the atmosphere, where an excited molecule will suffer many collisions in the course of its lifetime, there will be considerable scope for the spin conservation selection rule of non-radiative crossing to be overcome, so that the predissociating state does not need to have the same spin multiplicity as the excited state. Once the molecule is on the predissociating potential curve, it will dissociate to ground state atoms if it possesses sufficient energy.

One particular instance of predissociation in a diatomic molecule, of some consequence in the upper atmosphere as we shall see in Chapter 9, occurs in the

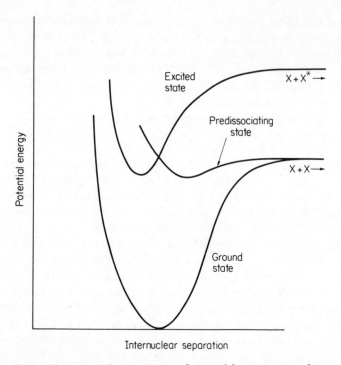

Figure 71. A typical arrangement of potential energy curves for a diatomic species which can give rise to predissociation.

$C^2\Pi$ state of nitric oxide. The lowest vibrational level of this state lies within 1 kJ mol^{-1} of $D_0^{\ominus}(NO)$ and it can be excited readily by irradiation of nitric oxide at 191.0 nm, when some photodecomposition of nitric oxide has been shown to occur. This has been interpreted as a predissociation of $NO(C^2\Pi)$ by the close lying $NO(a^4\Pi)$ potential curve and this quartet state dissociates directly to the ground state atoms, $N(^4S)$ and $O(^3P)$. The apparent deficit of energy (about 1 kJ mol^{-1}) is made up from the thermal (kinetic) energy of the system. Figure 72 shows the positions of the potential energy curves in this case.

In the case of polyatomic molecules, the strict analogies of the potential energy curves of diatomic molecules are potential energy surfaces in three or more dimensions. We shall not go into the resulting, rather complex, situation; nevertheless we can be sure that intersections of one surface by another become more probable as the number of atoms in the molecule, corresponding to the dimensions, increases. Therefore we may expect that predissociation phenomena will become more likely in polyatomic molecules compared to diatomic molecules. Moreover the selection rules governing crossings become less effective in absolute terms the larger is the molecule; the crossing process becomes known as intersystem crossing. Hence it becomes clear that many of the polyatomic molecules present in

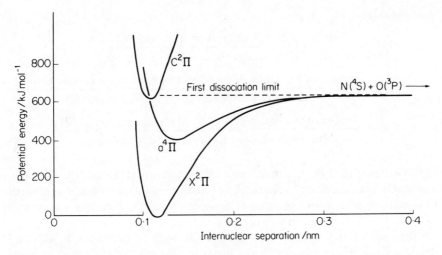

Figure 72. The arrangement of potential energy curves for nitric oxide which produces predissociation in the nitric oxide δ bands, $X^2 \Pi \rightarrow C^2 \Pi$.

the troposphere should be able to photodissociate in effect upon absorption of a suitable photon, giving rise to ground state products.

Returning to consider nitrogen dioxide, we should not now be surprised to find that effective predissociation is induced when the equivalent energy of an absorbed photon exceeds the lowest dissociation limit. That predissociation is the effective agency has long been known from a characteristic diffuseness in the rotational structure of the absorption spectrum under high resolution in the neighbourhood of 370 nm wavelength of nitrogen dioxide. There have been suggestions that the predissociating state is a weakly bound one with term symbol $^2A_1$ but the precise apposition of states in the vicinity of the first dissociation energy limit is uncertain as yet.

The Photodissociation of Polyatomic Molecules in the Atmosphere. The effective quantum yield for nitric oxide formation in the photolysis of nitrogen dioxide as a function of the wavelength of monochromatic radiation has been measured. This may also be taken as the quantum yield for $O(^3P)$ atom formation under atmospheric conditions of very low mixing ratios of nitrogen dioxide (in the p.p.h.m. range). Quantum yield is defined at a particular wavelength as:

$$\text{Quantum yield } (\Phi) = \frac{\text{Number of molecules transformed in a defined manner}}{\text{Number of photons absorbed by the species concerned}}$$

Φ may be defined for a product species or the original molecule. In the nitrogen dioxide photodissociation case the following quantum yields would be defined for

230

completeness:

$$\Phi_{NO_2} = \frac{\text{Number of NO}_2 \text{ molecules photodissociated}}{\text{Number of photons absorbed by NO}_2 \text{ molecules}}$$

$$\Phi_{NO} = \frac{\text{Number of NO molecules produced by photodissociation of NO}_2}{\text{Number of photons absorbed by NO}_2 \text{ molecules}}$$

$$\Phi_{O} = \frac{\text{Number of O atoms produced by photodissociation of NO}_2}{\text{Number of photons absorbed by NO}_2 \text{ molecules}}$$

In this instance all these quantum yields will have identical values. Figure 73 shows a plot of Φ as a function of wavelength as determined in a recent study, appropriate to conditions of low mixing ratios of nitrogen dioxide, i.e. secondary reaction of $O(^3P)$ with NO_2 is not taken into account.

Although there is a sharp fall from a value of Φ of near unity at wavelength 397.9 nm, which corresponds to the upper wavelength threshold for dissociation upon the basis of the energy of the absorbed photon itself, it is clear that there is effective photodissociation out to around 420 nm. The equivalent photon energy deficit for 420 nm radiation is around 16 kJ mol^{-1}. Whilst there is no full explanation for this at present, at least part of the deficit in the wavelength range 397.9 nm to 420 nm can be made up from the internal energy (rotation and

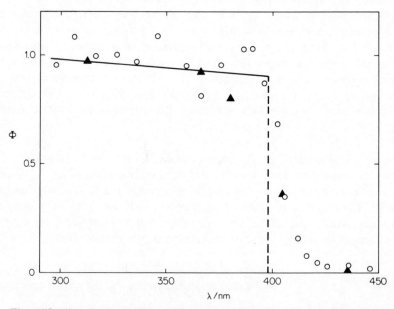

Figure 73. Plot of the quantum yield versus wavelength for the photodissociation of nitrogen dioxide at the levels encountered in the troposphere. From I. T. N. Jones and K. D. Bayes, *Journal of Chemical Physics*, **59** (1 November) 4840 (1973). Reproduced by permission of the American Institute of Physics.

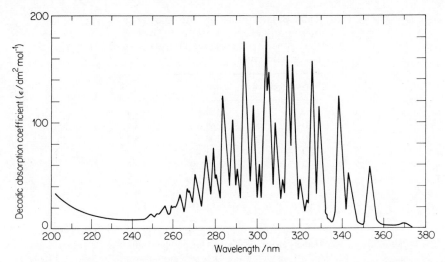

Figure 74. Decadic absorption coefficients for formaldehyde vapour as a function of wavelength. From J. G. Calvert and J. N. Pitts, Jr., *Photochemistry*, John Wiley & Sons Inc., New York, 1966, p. 368. Reproduced with permission of John Wiley & Sons Inc.

vibration) of nitrogen dioxide molecules. However any uncertainty in this spectral region turns out to be of little consequence in the estimation of the nitrogen dioxide photodissociation rate in the troposphere (next section); it is the maintainance of Φ with a value close to unity across the wavelength range 300 nm to 400 nm which is of prime importance in that.

Therefore we may conclude our discussion of nitrogen dioxide with a statement that the absorption of photons of wavelengths below 400 nm by this gas leads, through the agency of predissociation of high efficiency, to its photodissociation in the atmosphere with a 100% efficiency for practical purposes.

Another molecule of importance in the present context is formaldehyde, HCHO. With decreasing wavelength, absorption of radiation sets in at 372 nm and increases in strength thereafter, the spectrum showing pronounced absorption peaks as shown in Figure 74, which is a plot of the decadic absorption coefficient as a function of wavelength.

The initial act is the excitation of the ground state formaldehyde molecule ($^1A_1$) up to the excited state with the term symbol $^1A_2$. In this excited state the molecule can undergo an intersystem crossing onto another potential energy surface, which effects predissociation. In this instance there are two competing pathways for dissociation, represented by the equations:

$$HCHO + h\nu \longrightarrow H + HCO \qquad (8.5)$$
$$HCHO + h\nu \longrightarrow H_2 + CO \qquad (8.6)$$

The pathway (8.6) might be termed better as photodecomposition, since a structural rearrangement must be involved. For the pathway (8.5) a limiting

Table 8.4. Average product species quantum yields over 10-nm bands for the primary photochemical processes in formaldehyde

Wavelength/nm	Φ_H	Φ_{H_2}
290	0.81	0.19
300	0.66	0.34
310	0.52	0.48
320	0.40	0.60
330	0.29	0.71
340	0.18	0.82
350	0.09	0.91
360	0.01	0.99

From J. G. Calvert, J. A. Kerr, K. L. Demerjian, and R. D. McQuigg, 'Photolysis of formaldehyde as a hydrogen atom source in the lower atmosphere', *Science*, **175**, 751–752 (1972). Copyright 1972 by the American Association for the Advancement of Science and reproduced with permission.

wavelength (λ_L) can be deduced, defined simply as the longest wavelength at which the photon itself can supply all of the energy needed to effect the dissociation; this is 333 nm. It is not appropriate to quote a λ_L for the pathway (8.6) since the heats of formation for formaldehyde and carbon monoxide plus hydrogen are almost identical so that the overall change is virtually thermoneutral. In this photo-decomposition pathway, the energy derived from the absorbed photon is used only to effect activation to an excited state of formaldehyde in which such intramolecular rearrangement is feasible. Subsequently the original photon energy appears as thermal or internal energy of the products.

We may examine the variations of the quantum efficiencies Φ_H and Φ_{H_2}, manifestations of the partitioning between the two pathways, as functions of the wavelength of monochromatic radiation. Some recent data on these is tabulated in Table 8.4. It is observed that the quantum yields always total unity while Φ_H shows much the same form of fall-off above λ_L as can be seen in Figure 73. Again it can be postulated that internal energy modes of the molecules are involved. The cut-off of Φ_{H_2} above 372 nm simply reflects the influence of the Franck–Condon principle and the availability of a matching energy gap between potential surfaces, which prevent formaldehyde absorbing photons in the first place.

The nitric acid molecule, HNO_3, can in theory photodissociate along various pathways. These are listed below with the corresponding values of λ_L, the longest wavelength where an absorbed photon can contribute the minimum dissociation energy *per se*.

$$HNO_3 + h\nu \longrightarrow OH + NO_2, \quad \lambda_L = 598 \text{ nm} \tag{8.7}$$

$$HNO_3 + h\nu \longrightarrow HNO_2 + O(^3P), \quad \lambda_L = 401 \text{ nm} \tag{8.8}$$

$$HNO_3 + h\nu \longrightarrow H \quad + NO_3, \quad \lambda_L = 290 \text{ nm} \tag{8.9}$$

$$HNO_3 + h\nu \longrightarrow HNO_2 + O(^1D), \quad \lambda_L = 245 \text{ nm} \tag{8.10}$$

However nitric acid vapour is entirely colourless, which rules out absorption in the visible region of the spectrum. In fact the threshold for absorption is located at around 330 nm wavelength, well into the ultraviolet region of the spectrum. At shorter wavelengths continuous absorption is observed with the strength of absorption increasing rapidly with decreasing wavelength, as is shown in Figure 75, a plot of absorption cross-section versus wavelength.

It is now considered that the pathway represented by equation (8.7) is over-whelmingly the result of photodissociation down to 200 nm wavelength; a near unity quantum yield, Φ_{OH}, is applicable. None of the other pathways [(8.8), (8.9),

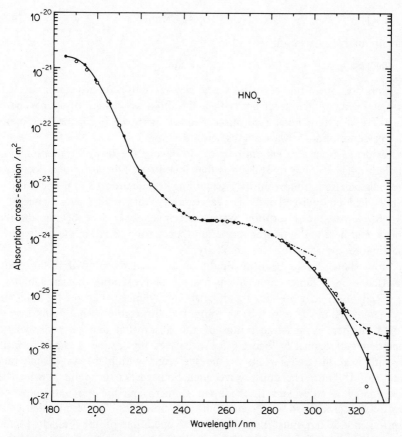

Figure 75. Absorption cross-section of nitric acid vapour as a function of wavelength in the middle ultraviolet spectral range. From F. Biaume, 'Nitric acid vapour absorption cross-section spectrum and its photodissociation in the stratosphere', *Journal of Photochemistry*, **2**, 145 (1973/74). Reproduced by permission of Elsevier Sequoia, S.A.

or (8.10)] appear to play any significant role in the photodissociation process above 200 nm wavelength.

This then sounds a cautionary note. Just because a particular photodissociation pathway is energetically possible, this does not mean that it will actually occur, even in polyatomic molecules. It is clear that the nitric acid molecule does not possess a suitable excited state potential energy surface to absorb photons in the wavelength range 330 nm to 598 nm in the first place. Moreover in the wavelength range 200 nm to 245 nm, even though the photon energies are sufficient to induce any of the four theoretical pathways for nitric acid photodissociation, the nature of the excited state potential energy surface and its intersections with other surfaces is such as to make only the first pathway (8.7) above be of significance.

Nitrous acid, HNO_2, is also a potential source of photodissociation in the atmosphere. The bond strength in the centre of the molecule (H–O–N–O) suggests the pathway represented by the equation:

$$HNO_2 + h\nu \longrightarrow OH + NO, \quad \lambda_L = 585 \text{ nm} \tag{8.11}$$

but an alternative exists in the equation:

$$HNO_2 + h\nu \longrightarrow H + NO_2, \quad \lambda_L = 366 \text{ nm} \tag{8.12}$$

The absorption spectrum of nitrous acid vapour only commences in the near ultraviolet region of the spectrum with a threshold wavelength of approximately 380 nm. The diffuse banded absorption spectrum is shown in Figure 76; it merges into a weak continuum at wavelengths below 303 nm.

The evidence from a recent study (Cox, 1974) of the effect of irradiation in the wavelength range 330 nm to 380 nm is that the nitrous acid absorption does result in photodissociation, almost entirely according to equation (8.11), and with the quantum yield of hydroxyl radicals near to unity; this result may be assumed to hold right across the absorption system. At most, only 10% of the absorbed photons could have led to photodissociation via equation (8.12) in the study above; we shall ignore this possibility for simplicity.

By now there should be little doubt in the reader's mind that the photodissociation of a complex molecule is more dependent upon the availability of electronically excited states in a particular energy range than on the minimum energy requirement for a particular route for dissociation. This point may be amplified further by reconsideration of the absorption and photodissociation characteristics of ozone. In Figure 6 it can be seen that ozone does absorb in the visible and near infrared regions of the spectrum, within the wavelength range 450 nm to 1180 nm. The upper wavelength limit marks the point at which the photon energy is equivalent to the bond strength and the products across this absorption are ground state oxygen atoms and molecules. The termination of the absorption at 450 nm must then reflect the operation of the Franck–Condon principle and the lack of availability of a suitable potential energy surface. At 350 nm, absorption sets in again and in the range of wavelengths 313 nm to 350 nm the products of photodissociation are a ground state oxygen atom and the first excited state of the oxygen molecule, $O_2(^1\Delta_g)$, with an excitation energy of

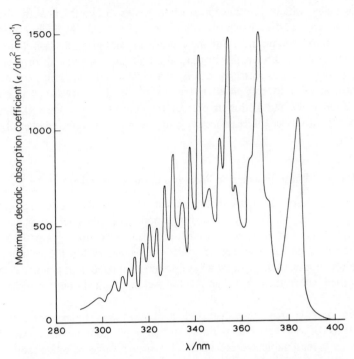

Figure 76. The absorption spectrum of nitrous acid vapour, with estimated maximum values of the decadic absorption coefficients as a function of wavelength. Spectrum reproduced in modified form by permission of the National Research Council of Canada from G. W. King and D. Moule, *Canadian Journal of Chemistry,* **40**, 2057–2065 (1962). Maximum decadic agsorption coefficients derived from P. L. Asquith and B. J. Tyler, *Chemical Communications,* 744 (1970).

94.3 kJ mol^{-1} with respect to the ground state. The value of λ_L for this photodissociation process is then 611.7 nm; again it is the availability of potential energy surfaces which enforces the true threshold at 350 nm. Below a wavelength of 313 nm the products of the photodissociation are the first electronically excited states of both the oxygen atom and oxygen molecule, $O(^1D)$ and $O_2(^1\Delta_g)$ respectively, for which process the value of λ_L is 310.4 nm. In this instance, therefore, it appears that there is a suitable potential energy surface available, so that as soon as the wavelength of the radiation approaches λ_L, the production of $O(^1D)$ atoms commences sharply to the exclusion of the production of the ground state atom.

In our above consideration of the photodissociation of the molecules NO_2, O_3, HCHO, HNO_2 and HNO_3, it will turn out that we have covered the primary sources of the induction of tropospheric chemistry. In the next section we shall see that these five species account for the overwhelming majority of acts of creation of atoms and radicals, the subsequent reactions of which constitute most of the chemistry in urban or industrial environments. This is not to say that there are no other molecules which can photodissociate under the influence of the solar radiation incident on the Earth's surface; examples would be higher aldehydes, nitro and carbonyl compounds, but the relative rates of photodissociation of these other molecules are much smaller in general on account of weak absorption across the penetrating solar spectrum and/or very low mixing ratios, and hence concentrations, even in the polluted regions of the troposphere. The book by Leighton, cited at the end of this chapter, gives a comprehensive account of most of these minor photodissociation processes and the reader is referred there for such information.

8.3 The Rates of Photodissociation Processes in the Troposphere

Consider a cube of air of unit volume (1 dm^3) within the troposphere close to the Earth's surface, with its upper face horizontal. A certain solar irradiance will be incident upon the top face while a reduced irradiance must leave through the bottom face if there has been absorption by molecules within the volume. In the first place we should wish to know what fraction of the incident irradiance is likely to be absorbed within the cube when it is located in a typical polluted region of the troposphere.

Nitrogen dioxide may be considered as the test case. Measurements of the mixing ratios of this gas in polluted urban environments have shown that 10 p.p.h.m. is a reasonable average value. Moreover nitrogen dioxide is an efficient

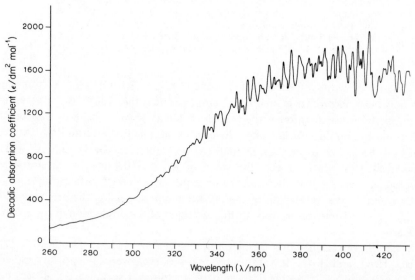

Figure 77. The decadic absorption coefficients of NO_2 gas in the tropospheric photodissociation spectral range. From T. C. Hall and F. E. Blacet, *Journal of Chemical Physics*, **20** (November), 1746 (1952). Reproduced by permission of the American Institute of Physics.

absorber of visible radiation, as is evidenced by pronounced brown colour of samples of this gas, but it has been found that, even when a considerable optical pathlength is viewed in a polluted atmosphere, there is no recognizable brown colouration. Hence we may take it that the extent of absorption within our 1 dm³ volume would be exceedingly small. In fact the solar irradiance reaches the ground in urban environments virtually undiminished by passage through the polluted layer close to the ground.

We may justify this view more quantitatively on the basis of the established absorption coefficients of nitrogen dioxide. Figure 77 shows the decadic absorption coefficients as a function of wavelength across the 300 nm to 450 nm range relevant to our context. It can be perceived that the maximum value of the decadic absorption coefficient, ϵ, is of the order of 1700 dm² mol⁻¹, which applies in the *Beer–Lambert* equation given as equation (2.2) earlier. Let us exemplify the use of this equation by calculating the attenuation of radiation of 400 nm wavelength on passage through a 1-km path containing a mixing ratio of 10 p.p.h.m. of nitrogen dioxide as the only effective absorber. The corresponding concentration of nitrogen dioxide, assuming that we are dealing with air under standard atmospheric conditions, will be $[NO_2] = 10^{-7} [M] = 4.2 \times 10^{-9}$ mol dm⁻³ and the pathlength L will be 10,000 dm. If $I_L(\lambda)$ is the irradiance at wavelength λ transmitted through the pathlength L and $I_0(\lambda)$ is the irradiance at the commencement of passage, then the relationship is as follows:

$$\frac{I_L(\lambda)}{I_0(\lambda)} = 10^{-\epsilon \cdot [NO_2] \cdot L} = 10^{-1700 \times 4.2 \times 10^{-9} \times 10,000} = 0.848$$

Hence the irradiance absorbed at the maximum in the nitrogen dioxide absorption spectrum over perhaps a pathlength corresponding to the maximum likely vertical development of the polluted layer, is only some 15% of the incident irradiance at entry.

It is clear that the fraction of the incident irradiance which will be absorbed within our cube will be exceedingly small. Under these circumstances we can apply what is called the 'weak absorption' form of the Beer–Lambert law, developed in a slightly different form in equation (3.9) from that which will be used here. If $I_A(\lambda)$ is the irradiance absorbed within the cube at a wavelength λ, then the weak absorption form of the Beer–Lambert equation is expressed as:

$$\frac{I_A(\lambda)}{I_0(\lambda)} = 1 - \frac{I_L(\lambda)}{I_0(\lambda)} = 1 - \exp(-2.303\epsilon \cdot [NO_2] \cdot L)$$

For weak absorption the exponential index may be expanded as a power series and all terms other than the first power can be ignored, so that we obtain:

$$\frac{I_A(\lambda)}{I_0(\lambda)} = 1 - (1 - 2.303\epsilon \cdot [NO_2] \cdot L) = 2.303\epsilon \cdot [NO_2] \cdot L$$

and hence the expression for the absorbed irradiance within a pathlength of

$L = 1$ dm is:

$$I_A(\lambda) = 2.303\epsilon \cdot I_0(\lambda) \cdot [NO_2] = j_\lambda [NO_2] \tag{8.13}$$

where j_λ is then the effective first order rate constant for the absorption of photons by nitrogen dioxide at wavelength λ and has units of reciprocal time.

Now our principal interest in the absorption of photons by any molecule in the troposphere is where this leads to photodissociation. If we define $R_p(\lambda)$ as the rate of photodissociation at the wavelength λ and derive the corresponding quantum yield from Figure 73, then we obtain the relationship:

$$R_p(\lambda) = \Phi(\lambda) \cdot j_\lambda \cdot [NO_2] \tag{8.14}$$

Equation (8.14) must then be integrated over the range of wavelengths which produce photodissociation of nitrogen dioxide in the troposphere. Figure 73 suggests an upper limit of 420 nm, while Table 2.4 suggests a lower limit of 300 nm for the integration. If R_p represents the total rate of photodissociation within the 1-dm^3 volume cube exposed to solar radiation near ground level, then we obtain:

$$R_p = \int_{300}^{420} \Phi(\lambda) \cdot j_\lambda \cdot d\lambda \times [NO_2] = J_{NO_2} \cdot [NO_2] \tag{8.15}$$

The parameter J_{NO_2} is the effective first order rate constant for photodissociation of NO_2 in the troposphere close to the ground; its units are reciprocal time. It is clearly a function of the integrated solar irradiance at ground level and will evidently then vary with solar angle.

If J_{NO_2} can be evaluated, then, for any mixing ratio of nitrogen dioxide in say an urban atmosphere, we can calculate the rate of photodissociation and hence the rate of production of reactive $O(^3P)$ atoms from this source.

Evaluation of J_{NO_2} for the Lower Troposphere

Evaluation of J_{NO_2} is hardly an original project and generally accepted values find wide usage in the literature of aerochemistry. However, it is important that we understand how such a parameter can be evaluated so that, say in another case of interest, we can do the calculation for ourselves. Accordingly we shall develop an elementary procedure of general applicability to the evaluation of the J parameter for any photodissociating species.

The integral of equation (8.15) cannot be evaluated in any straightforward mathematical way because $\Phi(\lambda)$ and j_λ are not usually expressible as amenable functions of λ. For example, j_λ incorporates the decadic absorption coefficient, ϵ, and the solar irradiance, $I_s(\lambda)$; glances at Figures 14, 15 and 77 should convince the reader of the impracticability of an exact integration procedure.

However the interpretation of the integral in graphical terms is that if $\Phi(\lambda) \cdot j_\lambda$ is plotted versus λ, then it corresponds to the area under the graph between the limiting wavelengths. This means that if average values of $\Phi(\lambda) \cdot j_\lambda$ can be derived over short wavelength integrals and the result is plotted against λ, producing a step-function graph, then the area under the graph will approximate to the value of the integral the more closely the shorter are the wavelength intervals. Compromising the fairly sharp variations of ϵ with changing wavelength evident in Figure 77 with the need to make the procedure practical, we should have to average $\Phi(\lambda) \cdot j_\lambda$ over intervals of 10 nm at most across the 120-nm wavelength range concerned to obtain a reasonably accurate result. In fact it turns out that, if we are to use the solar irradiance data of Table 2.4, then we are forced to use 10 nm as our averaging interval, since that is the averaging interval used there. Therefore our next task is to consider how we may evaluate the average values of the decadic absorption coefficients of nitrogen dioxide over 10-nm wavelength intervals centred on the decade wavelengths (300, 310, etc.).

Perhaps the simplest method of integrating an irregular variation is by weighing. A tracing of Figure 77 would be on too small a scale for easy manipulation. It is therefore to be recommended that a fivefold or so photographic enlargement is made and either that or a tracing made therefrom be used. On such a scale it is relatively easy to cut out the area under the absorption coefficient profile with a sharp pair of scissors. That done, the vertical lines at the limits of 295 nm and 425 nm should be *lightly* drawn in with a fine point and cut along. Similarly light vertical lines are drawn in at each 10 nm interval along the base scale, starting at 305 nm. The paper is then stripped along these lines, handling it as little as possible, and each strip is weighed accurately on a sensitive balance and the weight noted. It is advisable to take off one strip at a time to avoid later confusion on its identification; the wavelength to which it refers must not be written on the strip (pencil lead has weight!). Finally from the piece of paper taken off above the absorption coefficient profile, a calibrating piece should be cut out, conveniently with a 10 nm wavelength interval as base and a height corresponding to 1500 dm^2 mol^{-1} on the ordinate scale. Then the ratio of the weight of any of the other strips to the weight of this calibration strip gives the average absorption coefficient, ϵ, for nitrogen dioxide over the corresponding wavelength interval when multiplied by 1500 dm^2 mol^{-1} in the instance quoted.

Table 8.5 lists the averaged values of the absorption coefficients, ϵ, and the overhead sun values of the surface solar irradiance, $I_s(\lambda)$ derived from Table 2.4 for 10-nm wavelength intervals; the resultant j_λ values are derived from equation (8.13) as $2.303\epsilon \cdot I_s(\lambda) \cdot 10$, which is then the area of a rectangle of base equivalent to a wavelength interval of 10 nm and a height equivalent to the average value of the decadic absorption coefficient over the particular 10-nm wavelength range concerned. In ideality this would be exactly equal to the area topped by the irregular actual variation of the decadic absorption coefficient over the same range. Also listed in Table 8.5 are the average values of the quantum yield for photodissociation of nitrogen dioxide, $\Phi(\lambda)$, obtained from Figure 73; these are taken as unity from 295 to 395 nm, with values averaged over 10-nm intervals given for the wavelength range 395 to 425 nm.

The integral of equation (8.15) is then simply equal to the sum of the right hand column of this table. Strictly this should have appeared with units of reciprocal time only, but in the derivation of the sum we have mixed mol units from ϵ with

Table 8.5. Averaged values over 10-nm wavelength intervals for parameters in the calculation of ground-level J_{NO_2} for overhead Sun

Middle λ/nm	ϵ/dm^2 mol^{-1}	$10^{-17} \times I_s(\lambda)/$ photons dm^{-2} s^{-1}	$10^{-18} j_\lambda$	$\Phi(\lambda)$	$10^{-18} . \Phi(\lambda) . j_\lambda$
300	415	0.030	2.9	1.0	2.9
310	571	0.302	39.7	1.0	39.7
320	765	0.734	129	1.0	129
330	905	1.239	258	1.0	258
340	1089	1.358	341	1.0	341
350	1332	1.468	450	1.0	450
360	1453	1.514	507	1.0	507
370	1581	1.759	641	1.0	641
380	1612	1.747	649	1.0	649
390	1652	1.786	680	1.0	680
400	1700	2.422	948	0.70	664
410	1647	3.092	1173	0.33	387
420	1593	3.201	1174	0.09	106

Sum = 4856

photons from $I_s(\lambda)$ so that the sum has apparent units of photons mol^{-1} s^{-1} as it stands above. Hence it must be divided by the Avogadro number to obtain J_{NO_2} in the correct reciprocal time units. Thus Table 8.5 yields the result:

$$J_{NO_2} = 4.856 \times 10^{21}/6.023 \times 10^{23} = 8.1 \times 10^{-3} \text{ s}^{-1}$$
$$= 29 \text{ h}^{-1}$$

In view of the approximate nature of $I_s(\lambda)$ (for example it does not take into account atmospheric aerosol scattering or ground reflectivity) we should perhaps say that our calculation leads to a predicted value of J_{NO_2} of approximately 30 h^{-1} for overhead sun conditions.

To gain some feel for the significance of this value of J_{NO_2} let us consider a tropospheric region with the typical urban mixing ratio of nitrogen dioxide of 10 p.p.h.m. The rate of photodissociation of this species is then $R_p = 30 \times 10 = 300$ p.p.h.m. h^{-1}. In other words, if, as is usually the case, the nitrogen dioxide is regenerated by subsequent reactions as fast as it is photolysed, then it would be turned over 30 times in a one-hour period.

How rapidly does J_{NO_2} decrease with increasing solar angle? This question is relevant to the more northerly latitudes where the sun never achieves an overhead position. For the average British latitude of 53.5° N the solar angles at various times of the day during the summer solstice can be calculated from equation (2.6) and are tabulated in Table 8.6. Hence we should wish to know values of J_{NO_2} at least for $Z = 20°$ and $40°$.

As may be seen in Table 2.4, for increasing solar angles there is a relative weakening of the incident irradiance in the near ultraviolet region compared to that in the visible. However, at the same time there is the compensating factor of a

Table 8.6. Solar angles (Z) at local times during the summer solstice at latitude 53.5° N

Local time/hours	Z	
08.00	53°	36′
10.00	37°	33′
Noon	28°	50′
15.00	45°	04′
17.00	62°	30′

(These data are extended by symmetry about noon, e.g. Z is the same at 10.00 and 14.00)

Data calculated from equation (2.6).

pathlength increased by the factor sec Z [Section 2.1(iv)] through our imaginary cube on account of the non-perpendicular path of the rays to the upper face: these factors are sec 20(1.064) and sec 40(1.305) in the two instances being considered. The calculational details are then exactly similar to those given in Table 8.5, with the exception that we incorporate sec Z into equation (8.13) and we read off values of $I_s(\lambda)$ for $Z = 20°$ and $Z = 40°$ from Table 2.4. The results, expressed as a fraction of the overhead sun value $J_{NO_2}(Z = 0)$ are:

$$J_{NO_2}(Z = 20°) = 0.97\, J_{NO_2}(Z = 0°)$$
$$J_{NO_2}(Z = 40°) = 0.92\, J_{NO_2}(Z = 0°)$$

These results for non-zero solar angles contain the inbuilt error of an erroneous assumption that sky radiance is also incident at angle Z. A more correct calculation might assume a weighted angular incidence of sky radiation upon the imaginary cube. An approximate estimation suggests that the above result for $Z = 40°$ could be as much as 20% too high on this account after considering that, across the effective range of wavelengths of 300 nm to 420 nm, the data of Figure 16 show that sky radiance accounts for near half the total incident irradiance. Even so, for simplicity and without involving significant error for our present purposes, it will be assumed that J_{NO_2} has a constant value of 30 h^{-1}, independent of solar angles below $Z = 40°$.

Evaluation of J_{O_3} for the Lower Troposphere

Ozone is also present in polluted regions of the troposphere at around the 10-p.p.h.m. level. We are aware that this may be photodissociated to yield $O(^3P)$ atoms through absorptions in both the near ultraviolet and the visible spectral regions of the solar spectrum. It is therefore possible that this could constitute an additional, significant source of these atoms to the photodissociation of nitrogen dioxide discussed above, and as such could be responsible for the induction of the secondary chemistry of the troposphere. We therefore need to estimate J_{O_3} in order to compare it with J_{NO_2} and assess the relative source strengths.

At the outset it must be remembered that the solar radiation has already passed through the stratospheric ozone layer on its way down to the troposphere. In Section 2.1, and in particular in Table 2.1, we saw that the resultant absorption in the stratosphere was mainly responsible for the attenuation of ultraviolet radiation of wavelength below 300 nm, just the radiation which could have enhanced J_{O_3} in the troposphere had it penetrated the tropopause. The stratospheric ozone layer corresponds to the equivalent of a layer of pure ozone approximately 3 mm thick at the pressure and average temperature of the Earth's surface. The ozone level in the polluted stratosphere can be expressed on the same basis for perspective. In this way we may take the maximum possible upward extension of the polluted layer as 1 km. The mixing ratio of 10 p.p.h.m. corresponds to a partial pressure of ozone of 10^{-7} atm and, ignoring the lapse rate with altitude, we must think of the ozone content as having to be compressed 10 million times to produce the standard partial pressure of 1 atm. In other words, the thickness of the equivalent layer of pure ozone which would correspond to a polluted layer extending up 1 km in altitude with a uniform ozone mixing ratio of 10 p.p.h.m., is one ten-millionth of 1 km, i.e. 0.1 mm. Thus we find that the ozone content of the polluted troposphere is rather small in comparison with the stratospheric column of ozone above it. This might in itself suggest that the rate of photodissociation of ozone in the troposphere will be rather small in comparison with that of nitrogen dioxide; the effective radiation for the former has been subjected to a considerable attenuation in the stratosphere in contrast to the radiation in the wavelength range 350 nm to 400 nm, which Table 8.5 shows to contribute most to J_{NO_2}.

Nevertheless we must make our argument more quantitative by calculating J_{O_3} for the troposphere. The method adopted is exactly similar to that used for J_{NO_2}. We require average absorption coefficients for ozone over 10 nm wavelength intervals and also a knowledge of the quantum efficiency as a function of wavelength for the production of $O(^3P)$ atoms. For wavelengths below 313 nm the primary product is actually the electronically excited state $O(^1D)$, but we shall anticipate what is said in Chapter 9 on the fate of these, namely that they suffer very rapid physical quenching, mainly by nitrogen molecules, so that the effective product is $O(^3P)$ atoms.

The averaged values of the decadic absorption coefficients of ozone over 10-nm wavelength intervals have been tabulated in Table 2.1 up to a wavelength of 350 nm. In point of fact, the visible absorption spectrum (the so-called Chappius bands) is extremely weak compared to the main ultraviolet absorption spectrum and hardly attenuates incoming solar radiation at all. Table 8.7 is the analogue for ozone absorption of Table 8.5 for nitrogen dioxide absorption in the lower troposphere. The available evidence suggests a universal quantum efficiency of unity for oxygen atom production so that this is not tabulated. The Chappius band absorption is fairly symmetrical about 600 nm wavelength save for a very weak tail

Table 8.7. Average values over 10-nm intervals for parameters in the calculation of ground level J_{O_3} for overhead Sun

Middle λ/nm	ϵ/dm^2 mol^{-1}	$10^{-17} \times I_s(\lambda)/$ photons dm^{-2} s^{-1}	$10^{-18}j_\lambda$	$\Sigma 10^{-18}j_\lambda \cdot \Phi(\lambda)$
300	919	0.030	6.4	
310	269	0.302	18.7	
320	74.0	0.734	12.5	44.6
330	18.8	1.239	5.3	
340	4.5	1.358	1.4	
350	0.8	1.468	0.3	
450	0.5	4.079	0.5	
460	0.8	4.332	0.8	
470	1.0	4.375	1.1	
480	1.8	4.598	1.9	
490	2.2	4.433	2.3	
500	3.6	4.549	3.7	
510	4.2	4.515	4.4	
520	5.6	4.503	5.8	
530	7.3	4.637	7.8	111.8
540	8.2	4.592	8.7	
550	9.1	4.526	9.5	
560	10.9	4.552	11.4	
570	12.2	4.699	13.2	
580	11.8	4.777	12.9	
590	11.6	4.855	12.9	
600	13.3	4.853	14.9	

Absorption coefficient data read off graphs produced by M. Griggs, *Journal of Chemical Physics*, **49**, 859 (1968).

extending above 800 nm. Accordingly, for the level of accuracy which we require, it will be sufficient to work up from 450 nm to 600 nm wavelength and then to double the result to cover the entire band.

The approximation of doubling 111.8×10^{18} photons mol^{-1} s^{-1} to obtain the integrated absorption rate above 450 nm wavelength is strengthened by the fact that the maximum in the solar irradiance also occurs in the vicinity of 600 nm. Hence from Table 8.7 the estimated value of J_{O_3} comes out as 4.5×10^{-4} s^{-1} or 1.6 h^{-1}.

It is therefore apparent that J_{O_3} is of the order of one-twentieth of J_{NO_2}, so that under normal conditions in a polluted region of the troposphere it will be the photodissociation of nitrogen dioxide which will be the main source of atomic oxygen, and the production from ozone can be ignored without great error.

Evaluation of J_{HCHO} for the Lower Troposphere

At the present moment it is considered that photodissociation of formaldehyde is the major primary source of atomic hydrogen in polluted regions of the troposphere. It is a significant minor component of tropospheric regions polluted by internal combustion engine exhaust gases and it can account for up to 50% of the total aldehydic emissions from vehicle engines. As a result, mixing ratios of formaldehyde in heavily polluted areas like the Los Angeles basin can be typically of the order of 10 p.p.h.m., extending as high as 200 p.p.h.m.

We may perform our now familiar approach to the evaluation of the photodissociation rate coefficient for formaldehyde producing hydrogen atoms, $J_{HCHO}(H)$, and to the overall photodecomposition rate coefficient, J_{HCHO}. The profile of the decadic absorption coefficient, ϵ, for formaldehyde vapour as a function of wavelength was given as Figure 74 in the preceding section. This absorption spectrum gives rise to the average values of ϵ over 10 nm wavelength intervals shown in Table 8.8. Values of the quantum yield for the production of hydrogen atoms, $\Phi(H)$, have been tabulated earlier in Table 8.4, while the overall quantum yield in terms of the photodecomposition of formaldehyde is taken as unity at all wavelengths.

Summation of the two right hand columns yield the final results:

$$J_{HCHO} \quad = 9.9 \times 10^{-5} \text{ s}^{-1} = 0.36 \text{ h}^{-1}$$

$$J_{HCHO}(H) = 2.7 \times 10^{-5} \text{ s}^{-1} = 0.1 \text{ h}^{-1}$$

Under conditions of approximately-overhead sun, if the concentration of formaldehyde in the air is C_0 and no more formaldehyde is added to, or created in, that region of the troposphere, then photochemical decomposition will reduce the concentration to $C_0/2$ in a time equal to $\ln 2/J_{HCHO}$. This photochemical half-life of formaldehyde is then approximately 2 h.

Also for a typical formaldehyde vapour mixing ratio of 10 p.p.h.m. in a sunlit

Table 8.8 Average values over 10-nm wavelength intervals for parameters in the calculation of J_{HCHO} and $J_{HCHO}(H)$ at ground level for overhead Sun ($Z = 0$)

Middle λ/nm	ϵ/dm^2 mol^{-1}	$10^{-18} \cdot j_\lambda$	$10^{-18}! \cdot j_\lambda \cdot P_H(\lambda)$
290	83.3	2.34×10^{-4}	0
300	85.1	0.588	0.39
310	82.3	5.72	2.97
320	61.3	10.36	4.14
330	61.9	17.66	5.12
340	51.7	16.17	2.91
350	21.9	7.40	0.67
360	4.6	1.60	0.02

Absorption coefficients taken from J. G. Calvert, J. A. Kerr, K. L. Demerjian, and R. D. McQuigg, 'Photolysis of formaldehyde as a hydrogen atom source in the lower atmosphere', *Science*, 175, 751–752 (1972). Copyright 1972 by the American Association for the Advancement of Science and reproduced with permission.

urban atmosphere, we may expect a rate of production of primary hydrogen atoms of around 1 p.p.h.m. h^{-1}. This is rather small compared to the typical rate of formation of oxygen atoms from the photodissociation of nitrogen dioxide which is of the order of 300 p.p.h.m. h^{-1}. However we shall see shortly that such production rates do not necessarily reflect the relative importance of the species in question in the secondary chemistry of the troposphere.

Evaluation of J_{HNO_3} for the Lower Troposphere

In this evaluation, and in the one for nitrous acid vapour which follows, we shall be concerned implicitly with the primary photochemical production rate of another highly reactive species, the hydroxyl radical. In the preceding section it was said that the only effective photodissociation pathway for nitric acid under irradiation with near ultraviolet wavelengths gave rise to hydroxyl radicals with near-unity quantum efficiency.

Examination of Figure 75 and comparison with Figure 6 shows that the nitric acid vapour absorption spectrum is rather similar to the near ultraviolet absorption spectrum of ozone, except that the threshold wavelength of the former is somewhat shorter than the near 350 nm limit of the latter. Moreover the absorption coefficients of nitric acid vapour are always considerably less than those for ozone across the range of solar ultraviolet radiation which penetrates to the ground; at a wavelength of 300 nm for example, the decadic absorption coefficient for nitric acid vapour is of the order of 7 dm^2 mol^{-1} compared with a value of around 900 dm^2 mol^{-1} for ozone. It is therefore apparent that J_{HNO_3} will be at least two orders of magnitude less than even the near ultraviolet contribution to J_{O_3}, shown in Table 8.7, which means that J_{HNO_3} is less than 2×10^{-3} h^{-1}. As will be seen, such a value cannot be of any great significance in connection with the production of hydroxyl radicals in the troposphere.

Evaluation of J_{HNO_2} for the Lower Troposphere

The absorption spectrum of nitrous acid vapour shown in Figure 76 can be applied to obtain an upper limit for J_{HNO_2}, which is likely to be correct to within a factor of three. The uncertainty arises from the difficulty of estimating the concentration of the unstable nitrous acid vapour in an absorption experiment. Accordingly we are forced to use the upper limiting value of the decadic absorption coefficient obtained at the wavelength of maximum absorption, calibrate the rest of the spectrum on that basis, and then make some comment upon what the value of J_{HNO_2} is likely to be.

Table 8.9 shows the basis of the calculation. The quantum yield of hydroxyl radicals is taken to be unity across the entire absorption. The result derived from Table 8.9 is then that J_{HNO_2} has a maximum value of 2.8×10^{-3} s^{-1} or approximately 10 h^{-1}. The region of the spectrum covered by nitrous acid absorption is similar to that covered by the photodissociating part of the nitrogen dioxide absorption spectrum; in fact the maximum value of the absorption coefficient for nitrous acid, $\epsilon = 1500 \, dm^2 \, mol^{-1}$ at $\lambda = 368$ nm is of the same order as the decadic absorption coefficients of nitrogen dioxide in this region (Table 8.5). However, the banded nature of the nitrous acid spectrum produces average absorption coefficients across the 10-nm wavelength ranges which are around one-half of those of nitrogen dioxide in the central region of Table 8.9, with no effective contribution above 390 nm wavelength. Hence J_{HNO_2} comes out with a maximum value of about one-third of J_{NO_2} and it is likely that the actual value lies within a range lower again by a factor of three. The half-life of nitrous acid vapour in a sunlit urban atmosphere is therefore in the range of 4 to 12 min as opposed to nearer 1 to 2 min for nitrogen dioxide. However, the rate of generation of hydroxyl radicals from nitrous acid photolysis relative to the rate of generation of oxygen atoms from nitrogen dioxide photolysis, will depend upon the relative concentrations or mixing ratios of the two species in a particular region of the troposphere also.

Table 8.9. Average values over 10-nm wavelength intervals for parameters in the calculation of J_{HNO_2} at ground level for overhead Sun

Middle λ/nm	ϵ/dm^2 mol^{-1}	$10^{-17} \times I_s(\lambda)$/ protons dm^{-2} s^{-1}	$10^{-18} \cdot J_\lambda \cdot \Phi_{OH}(\lambda)$
300	127	0.030	8.8
310	236	0.302	16.4
320	351	0.734	59.3
330	539	1.239	154
340	729	1.358	228
350	774	1.468	262
360	659	1.514	230
370	882	1.759	357
380	625	1.747	252
390	240	1.786	98.7
		Sum =	1666

Nitric-acid vapour mixing ratios in sunlit urban atmospheres have been found to be as high as 10 p.p.h.m., but the typical level is much lower, probably nearer to 0.1 p.p.h.m. The typical levels of nitrous acid vapour are two orders of magnitude lower still, at around 0.001 p.p.h.m. Under these circumstances the primary photochemical rate of production of hydroxyl radicals will be of the order of 10^{-2} p.p.h.m. h^{-1}, originating overwhelmingly from nitrous acid. This, then, is around 1% of the typical rate of generation of hydrogen atoms by the photodissociation of formaldehyde, and only around one thirty-thousandth of the rate of production of oxygen atoms by the photodissociation of typical levels of nitrogen dioxide.

8.4 The Overall Chemistry of Photochemical Smog

The photodissociation processes discussed in the preceding sections make up an exceedingly small part of the overall chemical transformation scheme of a sunlit urban atmosphere. Nevertheless it is the input of solar energy at these few stages which drives the total chemical ensemble of many thousands of elementary reactions. The complexity of the situation has been revealed by laboratory studies, where simulated polluted atmospheres have been irradiated. The detailed results derived from the irradiation of but a single olefinic species mixed with nitric oxide and nitrogen dioxide, and at large dilution in air, have demanded a scheme of over 200 elementary reactions for explanation. The reader may refer back to the range of hydrocarbon species present in the air of Figure 42 to realize the complexity of the real situation. This fact then immediately suggests that, within the scope of this book, it will be impossible to develop a complete chemical model for tropospheric chemistry: it is doubtful if that will ever be produced. What we shall try to do, however, is to locate key stages of the overall mechanism and to analyse on the basis of available kinetic data, how it comes to be that some elementary reactions must be highly important while others can only be insignificant.

Tropospheric pollution control will be accomplished in the end by the limitation of the primary emissions of nitric oxide, carbon monoxide, 'unburnt' hydrocarbon species, and sulphur dioxide principally. However since this objective will demand a collosal financial expenditure for the necessary reformation of our present energy technology, it is necessary to understand how this can be justified in terms of the correct identification of the critical polluting species. Moreover, since even the best control procedures will involve some leakage of pollutants, a quantitative basis must be established for the evaluation of tolerable emission levels. Added to this, there is also a purely scientific interest to be satisfied on the score of how the various species injected into the troposphere interact to produce the, albeit undesirable, results.

Even before World War II, it had been observed that Los Angeles in the United States suffered from the phenomenon now generally known as 'photochemical smog'. At that time the area possessed something like a monopoly on the problem, but with the ever increasing propagation of the internal combustion engine and the increase in localized population densities, the general phenomenon is now observed in such diverse parts as Rotterdam in the Netherlands, Ankara in Turkey, Sydney in Australia, and many large conurbations. In fact, it has been reported that smog formation in Sao Paulo in Brazil has, on occasions, almost reached those levels at which the Los Angeles city authorities would evacuate the population! Of course the problem has intensified in Los Angeles itself to the extent where much of the current exhaust emission control legislation is specifically geared to the Californian situation.

The outward manifestations of photochemical smog are much reduced visibility, rather like fog but more associated with dry aerosol materials, coupled with the more insidious aspects of eye and bronchial irritation and plant damage (phytotoxication), associated with high levels of oxidant species such as ozone. Although ozone itself was blamed initially for the plant damage, experimental work has shown that all the typical plant injuries and symptoms could only be produced when automobile exhaust gases were present also. Evidence soon mounted that solar radiation was a third ingredient: the oxidant levels correlated strongly with incident solar irradiances both in diurnal variations and on integrated daily bases. It was also found that 'clean' samples of outside air would develop smog character-istics upon laboratory-simulated solar irradiation. Subsequent irradiations of mixtures of separate components then demonstrated the critical role of nitrogen dioxide: not only ozone but organic oxidant materials were also produced in experiments using low initial concentrations of nitrogen dioxide and single hydrocarbon species in air, together with other characteristic compounds such as aldehydes and organic nitro compounds. The levels so produced were similar to those detected in the 'natural' photochemical smog situations.

The level of sophistication which has now been attained has resulted in the identification of single phytotoxicant species, together with the development to a significant extent of computer simulation procedures for photochemical smog formation using kinetic data for the array of elementary reactions involved. Other efforts are currently being directed at the modes of formation of the aerosol particles which reduce the visibility. The climatic or meteorological conditions which favour the formation of photochemical smogs, and their combination with geographical features which makes Los Angeles so prone to the phenomenon, are now well understood.

It has also become evident in recent years that the original objective of the Clean Air Acts in Britain may have been substantially achieved in their original sense. Although the smoke and sulphur dioxide accumulations within British cities have been depressed to the levels where another occurrence like the reducing smog of London in December 1952 is unlikely, recent detections of high levels of ozone and dry aerosol particles suggest that the new problem of oxidizing or photochemical smog may be developing. It is ironic perhaps that the cleaner air in terms of smoke and induced cloud has given rise to more hours of sunlight in our cities and thus to photochemical consequences unthought of in the original concepts of the legislation.

The Typical Photochemical Smog Cycle

The classical exemplifying situation is naturally Los Angeles and Figure 78 shows the typical pattern of diurnal concentration variations of the main species during a sunlit day.

It can be seen that overnight there are background levels of nitrogen oxides, hydrocarbon species, and ozone, the last being at a very low level. Automobile activity begins at around 06.00 local time and has the initial effect of increasing the nitric oxide and hydrocarbon levels sharply, these being, as we have seen in Chapter 5, the primary species emitted at exhaust by internal combustion engines. It is anticipated that carbon monoxide levels would also increase sharply in the

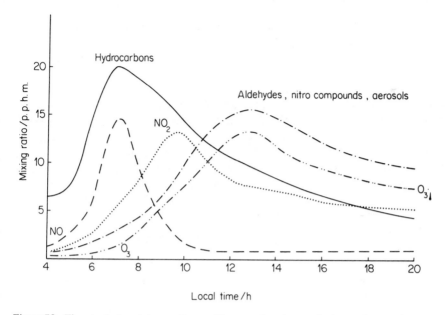

Figure 78. The typical mixing ratio profiles as functions of time of day in the photochemical smog cycle.

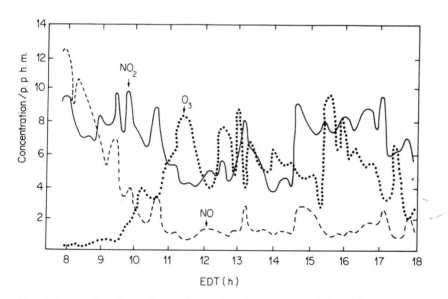

Figure 79. The diurnal variation of the concentrations of NO_2, NO, and O_3 measured for August 29, 1973 in Downtown Detroit, USA. From D. H. Stedman and J. O. Jackson, 'The photostationary state in photochemical smog', *International Journal of Chemical Kinetics*, Symposium No. 1, p. 496 (1975). Reproduced with permission of John Wiley & Sons Inc., New York.

250

early period of the day for the same reason. It is clear also that the emitted nitric oxide is converted to nitrogen dioxide fairly quickly, reaching a peak mixing ratio at around mid-morning. In turn the nitrogen dioxide level decreases through noon in favour of ozone, which reaches a peak mixing ratio shortly thereafter. Concurrently with the conversion of nitric oxide through nitrogen dioxide to ozone, the emitted hydrocarbons are oxidized to produce aldehydes amongst other partially oxidized species. At the same time there is a build-up of aerosol material, with a peak density being achieved around noon, and with a sharp rise in the density of larger size (above 2 μm diameter) particles paralleling the rise of ozone towards its peak. Figure 79 shows the actual diurnal variation of the nitric oxide–nitrogen dioxide–ozone part of the cycle as measured in Detroit. Although there are evidently random deviations from the idealized picture in Figure 78, the basic pattern is clearly developed.

The situation is even better illustrated by the variations of concentrations as a

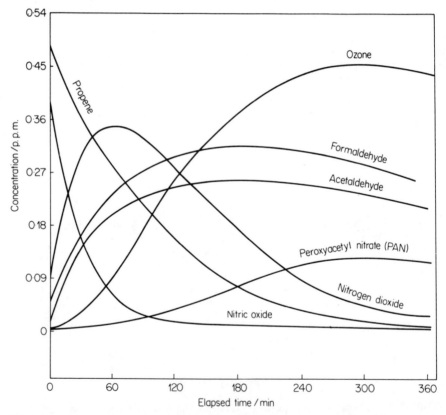

Figure 80. The concentration profiles during the laboratory irradiation of propylene–NO–NO$_2$–air mixtures. From J. N. Pitts, Jr., A. C. Lloyd and J. L. Sprung, 'Ecology, energy and economics', *Chemistry in Britain*, **11**, 247 (1975). Reproduced by permission of the Chemical Society.

function of time observed in artificial irradiations conducted with mixtures of a single hydrocarbon species with nitric oxide and nitrogen dioxide in air. A typical experimental result is shown in Figure 80. The rises in the concentrations of the partially oxidized organic species such as formaldehyde, acetaldehyde, and peroxyacetyl nitrate parallel the fall in the original propylene concentration, and reach maximum development when ozone concentrations are rising towards their peak. Of particular note is the observation that the peak in the level of peroxyacetyl nitrate appears almost contemporarily with the ozone peak.

Associated Meteorological Factors

As has already been remarked, meteorological conditions are highly important for the development of photochemical smog conditions. The critical feature appears to be the occurrence of a temperature inversion close to the ground, which traps a body of air and allows pollutant species to accumulate therein.

A *temperature inversion* is defined as the situation where the normal lapse rate (Section 3.2) in the lower part of the troposphere is inverted and the temperature actually increases with altitude. Such inversions normally extend to only a few hundred metres above the ground at most.

In Los Angeles, the commonest origin of the very frequent inversions (occurring on average some 320 days per annum) is the influx of comparatively cold air from the adjacent ocean, which pushes in below the warmer air above the land. The Los Angeles basin is enclosed on three sides by mountains and is open to the sea on the fourth. Therefore, once the cold lower layer is established, it will have a high degree of stability, particularly as winds tend to be light in the area. It is apparent that vertical mixing will become highly suppressed within the inverted layer. Consider a parcel of air near to the ground attempting to rise. As it ascends it will expand adiabatically, thereby cooling and becoming more dense; but in moving upwards the cool air passes into warmer and consequently less dense air; therefore it will tend to sink back to the ground.

It is therefore clear that, because of the inhibition of vertical mixing so produced, pollutant gases cannot be dispersed vertically or horizontally over vertical barriers of sufficient height and will therefore tend to accumulate below the inversion. On the other hand, horizontal mixing within the inverted layer will probably be unaffected to any large extent, and slight breezes might be expected to produce an approximately homogeneous mixture below the inversion. Therefore for simplicity we may consider the inverted layer as a homogeneous chemical reaction system, to a reasonable approximation, and develop the photochemical consequences without recourse to fluid dynamic considerations. Moreover we can treat the inverted layer as a closed system, at a further approximation, since vehicular activity subsides substantially during the high solar irradiance hours of the day.

Temperature inversions can occur under circumstances other than the typical Los Angeles mode. A so-called subsidence inversion is associated with a developing anticyclone or high pressure area. The descending air warms as it contracts adiabatically and the warming is often greater in the upper layers than in the lower layers close to the ground. Thus a persistent anticyclone may lead to a persistent temperature inversion below which pollutants can accumulate. This is aided by the usual windless conditions associated with anticyclonic weather, with the often cloudless skies promoting the photochemical conversions which follow. Subsidence inversions are more pertinent to the British situation where they are likely to coincide in the summer months with high incident solar irradiance. As was pointed out in Section 2.1 and illustrated by Figure 10, the integrated daily solar irradiance in the British midsummer is actually of the same order as, if not slightly greater than, that for Los Angeles at the same time by virtue of the longer hours of daylight in Northern latitudes. The development of photochemical smog in Britain is therefore principally governed by the incidence of temperature inversions. The main reason, therefore, for the absence of a problem of a similar order to that of Los Angeles lies in the comparative infrequency of temperature inversions in Britain; perhaps 5 days out of 100 in the summer period is an average British frequency.

A further circumstance of temperature inversion occurs where there is an enclosed valley in which cold air, because of its higher density, tends to collect. This then creates a strictly localized inversion and is sometimes made strikingly visible when mist forms within the cold layers near to the valley floor. One rather spectacular instance of such a localized inversion situation used to be found in the Hope Valley of Derbyshire. There a cement works had a chimney which, with a height of around 130 m, was not quite sufficient to penetrate the typical inverted layer. The unfortunate result was that the valley was frequently filled with fine white dust representing the accumulated emission from the stack. Eventually the height of the chimney was increased by a mere 10 m or so, which proved sufficient to penetrate the inversion on most occasions and the atmosphere of the valley cleared almost miraculously as a result.

The Generation of the Photoabsorbing Species and their Relative Significance

It is apparent that nitrogen dioxide, ozone, nitrous acid, and nitric acid are not primary emission products from combustion engines. Formaldehyde is a significant minor direct emission and may make up around half of the total aldehydic emission but formaldehyde is also generated as a secondary product of the partial oxidation of 'unburnt' hydrocarbon species in the air (see Figure 80). Hence at the outset of our consideration of the chemistry of the polluted troposphere, we must investigate the formation of these species which act as the main input vehicles for solar energy.

On the face of it, the conversion of nitric oxide into nitrogen dioxide might appear to have a simple origin. One of the most familiar examples of a visible, gas-phase, chemical reaction is the development of the brownish colouration of nitrogen dioxide when nitric oxide is released to air. The reaction concerned is actually third order and written as:

$$2\,NO + O_2 \longrightarrow 2\,NO_2 \tag{15}$$

The rate constant for this reaction at ambient temperatures is well established at around $k_{15} = 1.5 \times 10^4$ dm^6 mol^{-2} s^{-1}. In Figure 79 we see that the required atmospheric conversion rate is the conversion of the equivalent of some 10 p.p.h.m. of nitric oxide to nitrogen dioxide within a time scale of 2 h at most, when the average level of nitric oxide is of the order of 10 p.p.h.m.

A useful investigational technique for the assessment of whether the rate of a particular elementary reaction can be fast enough to account for a transformation within a given time is the Projection of Maximum Rates. In this we simply extrapolate the maximum (often the initial) rate over the allotted time. If that rate cannot account for the extent of conversion significantly, then the reaction concerned cannot play any major role therein. However the converse does not necessarily follow; it may be the case that reactant consumption lowers the actual rate well below the projected maximum.

The maximum rate of reaction (15) during the day is the initial rate since nitric oxide is consumed from left to right across Figure 79. A mixing ratio of 12 p.p.h.m. corresponds to a concentration $[NO] = 5 \times 10^{-9}$ mol dm^{-3} and the molecular oxygen concentration is $[O_2] = 8 \times 10^{-3}$ mol dm^{-3}. Hence the early morning rate of reaction (15) can be expressed as:

$$- d[NO]/dt = 1.5 \times 10^4 \cdot (5 \times 10^{-9})^2 \cdot 8 \times 10^{-3}$$
$$= 3 \times 10^{-15} \text{ mol dm}^{-3} \text{ s}^{-1}$$

since the reaction is second order in nitric oxide and first order in molecular oxygen. Converting to the incremental form to project this maximum rate and inserting the incremental time $\Delta t = 2$ h $= 7200$ s, yields the nitric oxide concentration decrement, $-\Delta[NO] = 2.2 \times 10^{-11}$ mol dm^{-3}, which corresponds to only 0.04 p.p.h.m., hardly matching the required decrement of 10 p.p.h.m. We may therefore state conclusively that reaction (15) is of no importance in the conversion of nitric oxide to nitrogen dioxide under typical atmospheric conditions. The underlying reason is that the rate of this reaction is dependent upon the square of the nitric oxide concentration, which is a very small term under atmospheric conditions. On the other hand, the rapid conversion observed in the demonstration referred to above stems from the much larger concentrations of nitric oxide involved there. For example, a mixing ratio of 0.1%, i.e. 100,000 p.p.h.m. can often be involved there, producing an initial rate of reaction (15) some one hundred million-times greater than that calculated above.

We can now appreciate that the reaction involved in the tropospheric conversion will almost certainly be first order in nitric oxide, i.e. rate proportional to $[NO]$ rather than $[NO]^2$. One possibility is therefore reaction between nitric oxide and ozone, the reaction mentioned in Section (8.1) as having a time scale for 1% conversion of typical and equal concentrations of each reactant of around 0.25 s.

$$NO + O_3 \longrightarrow NO_2 + O_2 \tag{14}$$

The problem attaching to this reaction in the early morning hours is evident in Figure 79: ozone levels are very low indeed before the nitric oxide has been substantially converted to nitrogen dioxide and, if we are to regard the polluted region as a closed system, then only a small conversion of nitric oxide could be achieved in any case on the basis of the unit stoichiometry of reaction (14). Moreover once the nitrogen dioxide is formed, the conclusion of Section 8.3 was that it can only have a photochemical lifetime of a few minutes before it is photodissociated to regenerate nitric oxide with effective first-order rate constant J_{NO_2}.

$$NO_2 + h\nu \longrightarrow NO + O(^3P)$$

The fate of the product oxygen atom must be overwhelmingly the three-body reaction with the abundant molecular oxygen to form ozone:

$$O(^3P) + O_2 + M \longrightarrow O_3 + M \tag{16}$$

Hence in an effectively closed system, the *net* change effected by this cycle in terms of nitric oxide to nitrogen dioxide conversion must be very small. However, this is not to say that these reactions are not proceeding individually at significant rates. As mentioned earlier, the effective value of the rate constant k_{14} is about 10^7 dm^3 mol^{-1} s^{-1}. If we take the early morning levels of nitric oxide and ozone as 12 p.p.h.m. and 0.2 p.p.h.m. respectively from Figure 79, the corresponding concentrations are $[NO] = 5 \times 10^{-9}$ mol dm^{-3} and $[O_3] = 8 \times 10^{-11}$ mol dm^{-3}. The predicted instantaneous rate of reaction (14) is then:

$$-d[NO]/dt = k_{14}[NO][O_3] = 10^7 \times (5 \times 10^{-9}) \times (8 \times 10^{-11})$$
$$= 4 \times 10^{-12} \text{ mol dm}^{-3} \text{ s}^{-1} = 0.01 \text{ p.p.h.m. s}^{-1}.$$

Extrapolation of this rate over an hourly period evidently produces a nitric oxide *turnover* well above the actual ambient level. In fact a photostationary state is established within the overall mechanism, where the rate of removal of nitric oxide by reaction (14) is exactly balanced by the rate of production by the photodissociation of nitrogen dioxide, i.e. $J_{NO_2}[NO_2] = k_{14}[NO][O_3]$. We may therefore define a photostationary state number, N_{ps}, as:

$$N_{ps} = J_{NO_2}[NO_2]/k_{14}[NO][O_3]$$

and this should be unity if the photostationary state is established. Figure 81 is a plot of values of N_{ps} derived from the data of Figure 79 for the observations in Detroit, and this shows clearly that N_{ps} achieves the value of unity very early in the day, with only random fluctuations evident thereafter.

Figure 81 also shows a plot of J_{NO_2} as measured with a chemical actinometer as a function of local time of day. A similar photostationary state must hold for ozone, representing balance between the rates of reactions (14) and (16), and for oxygen atoms, representing balance between photoproduction from nitrogen dioxide and the removal reaction (16).

A key to the mechanism which does effect the net conversion of nitric oxide to nitrogen dioxide was found in laboratory irradiations of simulated polluted atmospheres when carbon monoxide was present. Until that time it had been thought that carbon monoxide was an inert species from the chemical point of

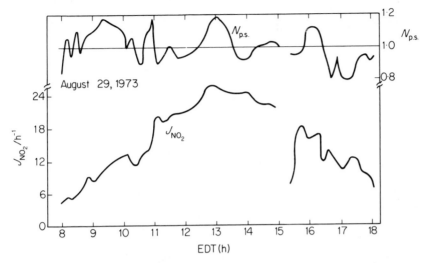

Figure 81. The variation of the measured value of J_{NO_2} and the photostationary state number, N_{ps}, on a diurnal basis for August 29, 1973, in Downtown Detroit, USA. From D. H. Stedman and J. O. Jackson, 'The photostationary state in photochemical smog', *International Journal of Chemical Kinetics*, Symposium No. 1, p. 496 (1975). Reproduced by permission of John Wiley & Sons Inc., New York.

view. Figure 82 shows the enhancement of the development of the simulated photochemical smog in the laboratory irradiations of isobutene–nitric oxide mixtures in air occasioned by the incorporation of 10,000 p.p.h.m. of carbon monoxide. There is only one reaction of carbon monoxide which has a rate high enough to be significant under these simulated tropospheric conditions, the reaction with hydroxyl radicals:

$$CO + OH \longrightarrow CO_2 + H \tag{13}$$

The value of the rate constant k_{14} is generally accepted to be around 9×10^7 dm^3 mol^{-1} s^{-1} for ambient temperatures. Accepting the occurrence of this reaction, we may consider the fate of the product hydrogen atom. The argument at this point is similar to that adopted in the assessment of the fate of the oxygen atom produced by nitrogen dioxide photodissociation, in that any three-body reaction involving molecular oxygen as specific coreactant must have a high rate in the troposphere on account of the abundance of this coreactant. Hence we would expect the predominant reaction removing hydrogen atoms in the troposphere to be represented by the equation:

$$H + O_2 + M \longrightarrow HO_2 + M \tag{17}$$

The rate constant for this reaction in air at ambient temperatures is of the order of 2×10^{10} dm^6 mol^{-2} s^{-1}, almost two orders of magnitude larger than k_{16} for the corresponding three-body reaction of oxygen atoms. Investigations of the rates of reaction of hydroperoxy, HO$_2$, radicals have shown that there is a rapid reaction

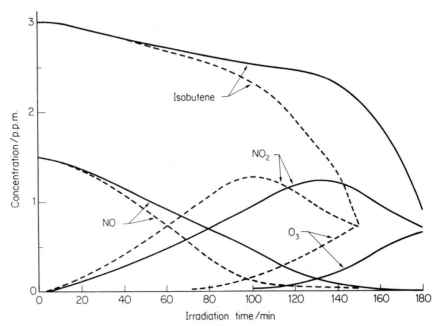

Figure 82. Enhancement of the rate of development of photochemical smog by carbon monoxide in a laboratory irradiation. Solid lines refer to 0 ppm of CO, dashed lines refer to 100 ppm of CO added. The relative humidity was 70% and the temperature was 301 K. The NO, NO_2, and O_3 curves were corrected for the rapid reaction between NO and O_3 that occurred during the transfer of the samples from the smog chamber to the measuring instruments. From K. Westberg, N. Cohen and K. W. Wilson, 'Carbon monoxide: Its role in photochemical smog formation', *Science,* **171** (12 March), 1014 (1971). Copyright 1971 by the American Association for the Advancement of Science and reproduced with permission.

with nitric oxide, represented as:

$$HO_2 + NO \longrightarrow OH + NO_2 \tag{18}$$

which occurs at more than one collision in one thousand of the reactant species. This reaction is therefore highly significant; not only does it close a cycle composed of reactions (13), (17), and (18), but it achieves nitric oxide to nitrogen conversion without involving ozone as a reactant. Our discussion of the implications of Figure 79 made this last point a necessary rider upon the situation in the real troposphere.

It may be imagined that the above cycle is started up by the photodissociation of formaldehyde, yielding a hydrogen atom directly, but there is also considerable evidence that the other fragment, the formyl (HCO) radical, provides a parallel impulse in reacting according to the equation:

$$HCO + O_2 \longrightarrow HO_2 + CO \tag{19}$$

The resultant hydroperoxy radical will enter the cycle through reaction (18).

The termination reactions of the chain cycle, which withdraw the chain carriers and do not give rise to propagating products, must involve principally hydroxyl radicals as reactants. That the termination steps compete principally with the propagation step (13) is perhaps not surprising: the rate constant k_{13} corresponds to reaction occurring between CO and OH effectively at only around one collision in about 2000 or so. Many other species, such as many of the hydrocarbons present in polluted atmospheres, react with hydroxyl radicals at much higher collisional efficiencies than this. For example, the alkane, n-butane, reacts with hydroxyl radicals at about one collision in 100, while olefinic and aldehydic species often approach reaction at every 10 collisions. Hence with an olefinic mixing ratio typically 3 p.p.h.m. in the polluted atmosphere, the mixing ratio of carbon monoxide required to make the rate of reaction (13) an order of magnitude larger than the rate of the competing olefinic termination step is of the order of 6000 p.p.h.m. Typical levels of carbon monoxide in polluted urban atmospheres are in the range 200 p.p.h.m. to 2000 p.p.h.m. Hence although computer simulations have borne out the postulate that the cycle represented by the equations (13), (17), and (18), makes a significant contribution to the conversion rate of nitric oxide to nitrogen dioxide in photochemical smog, it appears necessary to invoke further reactions to counter the relatively short chain length predicted above.

By analogy with the reaction of hydroperoxy radicals in reaction (18), it might be expected that peroxyalkyl (RO_2) and acylperoxy ($RCO.O_2$) radicals could also oxidize nitric oxide according to the equations:

$$RO_2 + NO \longrightarrow RO + NO_2$$

$$RCO \cdot O_2 + NO \longrightarrow RCOO + NO_2$$

Here R represents alkyl groups given by the general formula C_nH_{2n+1} ($n = 1,2,3, \ldots$). Thus $R = CH_3$, C_2H_5, etc., and in fact the hydroperoxy radical, HO_2, might be regarded simply as the first of the series with $n = 0$ and $R = H$, just as formaldehyde can be regarded as the first of the aldehydic series. The complexity of the set of specific cases of the above general reactions defies detailed examination here. It will suffice to say that, on the basis of present knowledge, these reactions are to be considered as also contributing significantly to the nitric oxide to nitrogen dioxide conversion in the photochemical smog situation. More will be said on the organic reactions in a later subsection.

In the course of our discussion of the conversion mechanism of nitric oxide to nitrogen dioxide, we have come across a significant source of hydroxyl radicals in the chain cycle based on reactions (13), (17), and (18), which means that any source of hydrogen atoms or hydroperoxy radicals can be regarded as a source of hydroxyl radicals. As was discussed in the preceding section, nitrous acid provides a direct photodissociation source of hydroxyl radicals, far stronger than nitric acid. We must therefore consider the formation mechanism of nitrous acid in polluted atmospheres.

The probable major source of nitrous acid vapour is the three-body reaction represented by the equation:

$$NO + NO_2 + H_2O \longrightarrow 2 HNO_2 \tag{20}$$

At present the rate constant for this reaction is uncertain, but a recent study (England and Corcoran, 1975) has suggested a value of $k_{20} = 1.5 \times 10^5$ dm^6 mol^{-2} s^{-1}. Under the midmorning conditions of Figure 79 there are around 10 p.p.h.m. each of nitric oxide and nitrogen dioxide, corresponding to concentrations of 4×10^{-9} mol dm^{-3}. An average humidity corresponds to around 1% of water vapour in the air, corresponding to a concentration of 4×10^{-4} mol dm^{-3}. Hence the instantaneous rate of nitrous acid formation at this time will be:

$$d[HNO_2]/dt = 2\,k_{20}[NO][NO_2][H_2O]$$
$$= 1.9 \times 10^{-15} \text{ mol dm}^{-3} \text{ s}^{-1} = 4.6 \times 10^{-6} \text{ p.p.h.m. s}^{-1}$$

Accepting that photodissociation is the overwhelming fate of nitrous acid vapour under these conditions, with J_{HNO_2} of the order of 2×10^{-3} s^{-1} (previous section), the stationary state mixing ratio of nitrous acid vapour is predicted to be around 2×10^{-3} p.p.h.m. Such low ambient levels of species like nitrous acid are largely beyond the capabilities of current measuring techniques; consequently calculations of the sort performed above are often the only recourse.

It is interesting to compare the potential source strength of hydroxyl radicals derived from the photodissociation of nitrous acid with that which could arise from the indirect route commencing with the photodissociation of formaldehyde. We have evaluated $J_{HCHO}(H)$ as 2.7×10^{-5} s^{-1} in the preceding section; this should be doubled to take into account hydroperoxy-radical formation from both hydrogen atoms and formyl radicals. For a typical mixing ratio of formaldehyde of 10 p.p.h.m. in a polluted atmosphere, we can therefore calculate a maximum hydroxyl-radical source strength from formaldehyde of around 5×10^{-4} p.p.h.m. s^{-1} ($=2J_{HCHO}(H) \cdot [HCHO]$), which assumes that all hydrogen atoms and hydroperoxy radicals react to generate hydroxyl radicals. This source strength is then two orders of magnitude larger than that deriving from nitrous acid photodissociation, and therefore appears to be the major initiation route of the chain cycle involving hydroxyl radicals.

However it remains to be pointed out that the study of reaction (20), referred to above, revealed the possibility of a heterogeneous component. Consequently the real rate of formation of nitrous acid, perhaps involving reaction on the surfaces of aerosols, in the polluted atmosphere could be considerably faster than suggested above.

In conclusion to this subsection, we may say that a strong case has been made to postulate that only two major species, nitrogen dioxide and formaldehyde, can be considered as absorbing solar radiation to serve as major source strengths of atoms and radicals under typical conditions in a polluted atmosphere. We may now go on to consider the subsequent chemistry in more detail.

The Development of Photochemical Smog

In this subsection we shall attempt to assess the importance of various reactions of photochemically generated species in inducing the chemical reactions which lead to the formation of photochemical smog. Our first step is to try to estimate the concentrations of atoms or radicals typically present in a sunlit urban atmosphere.

In the case of the atomic species, the steady state concentrations are extremely small and can only be estimated by the calculational approach.

The concentrations of atomic oxygen in the sunlit polluted regions of the troposphere are determined overwhelmingly by the rate of photodissociation of nitrogen dioxide and the rate of removal of oxygen atoms in the three-body formation of ozone, reaction (16). For air the third body (M) can be assumed to be 20% O_2 and 80% N_2 for practical purposes. Values of k_{16} for M = O_2 and M = N_2 are fairly closely defined and may be combined linearly to yield k_{16} (M = air) equal to 3×10^8 dm^6 mol^{-2} s^{-1}, probably accurate to within 20%. We have calculated J_{NO_2} to be approximately 8×10^{-3} s^{-1} for near overhead sun; this value is closely confirmed by the actinometric measurements shown in Figure 81. Applying the steady state approximation to oxygen atoms gives the equality:

$$J_{NO_2}[NO_2] = k_{16}[O][O_2][Air]$$

and, on inserting the numbers for the various quantities, the ratio of concentrations is obtained:

$$[O]/[NO_2] = 8 \times 10^{-8}$$

Since the observed mixing ratio of nitrogen dioxide around noon in Figure 79 is approximately 5 p.p.h.m., equivalent to a concentration of 2×10^{-9} mol dm^{-3}, the predicted oxygen atom mixing ratio is 4×10^{-7} p.p.h.m., equivalent to a concentration of 1.7×10^{-16} mol dm^{-3}. In the earlier part of the day in Figure 79, say at 08.00, the nitrogen dioxide mixing ratio is rather higher, at around 10 p.p.h.m., while from Figure 81 we may read off the corresponding value of J_{NO_2} as 1.3×10^{-3} s^{-1}. Insertion of this quantity into the above equation then leads to a predicted atomic oxygen mixing ratio of 1.3×10^{-7} p.p.h.m. Hence we may, with confidence, consider the range of 10^{-7} to 10^{-6} p.p.h.m. for oxygen atom mixing ratios to be appropriate to the critical period for the development of photochemcial smog in view of the well-established photostationary state and the simplicity of the predominant mechanism.

The calculation of the typical mixing ratios of the other reactive species is much more difficult, largely because of their involvement in chain cycles. It is usually necessary to develop a computer modelling technique or to compute the solutions to coupled differential equations in order to obtain the mixing ratios of species like hydrogen atoms or hydroperoxy radicals. This is beyond our present scope but the papers by Levy cited at the end of this chapter exemplify the computational method.

In fact, the concentrations of hydroxyl radicals present in sunlit urban atmospheres have proved recently to be amenable to measurement, as detailed in the papers by Wang and his coworkers cited in the list at the end of this chapter. So far this is the only case where such has been accomplished for any of the highly reactive species. It is worthwhile digressing slightly to consider some details of the detection procedure.

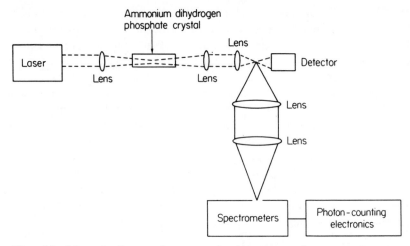

Figure 83. Schematic diagram of apparatus for the resonance fluorescence detection of hydroxyl radicals in air. From C. C. Wang and L. I. Davis, *Physical Review Letters,* **32** (18 February), 349 (1974). Reproduced by permission of the American Institute of Physics.

Figure 83 shows a schematic diagram of the *resonance fluorescence* apparatus used to measure the hydroxyl radical concentrations. The source of radiation used to excite the radical was a tunable laser, emitting a high intensity at 282.58 nm wavelength. The ammonium dihydrogen phosphate crystal serves to double the frequency of the original radiation output from the dye laser. The resultant second harmonic beam was brought to a focus in air on a spot 2 mm in diameter, and the resonantly scattered radiation perpendicular to the exciting beam was detected using photon-counting electronics coupled to a photomultiplier and tandem monochromators. Under these conditions the hydroxyl radicals within the focused spot are electronically excited from the ground state $OH(X^2\Pi)(v=0)$ to the first vibrationally excited level of the $OH(A^2\Sigma^+)$ state (denoted by OH* in the discussion to follow). In the excited state, the radicals have a radiative lifetime of around 750 ns but, at atmospheric pressure, vibrational relaxation is so rapid that the observed fluorescence originates entirely from $OH(A^2\Sigma^+)(v=0)$. Hence the wavelength of the detected radiation is in the hydroxyl (0,0) band around 309 nm, an advantage in that non-resonantly scattered radiation from the original laser beam gives no interference. In summary, the fluorescence mechanism can be written as:

$$OH \quad + h\nu(\text{laser}) \longrightarrow OH^*(v=1)$$
$$OH^*(v=1) + M \longrightarrow OH^*(v=0) + M$$
$$OH^*(v=0) \longrightarrow OH(v=0) + h\nu'$$

It is evident that the intensity of the fluorescence radiation detected will depend upon the number of hydroxyl radicals present in the focused spot, i.e. the concentration of hydroxyl radicals in the air. Figure 84 shows the clear development of the hydroxyl radical fluorescent band above and between spontaneous Raman scattering peaks of atmospheric components.

The apparent hydroxyl radical concentrations measured in Michigan at the Ford Motor Company research laboratories using the resonance fluorescence technique were overestimated initially because of ozone photodissociation within the focused spot. On the basis of our discussion of ozone photodissociation in Section 8.2, the absorption of the laser wavelength of 282.58 nm by ambient ozone will produce electronically excited oxygen atoms, $O(^1D)$, with unit quantum efficiency. Most of these are physically quenched to the ground state but a small fraction reacts with water vapour according to the equation:

$$O(^1D) + H_2O \longrightarrow 2\,OH \tag{21}$$

The rate constant for this reaction is close to the collision frequency. This 'photosynthesized' hydroxyl radical is then also excited by the incident laser radiation and consequently gives resonance fluorescence superimposed upon the

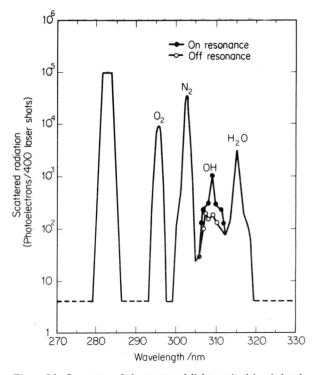

Figure 84. Spectrum of the scattered light excited in air by the incident laser radiation near a wavelength of 282.58 nm. The peaks at 295.6, 302.5, and 315.1 nm are due respectively to the spontaneous Raman scattering of oxygen, nitrogen, and water vapour in air. From C. C. Wang and L. I. Davis, *Physical Review Letters,* **32** (18 February), 349 (1974). Reproduced by permission of the American Institute of Physics.

resonance fluorescence of the ambient hydroxyl radicals. In confirmation of this point, there was a dependence of the apparent hydroxyl radical concentration upon the measured ozone and water vapour contents of the air. However, the extent of the ozone interference could be estimated and there was, particularly around midday, an excess resonant fluorescence signal above that which could have come from this source. The differences then indicated ambient hydroxyl radical concentrations in the range 10^{-13} to 10^{-12} mol dm^{-3}, corresponding to mixing ratios in the range 2×10^{-4} to 2×10^{-3} p.p.h.m. for the period around noon, quite variable from day to day. In this light we shall take a hydroxyl radical mixing ratio of 10^{-4} p.p.h.m. as a reasonable minimum value applicable to overhead (or nearly so) sun conditions in the relative-rate arguments to be advanced shortly. This value is then some two orders of magnitude larger than the typical noon mixing ratio of oxygen atoms deduced above, a view supported by the computer modelling results. The underlying reason must be that the major removal reactions for hydroxyl radicals involve coreactants which are relatively trace species, like carbon monoxide and olefinic hydrocarbons, as opposed to the oxygen atom removal reaction (16) based upon the major atmospheric component, molecular oxygen, as coreactant. The significance of this state of affairs will become apparent when the organic chemistry of photochemical smog is considered shortly.

It will turn out that the concentrations of other reactive species are of marginal importance only so that we need say little about them. However there is point in gaining an impression of the range of mixing ratios likely to be encountered in urban atmospheres for the more stable species. Accordingly Table 8.10 summarizes the results of many of the measurements which have been made, principally in the cities of the United States.

This table allows us to calculate ranges for the concentrations of some of the more reactive species. For example, if we assume that under high mixing ratio conditions of carbon monoxide most of the hydroxyl radicals are destroyed with CO as coreactant i.e., in reaction (13), then we can equate the rate of production

Table 8.10. Typical ranges of mixing ratios measured in photo-chemical smog-prone environments

Species	Mixing ratio range/p.p.h.m.
Carbon monoxide	200—2000
Nitric oxide	1—15
Nitrogen dioxide	4—20
Total hydrocarbons (excluding methane)	20—50
Alkanes (excluding methane)	11—26
Olefins	3—8
Aromatics	4—10
Acetylene	2—6
Aldehydes	5—25
Ozone	1—20

through the reaction of the hydroperoxy radical with nitric oxide, reaction (18), with the rate of destruction in reaction (13). Thus we obtain the equation:

$$k_{13} [CO] [OH] = k_{18} [HO_2] [NO]$$

In a specific case where, for instance, the carbon monoxide mixing ratio is 1000 p.p.h.m., the nitric oxide ratio is 10 p.p.h.m., and the values of the rate constants are taken as $k_{13} = 9 \times 10^7 \, dm^3 \, mol^{-1} \, s^{-1}$ and $k_{18} = 1 \times 10^8 \, dm^3 \, mol^{-1} \, s^{-1}$, which approximately average measured values, we obtain:

$$[HO_2]/[OH] = 90$$

Despite the degree of approximation made in obtaining this result and the likelihood that reactions other than (13) will contribute significantly to the removal of hydroxyl radicals, we can be sure that the ambient concentrations of hydroperoxy radical will be at least an order of magnitude larger than those of the hydroxyl radical. This will largely reflect the general chemical unreactivity of the hydroperoxy radical towards species like carbon monoxide, and there is no evidence of a significant reactivity towards organic molecules under atmospheric conditions. The main sink for hydroperoxy radicals in the troposphere appears, at present, to be identified as reaction with nitrogen oxides and the limitations imposed upon their rate of destruction on this account make for the relatively high ambient concentrations.

On the other hand, we should expect very low ambient concentrations of hydrogen atoms on account of their efficient destruction in reaction (17) to form hydroperoxy radicals. Equation of the rates of reactions (13) and (17) for typical conditions yield a predicted ratio $[H]/[OH]$ of the order of 10^{-11}. Therefore we may ignore hydrogen atom reactions other than the three-body formation of hydroperoxy radicals.

Organic Reactions in Photochemical Smog

We now make our approach towards gaining a basic understanding of the highly complex conversion of the original exhaust emission of 'unburnt' hydrocarbon species to the organic components, much smaller molecules, detected in the photochemical smog situation. Figure 80 provides a simulated example of the typical degradation for one hydrocarbon component.

The first step in this direction is to investigate the nature of the reactions which initiate the conversion. The experience gained from laboratory irradiations of simulated smog-forming mixtures based upon single organic components, has shown that it is generally the olefinic species which display high activity. Therefore, we shall concentrate upon the reactions of a general olefin, considering the rates of its reactions following its emission from exhaust into a sunlit urban atmosphere.

The first point to appreciate is that the chemical bonds within the general olefin are too strong to permit direct photodissociation or predissociation using the photon energies available in the incident solar spectrum at ground.

Of the reactive species present in the sunlit troposphere, three, $O(^3P)$, OH, and

O_3, stand out for main consideration. All are known to react with olefins at rates which could be significant under typical lower-tropospheric conditions.

Under the relatively high pressure conditions of the urban atmosphere, ground state oxygen atoms, $O(^3P)$, add across the double bond of olefins to give rise to the characteristic adducts of epoxides and aldehydes. For example in the case of l-butene the products are α-butene oxide and n-butyraldehyde, formed by the reactions represented:

$$O(^3P) + CH_2=CH \cdot CH_2 \cdot CH_3 \longrightarrow \quad H_2C-CH \cdot CH_2 \cdot CH_3$$
$$\diagdown \diagup$$
$$O$$

$$O(^3P) + CH_2=CH \cdot CH_2 \cdot CH_3 \longrightarrow \quad \underset{O}{\overset{H}{\diagdown}} C \cdot CH_2 \cdot CH_2 \cdot CH_3$$

A method often employed in the study of these reactions is the mercury-atom photosensitized decomposition of nitrous oxide in the presence of the olefin, and at large dilution in an inert gas like nitrogen. The mercury atom is excited into an electronically excited state (represented as Hg*) by irradiation with the resonance line at wavelength 253.7 nm; a fraction of the Hg* species transfers its excitation energy to nitrous oxide on collision causing it to dissociate. This process is represented by the mechanism:

$$Hg + h\nu \longrightarrow Hg^* + N_2O \longrightarrow Hg + N_2 + O(^3P)$$

Different olefins give rise to their own characteristic products; hence by using mixtures of olefins of defined composition in competition for the oxygen atoms, the relative yields of characteristic products reflect the relative rate constants for the corresponding addition reactions. Therefore with one absolute value of one of the rate constants obtained from studies in other types of system, the set of relative rate constants can be calibrated to the absolute basis.

A selection of the average values for rate constants of reactions of $O(^3P)$ atoms with organic substrate vapours for ambient temperature is given in Table 8.11.

Table 8.11 establishes the general point that, along a series like the olefins, the rate constant tends to increase with the complexity of the molecule. The second column gives the rate constants for other organic species likely to be found in polluted atmospheres; in most instances these are considerably lower than the rate constants for olefins in the first column. Diolefins follow the trend of the monoolefins while aromatic molecules like benzene are relatively unreactive for their complexity.

Although the main products of oxygen atom reaction with olefins are adducts, there is often a small degree of fragmentation to smaller molecules as well.

The ozonolysis of olefinic double bonds is a well-established reaction and inevitably results in the fragmentation of the molecule. As we have seen, ozone is present in the photochemical smog situation at relatively very much higher levels compared to the atomic or radical species, and hence the ozonolysis of olefins could well be of significance. The reactions are reasonably easy to study in the

Table 8.11. Approximate rate constants (in dm^3 mol^{-1} s^{-1}) for $O(^3P)$ + substrate gas-phase reactions at ambient temperatures

Olefinic substrates		Substrates	
Ethylene	4×10^8	Methane	6×10^3
Propylene	2×10^9	n-Butane	2×10^7
l-Butene	2×10^9	Benzene	2×10^7
cis-2-Butene	9×10^9	Toluene	5×10^7
trans-2-Butene	1×10^{10}	Methanol	8×10^6
2-Methyl-2-butene	3×10^{10}	Ethanol	3×10^7
Cyclohexene	1×10^{10}	Formaldehyde	9×10^7
Allene	2×10^9	Acetaldehyde	3×10^8
1,2-Butadiene	4×10^9	Ketene	3×10^8
3-Methyl-1,2-butadiene	2×10^{10}	Acrolein	2×10^8
2-Methyl-2,3-pentadiene	3×10^{10}	Methyl nitrite	2×10^6

Data collected and averaged from a variety of sources.

laboratory and the general procedure is to mix ozone and a substantial excess of the olefin at large dilution in an inert gas. The rate of the reaction is then followed by measuring the rate of decrease of ozone concentration as a function of time, for example, by the use of the attenuation by ozone absorption of the intensity transmitted through the reaction mixture from a weak source of ultraviolet radiation such as a mercury lamp. Table 8.12 lists a selection of the rate constants currently available for ambient temperatures.

Extensive product analyses for ozonolysis reactions have been made when the reaction conditions have not been too far removed from typical tropospheric conditions. The results of four such analyses are shown in Table 8.13 for the major products when the initial olefin:ozone ratio was 4:1 and this was diluted eightfold by molecular oxygen to give a total pressure of just over half a standard atmosphere.

Table 8.13 makes it clear that principal products are the aldehydes produced by oxidative fragmentation at the double bond. We see the appearance of products like

Table 8.12. Approximate rate constants (in dm^3 mol^{-1} s^{-1}) for O_3 + olefin gas-phase reactions at ambient temperatures

Substrate		Substrate	
Ethylene	1×10^3	2,3-Dimethyl-2-butene	8×10^5
Propylene	8×10^3	Tetramethylethylene	9×10^5
l-Butene	7×10^3	Cyclohexene	1×10^5
cis-2-Butene	1×10^5	1,3-Butadiene	5×10^3
trans- 2-Butene	2×10^5		

Data collected and averaged from a variety of sources.

Table 8.13. Main product distributions from the reactions of single olefins with ozone in the presence of molecular oxygen (Yields expressed as mols of product per mol of O_3 consumed.)

Propylene		*cis*-2-Butene	
Acetaldehyde	0.38	Acetaldehyde	1.02
Formic acid	0.34	Carbon dioxide	0.42
Carbon dioxide	0.32	Formic acid	0.20
Methanol	0.033	Methanol	0.152
cis-2-Pentene		1,3-Butadiene	
Acetaldehyde	0.56	Acrolein	0.43
Propionaldehyde	0.56	Carbon dioxide	0.24
Carbon dioxide	0.30	Formic acid	0.18
Formic acid	0.20	Ethylene	0.114
		Acetaldehyde	0.026

Reproduced by permission of the National Research Council of Canada from T. Vrbaski and R. J. Cvetanovic, 'A study of the products of the reactions of ozone with olefins in the vapor phase as determined by gas–liquid chromatography', *Canadian Journal of Chemistry*, **38**, 1063–1069 (1960).

formic acid and acrolein which could be highly irritant components of photochemical smog if the ozonolysis reactions are important.

The reactions of ozone with organic molecules other than unsaturated ones are generally too slow to have any possibility of significance under normal conditions.

The position has now been reached where a decision can be made upon whether the reactions of oxygen atoms or the reactions of ozone with olefins will be the more significant under typical photochemical smog-forming conditions. In the earlier part of this subsection it was suggested that typical mixing ratios of oxygen atoms would be limited to 10^{-6} p.p.h.m. while a typical ozone mixing ratio would be 5 p.p.h.m. (see Figure 79). The ratio of the rate of oxygen atom attack (R_O) on the olefin to the rate of the competing ozonolysis (R_{O_3}) will be simply expressed as:

$$R_O/R_{O_3} = k_O[O]/k_{O_3}[O_3]$$

where k_O and k_{O_3} are the corresponding rate constants and the substrate concentration is cancelled out on the right hand side of this equation. In Table 8.14 values of this ratio for the reactant levels specified above are listed, using rate constant values derived from Table 8.11 and 8.12.

This table makes it very clear that the reactions of atomic oxygen with olefins are rather unimportant in the development of photochemical smog since the corresponding reactions with ozone are at least an order of magnitude faster under any typical conditions.

Proceeding to the next stage of our assessment of the initiating steps in photochemical smog, we must consider whether there is any other reactive species which might achieve greater rates of reaction with the general olefin than does

Table 8.14. Values of the rate ratio (R_O/R_{O_3}) for various olefins for typical mixing ratios of $O(^3P)$ of 10^{-6} p.p.h.m. and of O_3 of 5.p.p.h.m. at ambient temperature

Olefin	$R_O/R_{O_3}(=2 \times 10^{-7} k_O/k_{O_3})$
Ethylene	0.08
Propylene	0.05
cis-2-Butene	0.018
trans-2-Butene	0.010
Cyclohexene	0.020

ozone, and perhaps at the same time can react more rapidly with other molecules than can oxygen atoms. We have remarked earlier that hydroxyl radicals can approach efficiencies of reaction with olefins of one collision in ten, and furthermore we have seen that the mixing ratios of hydroxyl radicals can be at least two orders of magnitude larger than those of oxygen atoms, with 10^{-4} p.p.h.m. a likely minimum under sunlit conditions. On these grounds we can discern a strong case for regarding the hydroxyl radical as the main initiating species, a view which may now be substantiated more quantitatively.

Rate constant data for hydroxyl radical reactions are currently becoming available for a wide variety of gaseous or vapour substrates. Many different experimental techniques have been developed which have provided both absolute and relative values of the rate constants. Perhaps it is worth describing briefly the basic principles of just one of the absolute techniques, the discharge flow system, which has achieved considerable success in this direction. The discharge flow system consists of a uniform diameter tube, down which reactant species are pumped in an inert carrier gas at constant linear and volume flowrate under conditions of relatively low total pressure (usually in the range 0.1 to 1 kN m^{-2}). The gas which is discharged (usually by passage through a microwave cavity) is hydrogen at large dilution in argon or helium and the resultant hydrogen atoms are drawn into the upstream end of the flow tube. There the hydroxyl radicals are generated by the addition of a flow of nitrogen dioxide through a jet inset into the flowtube, when the very rapid stoichiometric reaction is induced:

$$H + NO_2 \longrightarrow OH + NO \tag{22}$$

The substrate species may then be added through a second jet a short distance downstream and the flowrate of this is usually sufficient to create a large excess over the concentration of hydroxyl radicals in the flowtube. The consequent pseudo-first-order decay kinetics of hydroxyl radicals as a function of flowtime (linearly proportional to distance along the tube) can then be followed using one of a variety of detection techniques, including near-ultraviolet absorption spectrophotometry, electron spin resonance, or the measurement of the resonance fluorescence intensity as described before in connection with the detection of hydroxyl radicals in the atmosphere. An alternative method is to flow in the substrate to produce

Table 8.15. Approximate rate constants (in dm^3 mol^{-1} s^{-1}) for OH + substrate gas-phase reactions at ambient temperatures

Substrate		Substrate	
Carbon monoxide	9×10^7	Ethylene	1×10^9
Methane	5×10^6	Propylene	1×10^{10}
n-Butane	1×10^9	1-Butene	2×10^{10}
Cyclohexane	5×10^9	cis-2-Butene	4×10^{10}
Formaldehyde	9×10^9	trans-2-Butene	5×10^{10}
Acetaldehyde	9×10^9	2,3-Dimethyl-2-butene	9×10^{10}
Methanol	6×10^8	Tetramethylethylene	1×10^{11}
Ethanol	2×10^9	Benzene	1×10^9
Methyl nitrite	8×10^8	Toluene	4×10^9
Ammonia	1×10^8	Nitric acid	8×10^7
Hydrogen sulphide	2×10^9	Hydrogen	4×10^6

Data collected and averaged from a variety of sources.

concentrations much less than that of the hydroxyl radical; in this case the pseudo-first-order decay rate of the substrate may be determined using mass spectrometric detection methods. Both approaches have yielded absolute values for the second-order rate constants for the reaction between hydroxyl radicals and a substrate.

Table 8.15 lists representative values for the rate constants at ambient temperatures of reactions involving hydroxyl radicals as a reactant. The data encompass several inorganic gases for which the reaction with hydroxyl radicals is an important initiation step for the consumption in the troposphere, even well away from polluted regions.

Let us now compare the rates of hydroxyl radical reactions (R_{OH}) under the conditions of a mixing ratio of 10^{-4} p.p.h.m. of hydroxyl radicals with the rates of corresponding ozonolysis reactions (R_{O_3}) in the case of olefins, under conditions of a typical ozone mixing ratio of 5 p.p.h.m. Also of interest is the relative magnitude of R_{OH} under the above conditions compared with the rates of the corresponding reactions of oxygen atoms (R_O), taking the mixing ratio of atomic oxygen as the upper limit of 10^{-6} p.p.h.m., for species which do not react at significant rates with ozone. Table 8.16 shows values of the ratios R_{O_3}/R_{OH} or R_O/R_{OH} as appropriate.

The point is clearly made by the tabulated ratios of rates that it is to the hydroxyl radical that we must look for the overwhelming source of the initiation of the organic conversion reactions of photochemical smog. Perhaps the ozonolysis of olefins (particularly the more complex ones) contributes to this also in a minor way. However, consideration of the typical internal combustion engine exhaust composition of Table 5.12 suggests that olefins above C_4 are comparatively minor components compared to such as ethylene and propylene. Therefore we shall not move too far from the real situation if we consider, for simplicity, that the

Table 8.16. Values of the rate ratios (R_{O_3}/R_{OH}) and (R_O/R_{OH}) for typical mixing ratios of 10^{-6} p.p.h.m. (O), 10^{-4} p.p.h.m. (OH) and 5 p.p.h.m. (O_3) at ambient temperature

Substrate	R_{O_3}/R_{OH}	Substrate	R_O/R_{OH}
Ethylene	0.05	Methane	1×10^{-5}
Propylene	0.04	n-Butane	2×10^{-4}
1-Butene	0.02	Acetaldehyde	3×10^{-4}
cis-2-Butene	0.13	Benzene	2×10^{-4}
Tetramethylethylene	0.45	Methanol	1×10^{-4}

hydroxyl radical is the unique initiator of photochemical smog, principally through its reactions with olefins.

Before we go on to examine the secondary reactions following the initial reaction of hydroxyl radicals, we must substantiate the implication that electronically excited molecular oxygen species play no significant role. The two in question are $O_2(^1\Delta_g)$, with an excitation energy from the ground state, $O_2(^3\Sigma_g^-)$, of 94.3 kJ mol^{-1}, and $O_2(^1\Sigma_g^+)$, with an excitation energy of 157.1 kJ mol^{-1} from the ground state. Both of these are produced by energy transfer from donor molecules like nitrogen dioxide, sulphur dioxide, and organic species which can absorb solar radiation directly, as was discussed in Section 8.2. These states cannot be excited at any significant rate by the direct absorption of solar radiation since the transitions are made difficult by the contravention of the selection rules. The difficulty of the transitions is reflected in the high radiative lifetimes of these species, 1.08 h for $O_2(^1\Delta_g)$ and 7 s for $O_2(^1\Sigma_g^+)$. We shall not attempt to go into the detailed analysis of the production rates of these species in polluted atmospheres. An estimate of 1220 p.p.h.m. h^{-1} has been made (Kummler and Bortner, 1970) for the maximum production rate of $O_2(^1\Delta_g)$ under typical conditions in a polluted urban atmosphere. Once formed, the steady state concentration of $O_2(^1\Delta_g)$ will be determined by physical quenching by nitrogen and oxygen molecules, principally the latter for which the rate constant is around 10^3 dm^3 mol^{-1} s^{-1}. It then follows, on equating the maximum production rate with the removal rate, that the maximum steady state concentration of $O_2(^1\Delta_g)$ is of the order of 2×10^{-11} mol dm^{-3} or 0.04 p.p.h.m. This is then only two orders of magnitude larger that the steady state concentration of hydroxyl radicals proposed earlier for similar conditions. It is therefore apparent that the rate constants for *chemical* reactions of $O_2(^1\Delta_g)$ with, for example, olefins, would have to be of the order of 10^7 dm^3 mol^{-1} s^{-1} (cf. Table 8.15) to be of any significance in the organic cycle of photochemical smog. In fact the available kinetic data suggest that the rate constants for these *chemical* reactions (as opposed to physical quenching) are at least four orders of magnitude below this minimum requirement. On this basis $O_2(^1\Delta_g)$ can be ignored in the photochemical smog situation.

On much the same basis basis the other excited oxygen molecule, $O_2(^1\Sigma_g^+)$ can be eliminated from consideration also.

Finally in this subsection we must show that the projected maximum rates for hydroxyl radical—olefin reactions are sufficiently large to be effective within the time scale of the development of photochemical smog. The general rate requirement is that the total 'reactive' hydrocarbon mixing ratio must decrease at about 10 p.p.h.m. h^{-1} when the average level is about 30 p.p.h.m. From the data of Table 8.15 we may deduce a minimum value of 10^9 dm^3 mol^{-1} s^{-1} for the general reaction rate constant:

$$OH + \text{'Reactive' hydrocarbon} \longrightarrow \text{Primary products}$$

In converted form, this rate constant may be expressed as 1.5×10^3 $(p.p.h.m.)^{-1}$ h^{-1}. Hence for an average hydroxyl radical mixing ratio of 10^{-4} p.p.h.m., the projected rate of the general reaction is:

$$d[\text{'reactive' hydrocarbon}]/dt = 1.5 \times 10^3 \times 10^{-4} \times 30$$
$$= 4.5 \text{ p.p.h.m. } h^{-1}$$

This is of the correct order of magnitude. Since the rate constant which was applied was very much a minimum value for the olefin substrates of Table 8.15, and these may be considered as the main components of the 'reactive' hydrocarbon mixing ratio, this gives final credence to the view that hydroxyl radicals are the initiators of the organic part of photochemical smog. We shall now go on to consider the secondary chemistry which follows this primary act.

The Main Organic Cycles of Photochemical Smog

As was pointed out earlier, it will not be possible to detail any full mechanism for the routes to the main organic products because of the truly massive number of elementary reactions involved. All that we can attempt is to examine the last few stages leading towards characteristic product species, where the details of the chemical mechanism are reasonably well understood. The bulk of the reaction pathway, between the induction step and these last few steps, can only be speculated upon.

The initial interaction of the hydroxyl radical produces an adduct radical most probably in the case of an olefinic coreactant, with the radical adding on to one side of the double bond. The mechanism represented in Figure 85 may be a part of the ensuing reaction scheme for propylene.

Amidst the complexity of Figure 85 we may pick out one important general point. However complex the orginal molecule, eventually small radicals such as methyl (CH_3), methoxy (CH_3O), methylperoxy (CH_3OO), and acetyl (CH_3CO) appear, and it is reactions of these which generate many of the pollutants actually observed. It is to the reactions of these radicals that we direct our attentions, rather than to the mechanism by which they arise in the first place.

Reactions of Alkyl (R) radicals. Alkyl radicals such as methyl, ethyl, etc., when generated in the troposphere react overwhelmingly with molecular oxygen to

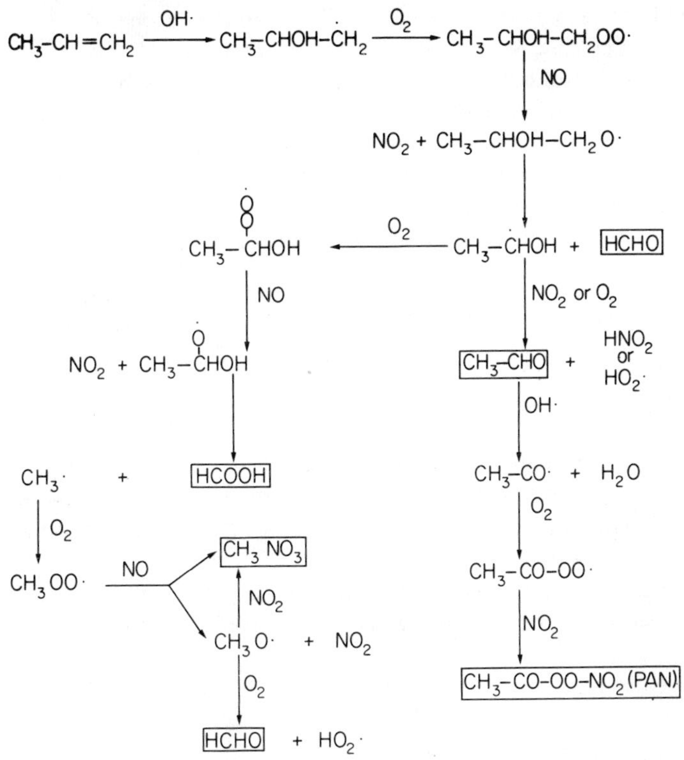

Figure 85. Speculative reaction scheme for propylene under photochemical smog-forming conditions. The boxed species are prominent components which have been detected.

produce the corresponding peroxy radical (RO_2). As was the case for hydrogen atoms and oxygen atoms, it is the abundance of molecular oxygen which promotes this state of affairs. For the methyl radical the predominant reaction is simply written as:

$$CH_3 + O_2 \longrightarrow CH_3O_2 \tag{23}$$

Rate-constant measurements for this reaction have produced values of around 10^9 dm^3 mol^{-1} s^{-1} for ambient tropospheric conditions. Hence the half-life for a methyl radical in the troposphere near to the ground is of the order of 10^{-7} s.

Reactions of Peroxyalkyl (RO_2) radicals. The reactions involving alkylperoxy radicals of significance in the photochemical smog situation are still somewhat contentious at the time of writing. Reaction with molecular oxygen, according to the equation:

$$RO_2 + O_2 \longrightarrow RO + O_3$$

can be discounted largely on the grounds that it must involve a substantial endothermicity because of the weakness of the bond in ozone. This must therefore be an exceedingly slow reaction at ambient temperatures.

In laboratory studies, the methylperoxy radical can be generated by the photodissociation of azomethane ($CH_3 \cdot N_2 \cdot CH_3$) on irradiation with wavelengths below 320 nm in the presence of oxygen; the initially formed methyl radicals form methylperoxy according to equation (23). It is considered that nitric oxide is the principal coreactant for alkylperoxy radicals in polluted regions of the troposphere. In principle there are three potential reaction pathways, written as the equation:

$$CH_3O_2 + NO \longrightarrow CH_3O + NO_2 \qquad (24)$$

$$CH_3O_2 + NO \longrightarrow CH_3O_2NO \text{ (Pernitrite)}$$

$$CH_3O_2 + NO \longrightarrow HCHO + HONO$$

The balance of currently available evidence (Pate, Finlayson, and Pitts, 1974) suggests that reaction (24) is the dominant channel. This, then, is also one of the reactions which contributes to the conversion of nitric oxide into nitrogen dioxide early in the smog cycle.

It is also likely that the peroxyalkyl radicals can react with nitrogen dioxide and possible channels for this reaction can be written as the equations:

$$CH_3O_2 + NO_2 \longrightarrow \underset{\text{(Pernitrate)}}{CH_3OONO_2} \overset{O_2}{\longrightarrow} \underset{\text{(Methyl nitrate)}}{CH_3ONO_2}$$

$$CH_3O_2 + NO_2 \longrightarrow HCHO + HNO_3$$

The first of these reactions and its analogues for higher peroxyalkyl radicals could contribute to the alkyl nitrate component of photochemical smogs, such as the methyl nitrate production shown in Figure 80, but there is also likely to be a contribution from the reaction of alkoxy radicals with nitrogen dioxide, considered below.

The alkyl nitrate products may be subject to photodissociation, according to the equation:

$$RONO_2 + h\nu \longrightarrow RO + NO_2$$

However the spectral absorption characteristics of alkyl nitrates are rather similar to those of nitric acid and, in view of the very small value of J_{HNO_3} estimated in Section 8.3, it would not be expected that such photodissociations would be of importance in the polluted troposphere.

Reactions of Alkoxy (RO) Radicals. Measurements of the relative rate constants for reactions of methoxy radicals have been obtained in recent work (Wiebe and coworkers, 1973) and these confirm what was supposed previously on the basis of the observed product distributions in laboratory irradiations of simulated polluted air. Methyl nitrite (CH_3ONO) was used as the source of the methoxy radicals; the vapour absorbs near ultraviolet radiation of wavelengths below 410 nm and, as a result, it is photodissociated according to the equation:

$$CH_3ONO + h\nu \longrightarrow CH_3O + NO$$

Accordingly, when this photodissociation was effected using monochromatic radiation of wavelength 366 nm in the presence of known proportions of nitric oxide, nitrogen dioxide, and molecular oxygen, the following reaction pathways of methoxy radicals were shown to occur:

$$CH_3O + O_2 \longrightarrow HCHO + HO_2 \tag{25}$$

$$CH_3O + NO \longrightarrow \begin{cases} HCHO + HNO\ (15\%) & (26) \\ CH_3ONO\ (85\%) & (27) \end{cases}$$

$$CH_3O + NO_2 \longrightarrow \begin{cases} HCHO + HONO\ (8\%) & (28) \\ CH_3ONO_2\ (92\%) & (29) \end{cases}$$

We may usefully adapt the measured rate-constant ratios to a typical urban atmosphere situation of 2×10^7 p.p.h.m. of O_2, 5 p.p.h.m. of NO, and 10 p.p.h.m. of NO_2. Under these conditions over 98% of the methoxy radicals will react with molecular oxygen in reaction (25), with reaction (29) accounting for most of the remainder.

Hence reaction (25) and its analogues for the higher alkoxy radicals may be considered as the major fate under normal tropospheric conditions. Nevertheless, alkyl nitrites, almost certainly the result of reactions like (27), have been detected in photochemical smogs.

Reactions of Acyl (RCO) radicals. Under tropospheric conditions there is little doubt that there overwhelming fate of acyl radicals is reaction with the abundant molecular oxygen to form peroxyacyl (RCO · OO) radicals, as is represented by the equation:

$$RCO + O_2 \longrightarrow RCO \cdot OO$$

The exception is the formyl radical, which appears to react mainly according to equation (19):

$$HCO + O_2 \longrightarrow HO_2 + CO \tag{19}$$

A definite product of the reactions of peroxyacyl radicals is the notorious component of photochemical smog, peroxyacyl nitrate (RCO · OO · NO$_2$), conventionally known as PAN. The origin of peroxyacetyl nitrate, the first of the series, has been made quite clear by its synthesis in the photodissociation of biacetyl ($CH_3CO \cdot COCH_3$) vapour in the presence of oxygen and nitrogen dioxide. The only realistic mechanism for this process is represented by the set of equations:

$$CH_3CO.COCH_3 + h\nu \longrightarrow CH_3CO \xrightarrow{\ O_2\ } CH_3CO.OO$$

$$\downarrow NO_2$$

$$CH_3-\underset{\underset{O}{\|}}{C}-O-O-NO_2$$

Peroxyacetyl nitrate is often simply referred to as PAN, it usually being the prinicipal component of the general PAN group of compounds. Now, considering that PAN is described as being a highly unstable, highly reactive, compound, the strong build-up of its level evident in the simulated irradiation of Figure 80 suggests a formation path of high efficiency. On this basis the potential reaction of peroxyacyl radicals with molecular oxygen must be extremely slow:

$$RCO \cdot OO + O_2 \longrightarrow RCOO + O_3$$

This, then, allows the combination of peroxyacyl radicals with nitrogen dioxide to become an efficient process under atmospheric conditions.

It seems that it is the juxtaposition of the nitro group with the peroxide link which creates the high chemical reactivity of PAN. It is the extension of this reactivity to interaction with biologically important materials such as enzymes which accounts for the plant-damaging capability within a few hours exposure at levels of PAN below 1 p.p.h.m. Under these conditions the PAN compounds are powerful eye and bronchial irritants; this is a considerably higher level of potency in this direction than is achieved by other noxious pollutants such as formaldehyde. Hence we may appreciate that here we are close to the nub of the photochemical smog problem.

The appearance of PAN and ozone together appears to be a good indication of a developing photochemical smog problem. Until a few years ago it was generally postulated that this was not a European problem of any consequence. However, oxidant levels (mainly ozone) exceeding 10 p.p.h.m. have been detected on several occasions since, both on the continent and in southern England. It is hardly surprising that there should be this apparently widespread development of ozone build-up. As was mentioned earlier, during the relatively infrequent periods of temperature inversion near to the ground in Britain, there is just as large an integrated daily solar irradiance as on a summer's day as in Los Angeles. This is coupled with a high level of combustive activity. The detection of PAN in southern England was achieved in rural surroundings (Harwell, Berkshire) on a sunny anticyclonic day, with a strong correlation with the ozone level (Penkett, Sandalls, and Lovelock, 1975). However the peak level of PAN was only of the order of 0.2 p.p.h.m., compared with the Los Angeles peak of some 5 p.p.h.m. At the same time the peak ozone levels observed in England have been up to 6 p.p.h.m., compared with over 20 p.p.h.m. in Los Angeles. The measurements at Harwell also detected a trace amount of the second member of the PAN series, peroxypropionyl nitrate (PPN).

8.5 The Photooxidation of Sulphur Dioxide

It will become clear that sulphur dioxide photooxidation is a more widespread phenomenon than photochemical smog as such. It is for this reason that these are dealt with in separate sections, although under some circumstances parts of the two processes can be coupled together. For example, in Britain on sunny days it is often found that high ozone levels are associated with sulphuric acid aerosol formation, with a developed correlation between the two phenomena. Moreover in laboratory irradiations of olefin–NO–NO_2–air mixtures, it is found that added sulphur dioxide is much more rapidly photooxidized than it is under the same conditions in clean air.

The first point to make for sulphur dioxide is that this molecule cannot be photodissociated in the troposphere. The bond strength is 565 kJ mol^{-1}, which is equivalent to the photon energy of radiation of wavelength 210 nm. None of the required radiation can penetrate the stratospheric ozone layer on any effective scale (see Table 2.1). Nevertheless sulphur dioxide gas absorbs across the near ultraviolet region of the spectrum. The absorption is banded, as is shown in the profile of decadic absorption coefficients as a function of wavelength in Figure 86.

There is a clear indication of two distinct absorption systems. The weak system, comprising the so-called Triplet Bands, commences in the vicinity of 400 nm wavelength, goes through a maximum with a decadic absorption coefficient of the order of 1 dm^2 mol^{-1} at 369 nm, and terminates near to 340 nm wavelength. The second and much stronger system, comprising the so-called Singlet Bands, has a threshold at a wavelength close to 330 nm and the absorption coefficients increase strongly across the peaks of the bands towards shorter wavelengths, reaching values of the order of 1000 dm^2 mol^{-1} in the vicinity of 310 nm. The singlet system then rather ressembles a truncated and more irregular analogue of the near ultraviolet absorption spectrum of ozone (see Figure 6 and Table 8.7) for which J_{O_3}(u.v. component) = 7.4×10^{-5} s^{-1}. Under tropospheric conditions we should therefore expect the first-order rate constant for the production of electronically excited sulphur dioxide molecules to be of around the same order for near-overhead sun.

We have already discussed the term symbols for the electronic states involved in sulphur dioxide excitation in Section 8.2. Figure 87 shows their energetic location and the transition processes in a schematic form. The transition probabilities in the two absorption systems are mirrored by the radiative lifetime of around 2×10^{-7} s for the 'easy' $SO_2(^1B_1 \rightarrow {}^1A_1)$ singlet transition compared to that of around 8×10^{-4} s for the 'difficult' $SO_2(^3B_1 \rightarrow {}^1A_1)$ triplet transition. Accordingly we may expect the triplet system to be ineffective from the point of view of the generation of electronically excited sulphur dioxide molecules in the atmosphere.

That the intersystem crossing process represented in Figure 87 does occur has been shown in laboratory experiments, where monochromatic light absorbed within the singlet system alone produced not only the expected fluorescence in the singlet system but also emission within the triplet system. This might be referred to as *phosphorescence* on account of the associated spin multiplicity change, by analogy with the true process of phosphorescence in large organic molecules. Under the relatively high pressure conditions of the troposphere, the singlet molecule, $SO_2(^1B_1)$, produced by absorption of the ultraviolet radiation, suffers many collisions with nitrogen and oxygen molecules within its radiative lifetime. As a result of relatively high physical quenching probabilities on a collisional basis, the predominant fate of $SO_2(^1B_1)$ in the troposphere is physical deactivation back to the ground state. However a small fraction of the collisions (some 3% on current

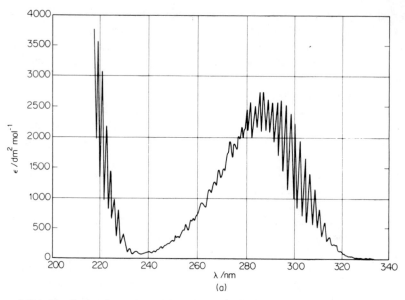

Figure 86(a). The singlet absorption system of sulphur dioxide in the middle ultraviolet spectral region, with decadic absorption coefficients (ϵ) as a function of wavelength, λ. From J. G. Calvert and J. N. Pitts, Jr., *Photochemistry*, John Wiley & Sons Inc., New York, 1966, p. 210. Reproduced with permission of John Wiley & Sons Inc.

estimates) is effective in inducing the intersystem crossing to $SO_2(^3B_1)$ and that species is likely to have some persistence on account of the spin multiplicity change demanded for its deactivation. Hence we may expect an effective first-order rate constant for the indirect photoexcitation of the triplet state of the order of 10^{-6} s^{-1} under overhead sun conditions (some 3% of the effective first-order rate constant for the excitation of the excited singlet state).

In mixtures of sulphur dioxide with large excesses of air in laboratory irradiation experiments (Cox, 1972) the quantum yield for the formation of sulphur trioxide (and hence sulphuric acid and sulphate aerosol) has been determined as 3×10^{-4} for clean atmospheric conditions. If we take the rate constant for absorption of solar radiation by sulphur dioxide to be of the order of 10^{-4} s^{-1} (as argued above by analogy with the ozone ultraviolet absorption rate constant), then, combined with the above quantum efficiency, this predicts an effective first-order rate constant for SO_2 to SO_3 conversion of the order of 10^{-4} h^{-1}. This corresponds to a conversion rate for sulphur dioxide photooxidation of the order of 0.01% per hour. We might call this the clean air rate; the degrees of approximation made in deriving it turn out to be of little consequence since it is rarely the case that the actual rate of removal of sulphur dioxide from the tropospheric air is not much larger than this. The clean air rate of photooxidation by itself predicts a half-life for sulphur dioxide approaching a year, whereas, as was remarked in Section 7.5, the average residence time of sulphur dioxide in the troposphere is only of the order of 4 days.

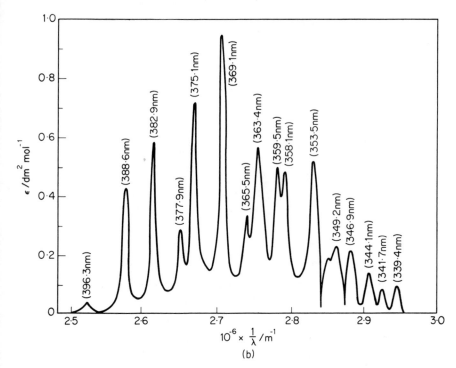

Figure 86(b). The triplet absorption system of sulphur dioxide in the near ultraviolet spectral region, with decadic absorption coefficients (ϵ) as a function of wavenumber ($1/\lambda$). The numbers above the spectrum are the wavelengths (λ) of the maxima. Reprinted with permission from H. W. Sidebottom and coworkers, *Journal of the American Chemical Society*, **93** (June), 2588 (1971). Copyright by the American Chemical Society.

The detailed mechanism of the clean-air photooxidation of sulphur dioxide is uncertain to the extent that while it is generally accepted that $SO_2(^3B_1)$ reacts with molecular oxygen to form sulphur trioxide, a possible additional contribution from the reaction of $SO_2(^1B_1)$ cannot be excluded. Moreover, whilst an SO_4 species is often considered to be an intermediate in the mechanism, no direct evidence for this has been obtained.

In polluted air, the rate of photooxidation of sulphur dioxide can be orders of magnitude faster than the clean-air value above. Many other species present therein can produce catalytic effects. For example, photooxidation rates approaching 10% per hour under simulated bright conditions have been measured in laboratory experiments, where initial mixing ratios of olefins and nitric oxide in the range 1 to 10 p.p.h.m. were introduced to the sulphur dioxide at large dilution in clean air.

As a first step towards assessing the nature of the catalytic effects, we might investigate whether reactions of the main reactive species, oxygen atoms, and hydroxyl radicals, with sulphur dioxide could make significant contributions to the overall rate of photooxidation. The two reactions concerned are represented by the equations (with the rates for ground level tropospheric conditions expressed in the

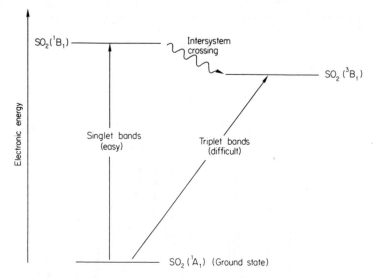

Figure 87. Relative energies of the electronic states of sulphur dioxide involved in photo-oxidation and radiation absorption.

second-order form):

$$O + SO_2 + M(air) \longrightarrow SO_3 + M \tag{30}$$

$$R_{30} = 1 \times 10^7 [O] [SO_2] \text{ mol dm}^{-3} \text{ s}^{-1}$$

$$OH + SO_2 + M(air) \longrightarrow HSO_3 + M \tag{31}$$

$$R_{31} = 4 \times 10^8 [OH] [SO_2] \text{ mol dm}^{-3} \text{ s}^{-1}$$

Subsequent to reaction (31), a possible course of reaction might be:

$$HSO_3 + O_2 \longrightarrow HSO_5$$

$$HSO_5 \longrightarrow HO_2 + SO_3$$

Thus sulphur trioxide could result from both reactions (30) and (31), the latter being a more speculative possibility at present.

In the preceding section we considered a mixing ratio for oxygen atoms of 10^{-6} p.p.h.m. (equivalent to a concentration of 4×10^{-16} mol dm^{-3}) to be a reasonable upper limit in a sunlit urban atmosphere. The incremental form of the rate of reaction (30) is then:

$$\Delta[SO_2]/\Delta t = 1 \times 10^7 \times 4 \times 10^{-16} [SO_2]$$

Hence for Δt taken as 1 h we will obtain:

$$\Delta[SO_2]/[SO_2] = 4 \times 10^{-9} \times 3600 = 1.5 \times 10^{-5}$$

Hence the predicted hourly extent of the conversion of SO_2 and SO_3 via reaction (30) is only 0.0015%, some six-times slower than the clean-air photooxidation rate.

Now, since the rate coefficient of reaction (31) is some 40-times that of reaction (30), and the typical hydroxyl radical concentration is some two orders of magnitude larger than that of oxygen atoms, as discussed in the last section, it is clear that photooxidation via reaction (31) could be three orders of magnitude faster than via reaction (30) and therefore overwhelm the clean air photooxidation rate. This must be tempered with the possibility that the peroxy species HSO_5 is not formed subsequently or, if it is, that other reaction routes which do not produce SO_3 compete significantly.

What is demonstrated by the above, therefore, is that it is not difficult to devise catalytic mechanisms for the photooxidation of sulphur dioxide of similar or greater significance than the clean-air mechanism. Further, there is considerable evidence binding the sulphur dioxide to sulphate aerosol conversion even more closely to the photochemical smog cycles of the preceding section. In laboratory irradiations it has been shown that the appearance of sulphate aerosol in olefin—NO—SO_2—air mixtures only comes after the almost complete oxidation of nitric oxide to nitrogen dioxide. In Figure 78 and 80 it may be seen that this is the condition which applies to ozone itself. It has been observed also that in general the time of maximum sulphate aerosol yield corresponds closely to the times of maximum yield of both ozone and PAN when the initial ratios of the concentrations are varied. The accelerative effects for sulphur dioxide photo-oxidation over the clean air rates are very large, but as yet there is no firm mechanism to explain the overall catalysis; it may only be said that some of the radicals (like hydroxyl) or intermediate species (like olefin—ozone adducts) must be able to interact oxidatively with sulphur dioxide.

The situation is made yet more complicated by the proven ability of particulate matter to catalyse the formation of sulphur trioxide. Soot can be effective in this respect and also many dry inorganic aerosols, composed of materials such as metallic oxides. Lead oxide particles have shown strong catalytic ability but the most effective material of all those tested so far appears to be ferric oxide. A typical capability of this latter material is conversion of 18 p.p.m. of sulphur dioxide completely to sulphate within a time scale of minutes. Such an instance may be referred to as a *synergic process*, where two species which are relatively unreactive in isolation act in concert to produce a highly reactive situation with potentially harmful consequences. This illustrates that any control measures proposed for pollution in general cases must take into account the synergic possibilities when separate sources of two emissions are close enough for eventual interaction.

It is obvious that the photooxidation of sulphur dioxide to sulphur trioxide and hence sulphuric acid by hydration is an immensely complicated reaction system even in the gas phase. Even then there is a further possibility that sulphur dioxide can be photooxidized in solution within the droplets present in clouds or fogs and also within the sulphuric acid aerosol particles themselves. More of this last aspect will be discussed in the next section in connection with chemically formed aerosols. In the particular case of an atmospheric fog containing droplets of an average diameter of $20\,\mu m$ with a content of manganese or iron ions, it has been demonstrated that typical tropospheric levels of sulphur dioxide can be oxidized at conversion rates approaching 60% per hour.

Residence times of sulphur dioxide in the troposphere have been derived empirically from extensive measurements at stations in many different locations throughout the world. The actual results show high variability, ranging from minutes to months, perhaps hardly surprising in view of the diversity of critical parameters discussed above. The average value of around 4 days is not only determined by photooxidation but also by direct solution into natural waters (see Section 7.5).

Another point to be made is that present dispersal techniques are not always as successful as might be desired. The emission of combustion gases from high chimneys can certainly alleviate potential sulphur dioxide level problems in the immediate vicinity of the source. However, evidence is now coming to light that this procedure merely transfers the problem to other localities, which can be quite remote from industrial activity. This point is borne out by the map shown as Figure 88, which shows the average levels of sulphur dioxide throughout the county of South Yorkshire.

Two points are of interest in connection with this map. The first is the method of measurement. Lichens are a group of primitive plants which are highly sensitive to sulphur dioxide poisoning with a varying degree of tolerance between species. It has been shown that a relatively simple scale based upon easily identified lichen species which grow on walls, asbestos roofs, and tree-bark, can be drawn up allowing assessment across the ranges shown in the figure ($100\,\mu g\,m^{-3}$ corresponds to 3.7 p.p.h.m.).

The second point to be drawn is the general relationship of the mapped levels to the location of principal sources. The levels are very high around Sheffield and Rotherham, as would be anticipated from the iron and steel industries based there. Other major centres of population and industry, like Doncaster, also show very high average levels. However, the agricultural belt running between Rotherham and Doncaster also shows fairly high levels of sulphur dioxide and indicates that pollution is not confined even closely to the places which produce it. There is considerable evidence also that the large power stations outside the area shown in the map, to the North, also contribute to the levels encountered. This then bears out the postulation that high chimneys are not the whole answer to pollution control.

On an ever wider scale, extensive measurements in southern Sweden, an agricultural area to a large extent, have shown that around 30% of the sulphate content of the soils can be attributed to the deposition of natural sulphur, a further 20% represents Swedish anthropogenic sulphur, while something of the order of 50% is anthropogenically-produced sulphur derived from outside Sweden; the direction of the prevailing winds must put Britain in line for part of the blame here. This, then, is the international dimension of pollution production; it is difficult to justify the poisoning of the soils of Sweden by foreign industrial activity.

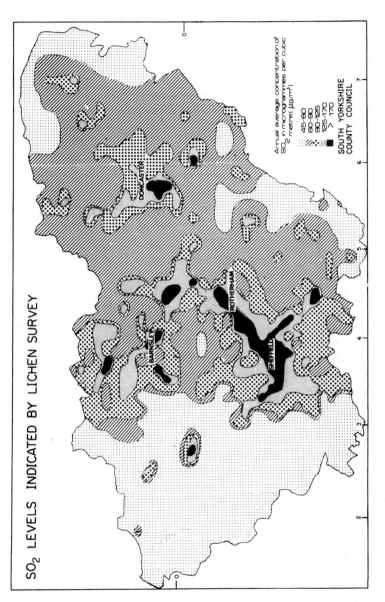

Figure 88. Annual average sulphur dioxide concentrations in air in South Yorkshire as indicated by a lichen survey. Reproduced by permission of South Yorkshire County Council from *Lichen Survey of Air Pollution 1975.*

8.6 Tropospheric Aerosols

Three general categories of aerosols may be distinguished for our present interest:

(i) Direct anthropogenic emissions, e.g. ash particles from power stations, lead halide particles from automobile exhausts.
(ii) Chemically formed aerosols, e.g. sulphuric acid from sulphur dioxide photo-oxidation, particulates formed in photochemical smog.
(iii) Natural aerosols, e.g. soil dust, evaporated marine spray.

The definition of an aerosol is rather imprecise and virtually any entity from an aggregation of only a few molecules to a substantial solid particle or liquid droplet can be so regarded, but in general the term aerosol is usually reserved for particles with diameters in the range 10^{-9} to 10^{-4} m. The lowest range (10^{-9} to 10^{-7} m diameter) are often referred to as Aitken particles, the next range (10^{-7} to 10^{-6} m) as large particles and above that as giant particles. Where aerosols are actually chemically formed within the atmosphere, new particles tend to grow through these size ranges to various extents involving a variety of mechanisms.

In the troposphere there are so-called 'haze layers' which extend up to a few kilometres altitude; there is a ceiling at approximately 3 km to aerosols originating over the oceans as sea-salt particles. It therefore appears unlikely that there can be much effective transport of these aerosols as such from the lower troposphere into the stratosphere.

We shall pick out three aspects of tropospheric aerosols for special attention. The inorganic elements present in an aerosol often give a good indication of its source. The organic content of an aerosol is, apart from soot, often the result of tropospheric photochemistry, particularly in the case of the aerosols associated with photochemical smog. The sulphate and nitrate contents of aerosols are frequently tied in again with photochemical conversions.

Direct Anthropogenic Emissions

Soot is the most obvious starting point. This is a highly variable entity, both in terms of particle size and structure and in terms of trace elemental composition. It is basically carbon with a graphite-like structure; in fact carbon constitutes something of the order of 50% of the total direct particulate emissions in urban atmospheres with up to three-quarters of this in the form of soot. However other trace elements are incorporated into the basic structure depending on the origin. For example, vanadium and nickel often occur in the soot formed from oil combustion. These metal atoms are incorporated in crude oil in chelated forms, commonly with porphyrin ligands. Vanadium can be present at levels in the range of 2 to 200 p.p.m. by weight in crude oil, with Venezuelan crude being particularly noted for its relatively high levels.

Soot particles in the troposphere furnish a disproportionately large surface area available for heterogeneous catalytic reactions relative to their weight, because of the porous structure. For instance, it has been shown (Novakov, Chang, and Harker, 1974) that sulphur dioxide oxidation in polluted atmospheres is significantly enhanced by such a catalytic action to the extent that this could be a major origin of the sulphate in urban regions.

Another significant aspect of directly injected particulate materials has been revealed by the trace element analysis of the fly ash produced by the combustion of coal in power stations. It is well known that many trace elements within high temperature combustion situations become mobilized in association with airborne particles. The surprising point to emerge from the investigations was that many of the most toxic elements were found at the highest concentrations in the smallest particles collected from the inside of the power plant stacks.

The importance of these observations becomes clear when the relationship between the size of an airborne particle and the extent to which it can penetrate the human breathing system is considered. Particulate materials with diameters exceeding $10 \, \mu m$ can rarely penetrate below the larynx, but particles with diameters below $6 \, \mu m$ can, in general, penetrate the bronchial passages and those of diameter less than $1 \, \mu m$ can usually penetrate right into the capillaries and alveoli of the lungs. The larger particles deposited within the upper parts of the respiratory system usually move to the stomach where the absorption efficiencies of metals and other trace elements into the bloodstream range up to only 15% or so, but for particles deposited on the inner surfaces of the lungs the absorption efficiencies for trace elements can be as high as 80%. The majority of living organisms possess little tolerance for the elements in question like arsenic, selenium, cadmium, chromium, and some of the other elements which are preferentially concentrated in the fly ash which would penetrate the lungs. Cyclonic particle collectors in current usage for the reduction of particulate emissions from power stations cannot prevent most of the smallest particles reaching the outside air.

Table 8.17 gives data from a single representative coal-fired power station in the United States, which demonstrate the concentration effect.

Under present operating conditions the $44-74 \, \mu m$ range would be retained in the plant; in the instance of Table 8.17 this represented about 23% of the particulate mass fraction, with just over 66% of the total being particles with diameters larger than $74 \, \mu m$. The other fractions would be airborne fly ash to an increasing extent down the table. The smallest diameter fraction, $2.1-3.3 \, \mu m$, would be almost entirely released to the atmosphere; it represents by mass an extremely small fraction, less than 0.01%, of the total particulates generated. The general enhancement of the set of toxic trace elements in the smaller size ranges is made very evident, emphazing the potential environmental health hazard in the vicinity of power stations.

Table 8.18 shows the major matrix element composition of coal fly ash determined in the same study as the data in Table 8.17.

The elements in Table 8.18 show less of a tendency for enhancement in the smaller particles than do those of Table 8.17.

The most attractive explanation for this phenomenon stems from the high volatilities of the trace elements or of some of their compounds, such as sulphides,

Table 8.17. Analytical data for the elements in typical coal fly ash (Concentrations expressed as $\mu g\ g^{-1}$.)

Particle diameter/μm	Pb	Sb	Cd	Se	As	Ni	Cr	Zn
44 −74	160	7	< 10	< 20	500	140	90	411
15 −20	520	19	< 10	< 30	300	200	170	720
4.7− 7.3	1500	34	18	16	1000	440	460	6600
2.1− 3.3	1500	37	26	19	1200	900	1500	15,000

From D. F. S. Natusch, J. R. Wallace, and C. A. Evans, Jr., 'Toxic trace elements: Preferential concentration in respirable particles', *Science,* **183** (18 January), 202−204 (1974). Copyright 1974 by the American Association for the Advancement of Science and reproduced with permission.

at the combustion temperatures. If condensation, adsorption, or absorption rates in the subsequent cooling region are dependent upon the surface area of existing particles, the higher surface-area to mass ratio of the smaller particles would be expected to enhance the mass fractions of the volatile trace elements therein. The major matrix elements of Table 8.18 would be expected to exist in the least volatile forms and accordingly compose the particles upon which the volatile species condense.

In this connection we may return to the fluidized-bed combustor concept described in Section 6.3 to emphasize that the much lower combustion temperatures achieved in that way should lead to some considerable abatement of the problem.

The lead incorporated into petrol to inhibit knock emerges at exhaust substantially in the form of lead halides. The explanation of this lies in the additions of ethylene dibromide and ethylene dichloride in conjunction with tetraethyl lead which, by itself, would produce involatile lead compounds, such as

Table 8.18. Major matrix elemental composition in typical coal fly ash (percentages by weight)

Particle diameter/μm	Al	Si	Fe	K	Ca	S
44 −74	9.4	18	18	1.2	5.4	1.3
10 −15	9.8	19	6.6	4.0	4.0	4.4
4.7− 7.3	16.2	27	12	4.2	4.2	7.9
2.1− 3.3	21	35	17	5.0	5.0	25.0

From D. F. S. Natusch, J. R. Wallace, and C. A. Evans, Jr., 'Toxic trace elements: Preferential concentration in respirable particles', *Science,* **183** (18 January), 202−204 (1974). Copyright 1974 by the American Association for the Advancement of Science and reproduced with permission.

oxides, and hence deposits on spark plugs. The formation of volatile lead halides such as PbBrCl mitigates the potential problem. Investigations of the lead particle size distribution have shown a maximum for particles of diameters around 0.25 μm, with about 25 percentile points between 0.15 and 0.45 μm. These fine lead particles can then penetrate to the very inner parts of the lungs, as discussed above, and constitute a considerable health hazard in view of the undesirability of enhanced lead levels in the bloodstream. A typical lead-particle level in the atmosphere in the vicinity of high road traffic densities can be in the range of 1 to 3 μg of lead per cubic metre of air, whereas the background level in regions remote from heavy traffic densities will frequently be below 0.1 μg m^{-3}.

In the atmosphere the fine lead halide particles can have moderately long residence times and in the course of their existence they can be photodecomposed by near ultraviolet radiation to generate chlorine atoms, according to the equation:

$$PbBrCl + h\nu \longrightarrow PbBr + Cl$$

Although there is no evidence as yet for significant rates of reactions induced by these chlorine atoms, it is probable that they inevitably produce the peroxy radical ClOO by reaction with the abundant molecular oxygen. There are strong possibilities for further reactions; for instance the peroxychlorine radical is known to be a strong oxidizing agent for both nitric oxide and sulphur dioxide, and could therefore enter into the photooxidation cycles discussed in the preceding section.

Chemically Formed Aerosols

Under this heading we shall discuss those aerosol particles which are formed within the troposphere as a result of photochemical reactions of initially gaseous components.

The most obvious end-product of photochemical smog cycles is the aerosol material which reduces visibility. An intensive investigation of the nature and composition of the particles under such conditions was conducted in Pasadena, California, in September 1969. The results of this were published as a collection of articles in the April 1972 issue of the *Journal of Colloid and Interface Science*. Principal interest in the chemical composition of the aerosol particles attaches to the various size ranges as a function of time of day (Novakov and coworkers, 1972). Nitrogen appeared in the forms of nitrate ions, ammonium ions, amino-nitrogen and pyridino-nitrogen. The latter two had the largest concentrations both in the larger diameter (2–5 μm) and in the smaller diameter (0.6–2 μm) particles which were analysed. However, it was considered to be more likely that these forms of nitrogen originated from compounds with amino and pyridino groups used as petrol additives rather than by synthesis during the course of the various photochemical cycles. There was clear evidence that initially smaller particles gradually coagulated into larger particles; the diurnal peaks in the densities of smaller particles per unit air mass appeared before many of the larger particles were detected. However a fairly steep rise in the density of larger particles in the air followed the peaks in the smaller particles.

Table 8.19. Compositional data for the Pasadena aerosol around noon (Concentrations expressed in μg N m^{-3}.)

Nitrate-N	Ammonium-N	Amino-N	Pyridino-N	Sulphite-S	Sulphate-S
0.4	1.8	3.4	5.3	5.4	3.9

Data read off graphs in T. Novakov, P. K. Mueller, A. E. Alcocer, and J. W. Otvos, 'Chemical composition of Pasadena aerosol by particle size and time of day. III. Chemical states of nitrogen and sulfur by photoelectron spectroscopy', *Journal of Colloid and Interface Science,* **39**, 225–234 (1972).

Sulphur was detected as sulphate and sulphite, with the former ion concentrated in the larger particles. The diurnal patterns were similar to those of the nitrogen species and could be explained in terms of absorption of sulphur dioxide into existing particles followed by conversion to sulphate. Table 8.19 shows a typical composition for the particles with diameters between 0.6 and 5 μm for local noon when densities were near to their peaks. As is illustrated by the data, the ammonium ion concentrations were generally insufficient to allow the nitrate and sulphate to be present only in the form of ammonium salts; it is likely that the corresponding acids and perhaps other salts also existed. The range of particle sizes encompassed by the 0.6 to 5 μm diameters accounted for some 50% of the total particulate matter. Lead was also a component of the aerosol at levels corresponding to lead-to-sulphur ratios of around 0.025 in terms of atoms, but variable by a factor of up to 2 during the day.

In the British aerosol experience there is more interest in the sulphate aerosols which occur in association with high ozone levels. Ammonium sulphate frequently accounts for more than 60% of the water-soluble particulate material in the British troposphere and it seems likely that this salt, in combination with the parent sulphuric acid aerosol, is the major source of typical haze conditions. We have discussed in Section 7.4 the strong source of tropospheric ammonia provided by animal grazing activity within general agricultural working of the land and this may be considered as the major source of the ammonium ion in the particulate material. As we saw in Section 7.5, and more particularly in Figure 66, sulphur dioxide is emitted by anthropogenic and natural sources in almost equal measure over land. However, the point made by Figure 88 is that it is the localized nature of the anthropogenic sources which gives rise to the high levels of sulphur dioxide either within or in the vicinity of centres of population. Hence it is likely that a predominant fraction of the sulphur in the British sulphate aerosol is derived from anthropogenic emissions. Ammonia tends to have a rather shorter lifetime in the troposphere than sulphur dioxide, in all probability due to its high water solubility and general chemical reactivity, even if both lifetimes are of the order of days. Therefore we might expect ammonium sulphate to be a more substantial component of aerosols in rural rather than urban areas. However the ammonia levels in any location are so variable with time that it is difficult to discern this trend in the few analytical data available.

The bulk of the formation of ammonium sulphate is likely to take place by way of a mechanism of solution and chemical reaction of the gases in existing sulphuric acid aerosol droplets. The reaction of sulphur dioxide and ammonia in humid air has been studied in laboratory systems (Hartley and Matteson, 1975). The important conditions are those where gaseous sulphur dioxide is in considerable excess over gaseous ammonia, the usual condition encountered in the troposphere. However, there are relatively infrequent periods where ammonia mixing ratios peak to reverse the situation. A reasonable mechanism proposed the formation of an initial adduct species, according to the equation:

$$NH_3(g) + SO_2(g) \longrightarrow NH_3 \cdot SO_2(g)$$

It is then envisaged that the gas phase adduct can condense out readily to a solid product represented as $NH_3 \cdot SO_2(s)$. Such a yellowish solid has been produced in the reaction of anhydrous ammonia and sulphur dioxide at 263 K (Scott, Lamb, and Duffy, 1969). Also produced was a white 2:1 compound, $(NH_3)_2SO_2(s)$, the formation of which may be ascribed to the reaction represented by the equation:

$$NH_3 \cdot SO_2(g) + NH_3 \longrightarrow (NH_3)_2SO_2(s)$$

In the laboratory studies of the reaction in humid air, it has been observed that the yield of solid product was independent of the oxygen concentration but strongly dependent upon the water vapour concentration. A mechanism which could explain these observations is represented by the equations:

$$(NH_3)_2SO_2(s) \quad + \tfrac{1}{2}O_2 \longrightarrow NH_4.SO_3.NH_2(s)$$
$$NH_4.SO_3.NH_2(s) + H_2O(g) \longrightarrow (NH_4)_2SO_4(s)$$

in which the second step is rate determining for the subsequent oxidation to sulphate. The position is by no means clear, however. Other laboratory work has suggested that ammonium sulphite is the product only when ammonia is in excess over sulphur dioxide, with the reverse situation yielding ammonium pyrosulphite $[(NH_4)_2S_2O_5(s)]$. There is also evidence that the ammonium ions generated in liquid aerosol droplets of sulphuric acid by absorption of ammonia lead to a considerably enhanced rate of oxidation of dissolved sulphur dioxide therein. This may reflect something of a buffering action, since it is established that the rate and extent of sulphur dioxide—sulphite oxidation in the liquid phase is strongly affected by the acidity, decreasing with decreasing pH. Reference back to Table 7.4 shows that the liquid-phase resistance to transfer of sulphur dioxide from the air to the liquid phase goes up sharply in the same direction, which may be a contributory factor.

A heterogeneous mechanism certainly seems to be required to explain the production of the larger diameter particles, both in the troposphere, and in the stratospheric sulphate layer to be discussed in the next chapter. In laboratory simulations with traces of ammonia present in irradiated air containing water vapour and sulphur dioxide at low levels, a very rapid production of both embryonic nuclei and larger particulates is observed. These larger aerosols exist at much larger densities than can be explained by simple coagulation from small embryos. The implication is that catalytic oxidation and resultant accretion must occur within the embryonic particles themselves. The most attractive mechanism considers the salt embryonic particle to the formed through two steps incorporating ammonia into an initial Aitken particle of sulphuric acid, represented by the equation:

$$NH_3 + H_2SO_4.nH_2O \longrightarrow NH_4^+ HSO_4^-.nH_2O$$

$$NH_3 + NH_4^+ HSO_4^-.nH_2O \longrightarrow (NH_4^+)_2 SO_4^{2-}.nH_2O$$

This provides a medium wherein the ammonium ion induces a rapid catalytic oxidation of sulphur dioxide and consequent growth in the size of the particle. One step of this process of growth could be represented by the equation:

$$2 SO_2 + 2 H_2O + O_2 \xrightarrow{\text{embryo } NH_4^+} 2 H_2SO_4(l)$$

As mentioned above it is the ability of the ammonium ion to buffer the medium through the hydrolysis represented by the equation:

$$NH_4^+ + H_2O \rightleftharpoons NH_3 + H_3O^+$$

which may be the critical factor in this; this keeps the pH up so that sulphur dioxide can enter the solution to form HSO_3^-. As the particle grows, presumably it scavenges more ammonia from the surrounding air; the availability of ammonia may be a limiting growth factor. If ammonia gas molecules impinge upon the surface of such an acid droplet, then reactive absorption is highly likely. The usual excess of sulphur dioxide mixing ratios over those of ammonia must encourage the prior conversion to sulphate before alliance with the ammonium ion.

8.7 Concluding Remarks

In this chapter we have established that the photochemistry of the troposphere, particularly in urban regions, is driven overwhelmingly by the radiant solar energy absorbed by a small number of different molecules. We have reviewed the possibilities and have come to identify nitrogen dioxide, formaldehyde, and sulphur dioxide as the significant species in this respect. We have identified the hydroxyl radical as the prime initiator of photochemical smog situations, with olefins as the predominant coreactants. The origin of the most noxious organic pollutants has been seen to be associated with the degradation of larger organic molecules to small radicals such as methyl and acetyl which go on to react with molecular oxygen and nitrogen oxides.

We have seen that the fate of the sulphur dioxide released to the troposphere through anthropogenic activity is in part oxidation to sulphate aerosols, with sulphuric acid and ammonium sulphate as major components. The average time scale of days for the photooxidation of sulphur dioxide implies that much of the sulphur returns to the ground in regions substantially removed from the source.

Hence, while photochemical smog is very much a problem localized within the region where the pollutants are emitted, the sulphur pollution problem has more international undertones. It has become clear that more efficient dispersal of sulphur dioxide at source cannot be regarded as an acceptable long term solution, since that merely transfers the problem to another region or country, the problem intensified by the fact that the anthropogenic term in the sulphur cycle is of the same order of magnitude as the natural terms (Section 7.2). On the other hand, the photochemical smog problem develops on a rather shorter time scale and also under conditions which prevent the dispersal of the anthropogenic emissions. Once dispersed, the species such as nitrogen dioxide and olefins are 'lost' in the larger terms of the natural cycles discussed in Chapter 7, and there is little transmission of the problem over distances of more than the order of tens of miles.

BIBLIOGRAPHY

1. *Photochemistry of Air Pollution*, by P. A. Leighton, Academic Press, New York, 1961.
2. 'Aerochemistry of Air Pollution', by R. S. Berry and P. A. Lehman, *Annual Reviews of Physical Chemistry*, 22, 47–84 (1972).
3. 'Photochemistry of the Lower Troposphere', by H. Levy, *Planetary Space Science*, 20, 919–935 (1972); 21, 575–591 (1973).
4. 'The Mechanism of Photochemical Smog Formation', by K. L. Demerjian, J. A. Kerr, and J. G. Calvert. Chapter 1 in *Advances in Environmental Science*, Volume 4, Eds. J. N. Pitts and R. L. Metcalfe, Wiley Interscience, New York, 1973, pp. 1–262.
5. 'Ecology, Energy and Economics', by J. N. Pitts, A. C. Lloyd, and J. L. Sprung, *Chemistry in Britain*, 11, 247–256 (1975).
6a. 'Measurement of Hydroxyl Concentrations in Air using a Tunable UV Laser Beam', by C. C. Wang and L. I. Davis, *Physical Review Letters*, 32, 349–352 (1974).
6b. 'Laser-induced Dissociation of Ozone and Resonance Fluorescence of OH in Ambient Air', by C. C. Wang, L. I. Davis, C. H. Wu, and S. Japar, *Applied Physics Letters*, 28, 14–16 (1976).
7. 'Quantum Yields for the Photo-oxidation of Sulphur Dioxide in the First Allowed Absorption Region', by R. A. Cox, *Journal of Physical Chemistry*, 76, 814–820 (1972).
8. 'Photochemical Ozone and Sulphuric Acid Aerosol Formation in The Atmosphere over Southern England', D. H. F. Atkins, R. A. Cox, and A. E. J. Eggleton, *Nature*, 235, 372–376 (1972).
9. 'Atmospheric Aerosol Formation by Chemical Reactions', by G. M. Hidy and C. S. Burton, *International Journal of Chemical Kinetics* 7, 509–541 (1975).

Chapter 9

Neutral Chemistry of the Upper Atmosphere

In this chapter we shall be considering the reactions of atoms, radicals, and molecules, principally in the stratosphere (11–50 km) and the mesophere (50–85 km altitude), but excluding the ions which begin to appear above 70 km altitude. A brief overview of the reactions of charged species will appear as Chapter 10.

In Chapter 3, we have considered one outstanding phenomenon of the stratosphere in some detail already (Section 3.3), namely the ozone layer. It is the absorption of ultraviolet radiation by ozone which produces the increase in temperature with altitude in the stratosphere. In view of what has been said on the topic of the photodissociation of ozone in Chapter 8, we now appreciate that the absorption of energy into the stratosphere through the agency of the ozone layer must inevitably lead to a complex chemistry following the production of $O(^3P)$, $O(^1D)$ and $O_2(^1\Delta_g)$ in this way.

We can also appreciate at the outset that many other species are involved also. In Chapter 7 it was suggested that molecules like methane, carbon monoxide, and nitrous oxide, generated at the Earth's surface, could eventually diffuse up through the tropopause into the stratsophere where they are more subject to reaction. In each of these instances sharp decreases in the mixing ratios have been detected in the vicinity of the tropopause, as illustrated for carbon monoxide in Figure 21. However we have to remember that the temperature at the tropopause is only around 220 K which creates a highly effective barrier to the upward transport of condensable species like water vapour; in fact the stratosphere is very dry with only a few parts per million of water vapour in reflection of this. In this connection there has been concern recently with regard to the totally man-made chloro-fluorocarbon molecules such as $CClF_3$ (normal boiling point = 192 K) and CCl_2F_2 (normal boiling point = 243 K). These have found rapidly increasing usage over the past 30 years as refrigerant fluids and aerosol propellants (they are often referred to by the patent name 'Freons'). These molecules are highly inert in the troposphere and it is evident that they must be gradually transported into the stratosphere. There they are subject to photodissociation by the shorter wavelength ultraviolet radiation and, as we shall see later, the product chlorine atoms pose a considerable threat to the existence of the ozone layer with its present equivalent thickness. The nature of this and other threats can only be understood if the origin of the ozone layer is understood in the first place.

Our first area of interest concerns how the gross structure and conditions of the upper atmosphere affect the rates of chemical processes and the extent of penetration of short-wavelength radiation.

9.1 The Upper Atmosphere as a Chemical Reactor

What are the critical features for chemistry in the upper atmosphere? The first part of the answer is that pressure and hence total species concentration, [M], are low compared to the troposphere. In the *U.S. Standard Atmosphere* model, which applies to an idealized annual mean for mid-latitudes in the mid-solar cycle period, the various parameters of significance as functions of altitude are shown in Table 9.1.

It is in the region below an altitude of 100 km that the overwhelming bulk of the chemistry involving only neutral species occurs.

In Table 9.1 we can see straight away that even at an altitude of 20 km the concentration of all molecules, [M], is only some 7% of that at the surface; as a result the number of collisions which a molecule suffers in a given period of time is reduced by more than an order of magnitude. On this basis alone we must consider that electronically excited species like $O(^1D)$, which are mainly destroyed by physical quenching in collisions, will have enhanced persistence compared to that in the troposphere. Such persistence will be increased by a further three orders of magnitude at an altitude of 70 km, where species travel a mean free path approaching 1 mm between collisions. It is therefore apparent that chemical reactions of electronically excited atoms and molecules could be of significance in the upper atmosphere, in contrast to the situation in the troposphere.

Moreover, even ground state atoms and radicals will be expected to have much longer average lifetimes in the upper atmosphere. For example, the main removal reaction of oxygen atoms, $O(^3P)$, will be the three-body recombination with molecular oxygen to form ozone, just as in the troposphere. This reaction is represented by the equation:

$$O + O_2 + M \longrightarrow O_3 + M \tag{16}$$

and it will follow a rate given by the equation:

$$-d[O]/dt = k_{16}[O][O_2][M] \tag{9.1}$$

This rate would be expected to be nearly 200-times slower at 20 km than at ground level on account of the dependence of the rate on the product $[O_2][M]$ alone. At an altitude of 70 km, the rate on this basis would be almost 200 million-times slower than at ground level. Since concentrations of reactants which preserve the same mixing ratio with increasing altitude will decrease in proportion to [M], bimolecular reaction rates for atoms involving such a species as coreactant will decrease less rapidly with increasing altitude than the rates of three body reactions like reaction (16). Therefore we may expect bimolecular reactions of atoms to play

Table 9.1. Parameters of the U.S. Standard Atmosphere from 0 to 100 km

Altitude/km	[M]/mol dm^{-3}	Pressure/atm	Collision frequency/collisions s^{-1}	Mean free path/dm	T/K
0	0.042	1.00	6.9×10^9	6.63×10^{-7}	288
10	0.0143	0.262	2.1×10^9	1.97×10^{-6}	223
20	0.00307	0.0546	4.3×10^8	9.1×10^{-6}	217
30	6.36×10^{-4}	0.0118	9.2×10^7	4.4×10^{-5}	227
40	1.38×10^{-4}	2.83×10^{-3}	2.1×10^7	2.0×10^{-4}	250
50	3.55×10^{-5}	7.87×10^{-4}	5.6×10^6	7.9×10^{-4}	271
60	1.06×10^{-5}	2.22×10^{-4}	1.6×10^6	2.7×10^{-3}	256
70	3.02×10^{-6}	5.45×10^{-5}	4.3×10^5	9.3×10^{-3}	220
80	6.90×10^{-7}	1.02×10^{-5}	8.9×10^4	0.0407	181
90	1.09×10^{-7}	1.62×10^{-6}	1.4×10^4	0.256	181
100	1.72×10^{-8}	2.97×10^{-7}	2400	1.63	210

Data source: *The U.S. Standard Atmosphere Tables*, U.S. Government Printing Office, Washington, D.C., 1962.

Table 9.2. Values of the Arrhenius factor, $\exp(-E/RT)$, as a function of E and T

T/K	$E/kJ\ mol^{-1}$					
	1	5	10	20	30	40
288	0.66	0.12	0.015	2.4×10^{-4}	3.6×10^{-6}	5.6×10^{-8}
250	0.62	0.09	8.1×10^{-3}	6.6×10^{-5}	5.4×10^{-7}	4.4×10^{-9}
220	0.58	0.064	4.2×10^{-3}	1.8×10^{-5}	7.5×10^{-8}	3.2×10^{-10}
180	0.51	0.035	1.3×10^{-3}	1.6×10^{-6}	2.0×10^{-9}	2.5×10^{-12}

a rather larger part in the chemistry of the upper atmosphere than was the case in the troposphere. At the same time, the steady-state concentrations under sunlit conditions for atoms and radicals will be relatively much larger compared with those of stable molecules in the upper atmosphere as opposed to the troposphere because of the much reduced rates of their removal. This state of affairs will be reinforced by the generally higher rates of photodissociation, which we shall consider shortly.

An additional factor to be taken into account, particularly in the lower stratosphere and in the upper mesosphere, is the effect of low temperatures on reaction rate constants. In Table 8.3 the point was made that bimolecular reactions involving atomic or radical reactants and which are exothermic tend to have relatively low Arrhenius activation energies (E). In most cases these may be considered to be less than 40 kJ mol^{-1}. In Table 9.2 the variation of the Arrhenius factor, $\exp(-E/RT)$, as a joint function of values of E and the temperatures encountered in the upper atmosphere is shown. We may illustrate the significance of the data in this table by referring to two reactions of importance in the upper atmosphere.

The reaction represented by the equation:

$$O(^3P) + OH \longrightarrow O_2 + H \qquad (-1)$$

is known to have a small value of E, less than 1 kJ mol^{-1}. At 288 K, the rate constant of this reaction is of the order of 2×10^{10} dm^3 mol^{-1} s^{-1}, corresponding to reaction of the two species at around one collision in ten. Examination of the left-hand column of Table 9.2 then shows that the rate constant of this reaction, k_{-1}, will be virtually invariant as a function of altitude.

On the other hand, the reaction between $O(^3P)$ atoms and ozone represented by the equation:

$$O + O_3 \longrightarrow 2O_2 \qquad (32)$$

has a temperature coefficient operating on the rate constant corresponding to a value of E of approximately 20 kJ mol^{-1}. Therefore we expect k_{32}, corresponding to an efficiency of reaction of the order of one collision in 5000 between oxygen atoms and ozone molecules at 288 K, to have decreased by a factor of just over 13 on ascending to 20 km altitude, and to have decreased by over two orders of magnitude for ascent to 80 km altitude. As a rough rule of thumb, we may consider that rate constants of reactions with Arrhenius activation energies greater than $10\ \text{kJ mol}^{-1}$ will vary significantly with altitude.

Three-body reaction rate constants, like k_{16}, generally have negative temperature coefficients corresponding to negative Arrhenius activation energies usually of the order of $-2\ \text{kJ mol}^{-1}$. This means that these rate constants will increase slightly as the temperature decreases but not by more than a factor of two over the range of temperatures encountered in the altitude range of 0 to 100 km. Taken in conjunction with the drastic decrease in the *rate* of three-body reactions with increasing altitude occasioned by the fall in [M], such a variation of the rate constant is virtually insignificant. We may choose to ignore variations in three-body rate constants as a function of altitude to a good approximation.

We have now dealt with the effects of physical conditions in the upper atmosphere mainly upon the reactions responsible for the consumption or interconversion of atoms and radicals. We must now consider the photodissociation processes responsible for the primary generation of these species, the other element determining the steady-state concentrations.

It has been implied in Section 3.3 that the principal absorbing molecule for radiation with wavelengths above 200 nm is ozone. In Figure 6 it can be seen that for radiation of wavelengths below 200 nm, the other principal upper atmospheric absorber, molecular oxygen, enters into the picture. The absorption by oxygen is mainly in the Schumann–Runge system and, as may be seen in Figure 70, this is a banded spectrum down to a threshold wavelength of 175 nm. Below this limit the absorption is continuous and results in photodissociation into the first electronically excited state of the oxygen atom, $O(^1D)$, and a ground state oxygen atom, $O(^3P)$. Hence this will be a further source of atomic oxygen in the upper atmosphere.

The distribution of ozone and molecular oxygen as functions of altitude are of course very different. Figure 23, taken in conjunction with Figure 26, shows that the ozone content of the upper atmosphere is overwhelmingly located in the lower stratosphere, and Figure 21 emphasizes the sharp increase in the ozone concentration just above the tropopause. Table 3.1 demonstrates that only a very small fraction of the total ozone layer extends above an altitude of 50 km. On the other hand, molecular oxygen preserves a constant mixing ratio for practical purposes some way up into the thermosphere and its concentration is given by $[O_2] = 0.2095\,[M]$. It therefore becomes apparent that, in the regions above 50 km altitude, absorption by oxygen is the principal source of attenuation of the solar irradiance with wavelength below 200 nm. The reinforcement of molecular oxygen, and ozone absorptions which might have been expected for radiation of wavelengths below 200 nm on the basis of Figure 6, hardly occurs because molecular oxygen has a significant column density well above the upper limiting altitude for effective attenuation by ozone. Let us now extend Table 3.2 into the spectral region of effective absorption by molecular oxygen below 200 nm wavelength.

The first requirement for the calculation is to obtain information on the total

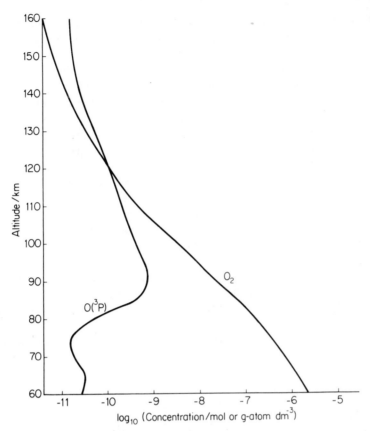

Figure 89. The concentration profiles versus altitude of $O(^3P)$ atoms and molecular oxygen from 60 km to 160 km at midday as computed by Hunt. From B. G. Hunt, 'Photochemical heating of the mesosphere and lower thermosphere', *Tellus*, **XXIV**, 48 (1972). Reproduced by permission of the Swedish Geophysical Society.

number of oxygen molecules within a vertical column of unit base area, i.e. the column density, above say 40 km altitude, and to partition this into fractions contained within certain ranges of altitude, in analogy to Table 3.1 for ozone. Figure 89 shows the concentration profile of molecular oxygen as a function of altitude in the range 60 to 160 km, and, for comparison, the concentration profile of ground-state atomic oxygen.

The increasing ratio of atomic oxygen to molecular oxygen concentrations with increasing altitude reflects the increasing content of wavelengths below 200 nm in the solar irradiance in the ascent. In fact, as shown in the figure, the atomic oxygen concentration exceeds the molecular oxygen concentration above an altitude of 120 km for the overhead sun conditions to which it refers. It is to be noted, however, that at 25 km altitude the maximum ozone concentration is only of the order of 2×10^{-5} of that of molecular oxygen.

The column density of molecular oxygen above a defined altitude has been measured by the extent of attenuation in the Schumann—Runge absorption continuum of the far ultraviolet spectrum of a 'hot' star, i.e. a star emitting a substantial black-body component of thermal radiation in the far ultraviolet region of the spectrum. This method relies upon the use of a rocket-borne telescope and scanning spectrometer. From the observed attenuation in the wavelength region of 130 to 175 nm where molecular oxygen is the only effective atmospheric absorber, the Beer—Lambert law, in conjunction with the established absorption coefficients, allows evaluation of the column density of molecular oxygen above the altitude where the observation is made. The disposition of the oxygen molecules in the column is of no consequence since they are effectively integrated in the extent of attenuation. The result can be usefully expressed as an equivalent thickness of pure molecular oxygen at atmospheric pressure (standard) and 288 K. This assumes that the absorption coefficients as a function of wavelength with the absorption continuum are independent of temperature, which is approximately true. In one such experiment (Opal and Moos, 1969) the column density of molecular oxygen above 100 km altitude was found to be 1.0×10^{-4} mol dm^{-2}. This corresponds to an equivalent thickness of 0.25 mm at 1 standard atmosphere and 288 K.

The average absorption coefficients over 10-nm wavelength intervals can be derived from Figures 6, 70, and 90 below, the last showing the detail in a second banded region between wavelengths of 105 and 135 nm. The maximum absorption

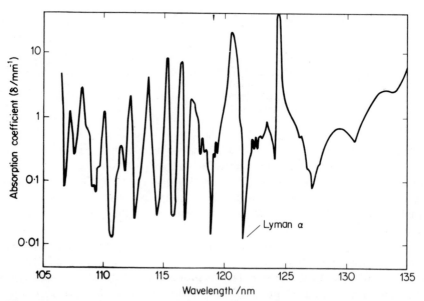

Figure 90. Decadic absorption coefficient of molecular oxygen as a function of wavelength in the region 105 to 135 nm. From K. Watanabe, E. C. Y. Inn and M. Zelikoff, *Journal of Chemical Physics*, **21** (June), 1030 (1953). Reproduced by permission of the American Institute of Physics.

coefficient in the continuum is $17\,\text{mm}^{-1}$ at wavelengths around 140 nm, applicable to the Beer—Lambert law expressed in the form:

$$\frac{I_l(\lambda)}{I_t(\lambda)} = 10^{-\delta \cdot l}$$

where $I_t(\lambda)$ is the extraterrestrial photon flux density and $I_l(\lambda)$ is the residual photon flux density after passage through an equivalent thickness of molecular oxygen of l, both at wavelength λ. δ is the appropriate absorption coefficient, expressed in the reciprocal units of l. Using $l = 0.25\,\text{mm}$ applicable to 100 km altitude, then for wavelengths in the vicinity of 140 nm, we can expect an attenuation of solar radiance by a factor of the order of 10^{-5}. Clearly we may regard such penetrating solar irradiance as negligible. However in Figure 70 and 90 we can see that the absorption coefficients are much smaller both above 170 nm and below 130 nm wavelengths, so that we have to investigate the extent of penetration of solar radiation in these spectral regions. In fact it happens that there is a 'window' (i.e. a spectral region of relatively low absorption coefficients) in the molecular absorption of oxygen located around the so-called Lyman α line of atomic hydrogen at wavelength 121.57 nm. Since atomic hydrogen is rather abundant in the outer layers of the Sun, this Lyman α emission causes a strong peak in the solar radiance spectrum (see Table 3.4), so that penetration of this radiation to altitudes of 100 km and below is of high potential significance; the photon equivalent energy is a massive $984.7\,\text{kJ}\,\text{mol}^{-1}$. A detailed set of absorption coefficients of molecular oxygen around the Lyman α wavelength is shown in Figure 91.

At the wavelength of 121.57 nm, the value of δ is only $0.010\,\text{mm}^{-1}$ so that at 100 km altitude an attenuation of less than 1% of the solar irradiance in the Lyman α line is expected. From the data of Table 3.4 we therefore expect an incident radiation energy density of the order of $9\,\text{mW}\,\text{m}^{-2}$ in the 10 nm wavelength band centred on 120 nm, but mainly concentrated at the Lyman α wavelength; this corresponds to a photon flux density of around 5×10^{13} photons $\text{dm}^{-2}\,\text{s}^{-1}$ on the basis that there is no other source of attenuation. We shall see in Chapter 10 that Lyman α radiation is an important source of ionization in the upper atmosphere through strong absorption by nitric oxide.

Table 9.3 lists calculated attenuation factors, averaged over 10—nm wavelength intervals, for the solar photon flux densities penetrating to 100 km altitude under overhead Sun conditions as a function of the middle wavelength, on the basis that ground-state molecular oxygen is the only effective absorber. Also listed are the photon flux densities, based upon the solar irradiance data of Table 3.4.

The critical feature which emerges from Table 9.3 is that, at the altitude of 100 km, the Lyman α irradiance is the most significant component of the penetrating solar irradiance below 180 nm wavelength. In fact the irradiance in the wavelength range 115 to 125 nm is comparable in magnitude with that between the wavelengths of 175 and 185 nm: the former is the result of a combination of the abnormally high Lyman α radiance from the Sun and the window in the molecular oxygen absorption spectrum, while the latter reflects mainly the weakness of the

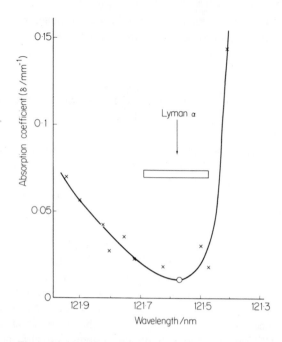

Figure 91. Decadic absorption coefficients of molecular oxygen in the neighbourhood of the Lyman α line (121.57 nm). The rectangle indicates the width of the effective atmospheric window. The circle point is at the Lyman α wavelength. From Po Lee, *Journal of the Optical Society of America,* **45** (September), 706 (1955). Reproduced by permission of the American Institute of Physics.

Schumann–Runge absorption in the banded region at the threshold of the continuum towards lower wavelengths. Hence from the point of view of the penetration of significant photon flux densities to altitudes below 100 km, we may reasonably ignore all wavelengths below 170 nm, save for the Lyman α line at 121.57 nm.

It is clear that, on descending towards lower altitudes, the high absorption coefficients of ozone in the Lyman α region and in the wavelength region above 170 nm wavelength, will combine with the weaker absorption of molecular oxygen in these spectral regions to attenuate the photon fluxes (see Figure 6). We may put this on a quantitative basis in order to assess the relative extents of the attenuation effected by each molecular species. In this we shall use an average total ozone-column equivalent thickness of 3 mm and the fractions of that column in various altitude ranges given in Table 3.1. Hence 0.057 of the 3-mm equivalent thickness of ozone lies above 40 km altitude. The equivalent thickness of the oxygen column between altitudes of 40 km and 100 km may be obtained by graphical integration of the concentration profile shown in Figure 89, extrapolated from 60 km to 40 km using $[O_2] = 0.2095[M]$ and $[M]$ data from Table 9.1. This is

Table 9.3. Attenuation factors and photon flux densities at 100 km altitude and overhead Sun as a function of wavelength

Middle wavelength/nm	Attenuation factor (averaged over 10 nm)	Photon flux density/ photons dm$^{-2} \cdot$ s^{-1} (10 nm)$^{-1}$
120	Negligible	5.5 x 10^{13}
130	0.53	2 x 10^{11}
140	2 x 10^{-5}	4 x 10^{7}
150	2 x 10^{-5}	1 x 10^{8}
160	3.6 x 10^{-4}	7 x 10^{9}
170	0.0086	5 x 10^{11}
180	0.95	1.1 x 10^{14}
190	Negligible	2.6 x 10^{14}

calculated as 5300 mm, much larger than the 0.25 mm equivalent thickness above 100 km altitude as expected.

Consider first the Lyman α flux, for which the oxygen absorption coefficient is $\delta = 0.010$ mm^{-1}. The attenuation of the extraterrestrial Lyman α flux is evidently over 50 orders of magnitude at an altitude of 40 km from oxygen absorption alone and this radiation is of negligible importance. The attenuation of the Lyman α flux as a function of altitude has been measured using a rocket-borne nitric oxide ionization chamber; this was sensitive to solar radiation in the wavelength range 105 to 135 nm, which is dominated by the radiation in the Lyman α line. Figure 92 shows the resultant profile of the extent of absorption versus altitude. As may be seen in the figure, over half the Lyman α flux has been attenuated at around 80 km altitude.

Next let us consider radiation of wavelength 200 nm, where the weak absorption by molecular oxygen, shown in Figure 6, arises from a different transition than does that which produces the Schumann–Runge system, and is known as the Herzberg continuum. The absorption coefficient at the wavelength of 200 nm is $\delta = 1.1 \times 10^{-5}$ mm^{-1} and for the 5300-mm equivalent thickness of molecular oxygen above the altitude of 40 km, it is predicted that the attenuation by this is some 13%. At the same time, the ozone column above 40 km altitude has an equivalent thickness of 0.17 mm and at 200 nm wavelength the absorption coefficient is $\delta = 0.28$ mm^{-1} and a minimum in this spectral region (see Table 2.1). Hence at an altitude of 40 km the attenuation of this radiation by the ozone column is some 10%. Overall, therefore, we expect that over 70% of the extraterrestrial solar irradiance at wavelengths in the vicinity of 200 nm will penetrate down to 40 km altitude. This will be a maximum in this spectral region. At lower wavelengths the oxygen absorption coefficients rise sharply, and at higher wavelengths the ozone absorption coefficients increase sharply as is shown in Table 2.1; hence there is a window in the Earth's atmosphere for radiation in the ultraviolet region.

The critical features of the upper atmosphere viewed from the chemical reaction standpoint are the significant penetration of high energy ultraviolet radiation (particularly above the main part of the ozone layer), which greatly enhances the

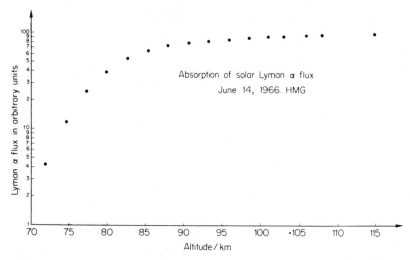

Figure 92. The penetration of solar Lyman α radiation as a function of altitude as determined by a rocket-borne nitric oxide photoionization detector. From J.-A. Quessette, *Journal of Geophysical Research*, **75**, 840 (1970). Copyrighted by American Geophysical Union and reproduced with permission.

potential for the generation of reactive species through photodissociation, and the physical conditions which promote bimolecular reactions rate at the expense of three body reaction rates.

9.2 Primary Photochemistry of the Upper Atmosphere

As would be anticipated on the basis of what was said in Section 8.2, the higher photon energies available in the upper atmosphere greatly increase the likelihood of the generation of electronically excited species. Moreover we have remarked in the preceding section upon the greater persistence expected for these species. Hence not only should there be rather larger photodissociation rate coefficients for molecules in the upper atmosphere exposed to substantial ultraviolet irradiances, but chemical reactions of electronically excited atoms in particular should make a significant contribution to the overall chemistry.

Let us examine how photodissociation rate coefficients, J, will be expected to vary as a function of altitude through as set of relevant case studies.

Case I – Nitric Acid Vapour

According to our earlier consideration of this photodissociation in Section 8.2, only the process represented by the equation below is effective over the ultraviolet range down to a wavelength of 200 nm.

$$HNO_3 + h\nu(\lambda < 330 \text{ nm}) \longrightarrow OH + NO_2$$

The absorption spectrum is shown as Figure 75. It is evident from the increasing absorption cross-sections to shorter wavelengths that J_{HNO_3} will be much larger in the stratosphere and above than it was in the troposphere. Figure 93 shows the profile of $j_\lambda \cdot \Phi_\lambda$ elements averaged over short wavelength intervals as a function of wavelength for various altitudes.

This figure illustrates very clearly the effective penetration of radiation of wavelengths in the range 180 to 220 nm almost to the base of the stratosphere, on account of the window between the molecular oxygen and ozone absorptions. It is also evident that J_{HNO_3} will increase substantially from an altitude of 20 km to 40 km due to the decreasing effect of the ozone column above in removing the middle wavelengths in the figure. The difference between the extraterrestrial and 40 km altitude values of $j_\lambda \cdot \Phi_\lambda$ around 200 nm wavelength largely reflects the weak molecular oxygen absorption across this region.

Figure 94 shows plots of J_{HNO_3}, separated into contributions from the two spectral ranges divided by the wavelength 250 nm. The point is emphasized that in the upper stratosphere it is mainly the shorter wavelength flux which produces the main component of J_{HNO_3}. On the other hand, in the lower stratosphere the value of J_{HNO_3} almost totally reflects the higher wavelength range contribution and has moved very close to its tropospheric low value.

All of the above data refer to overhead-Sun conditions. For non-zero solar angles (Z), the greater pathlengths through the atmosphere above a particular altitude will

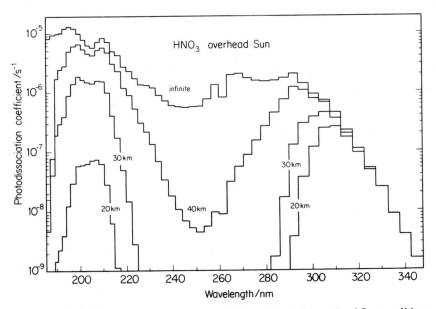

Figure 93. Nitric acid photodissociation coefficients ($j_\lambda \Phi_\lambda$) for overhead Sun conditions as a function of wavelength for indicated altitudes: infinite may be considered to apply above 60 km. From F. Biaume, 'Absorption cross-section spectrum and its photodissociation in the stratosphere', *Journal of Photochemistry*, **2**, 147 (1973/74). Reproduced by permission of Elsevier Sequoia S.A.

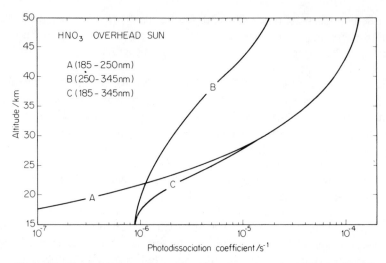

Figure 94. Plots of J_{HNO_3} (curve C) and its longer wavelength (curve B) and shorter wavelength (curve A) components as functions of altitude for overhead Sun conditions. From F. Biaume, 'Absorption cross-section spectrum and its photodissociation in the stratosphere', *Journal of Photochemistry*, **2**, 147 (1973/74). Reproduced by permission of Elsevier Sequoia S.A.

lead to increased attenuation of solar irradiance, which may be assessed by incorporation of a sec Z factor as described in Section 2.1. At the same time, the pathlength through an imaginary, vertically aligned, cubic volume will be increased by the factor sec Z, leading to a partial compensation of the J value. However, the net effect on J_{HNO_3} is a decrease as the solar angle increases; for example with Z taken as $60°$, J_{HNO_3} is approximately halved compared to the overhead sun values.

In the case of nitric acid vapour, the species produced by photodissociation in the stratosphere are the same electronic ground-state products as are produced by photodissociation in the troposphere.

Case II – Ozone

A more complex case is the photodissociation of ozone in the stratosphere, where the production of the electronically excited atom, $O(^1D)$ and molecule $O_2(^1\Delta_g)$, will occur at a much greater rate than in the troposphere. The wavelength threshold for the atomic species is 313 nm, while the electronically excited molecule is produced by the absorption of photons with wavelengths below 350 nm.

Let us proceed to calculate the rate of production of $O(^1D)$ atoms, the rate of removal and hence the steady-state concentration at a representative stratospheric altitude, say 25 km. The first stage entails derivation of the solar irradiance at 25 km altitude as a function of wavelengths; we shall specify overhead-Sun conditions for simplicity. According to Table 3.1, a fraction of 0.579 of the ozone layer, taken to be an equivalent thickness of 3 mm in total, lies above 25 km altitude. On the basis of the average absorption coefficients given in Table 2.1 and the average solar irradiances over 10-nm wavelength intervals given in Tables 2.4, 3.2, and 9.3, the ozone-attenuated photon flux densities (I') and net photon flux densities (I), incorporating the weak absorption by molecular oxygen, at 25 km altitude are as shown in Table 9.4.

The conversion of I' into I in the above table demands that the molecular oxygen column density above 25 km altitude is calculated. A graphical integration procedure may be adopted, using the relationship $[O_2] = 0.2095[M]$ in conjunction with the U.S. Standard Atmosphere tabulated data, to obtain the portion of the column between 25 and 40 km altitude. The result is an equivalent thickness of 37,800 mm, to which is to be added the 5300 mm equivalent thickness for the column above 40 km altitude calculated earlier. The attenuating effects of such a total equivalent thickness of molecular oxygen of 43,100 mm appear in the ultraviolet window region around the wavelength of 200 nm. The result is depression of I' to I by nearly an order of magnitude and a sharp cut-off at wavelengths of 190 nm or below.

It is clear from the I data of Table 9.4 that as far as $O(^1D)$ production in the middle stratosphere is concerned, it is overwhelmingly radiation in the narrow band between wavelengths of 285 to 313 nm which must be considered further.

The next stage of the calculation is the evaluation of the rate of absorption of the photons by the ozone within a pathlength of 1 dm through a cubic volume of 1 dm sides, vertically aligned at 25 km altitude. We may be sure that absorption within such a relatively short pathlength will accord to the weak absorption approximation, given by equation (8.13), adapted to apply to ozone, viz.

Table 9.4. Solar photon flux densities at an altitude of 25 km for overhead Sun (I', ozone attenuation only; I, attenuation by O_3 and O_2)

Middle wavelength/nm	I'/photons dm^{-2} s^{-1} (10 nm)$^{-1}$	I/photons dm^{-2} s^{-1} (10 nm)$^{-1}$
310[a]	5.22×10^{16}	5.22×10^{16}
300	1.51×10^{16}	1.51×10^{16}
290	1.60×10^{14}	1.60×10^{14}
280	2.04×10^{8}	2.04×10^{8}
270		
260		
250	Nil[b]	Nil[b]
240		
230		
220	2.10×10^{12}	0.37×10^{12}
210	1.72×10^{14}	0.195×10^{14}
200	3.50×10^{14}	0.164×10^{14}
190	5.9×10^{13}	Nil[b]

[a] Limited to the wavelength range 305–313 nm.
[b] Nil used for photon flux densities less than 10^8 photons dm^{-2} s^{-1} (10 nm)$^{-1}$.

$$I_A(\lambda) = 2.303\,\epsilon \cdot I_0(\lambda) \cdot [O_3]$$
$$= j_\lambda [O_3] \tag{9.2}$$

Here I_A (λ) is the rate of absorption of photons of wavelength λ, where ϵ is the decadic absorption coefficient and when $I_0(\lambda)$ is the photon flux density incident upon the top face of the cube, also with wavelength λ, which appears as I in Table 9.4 as average values across 10-nm wavelength intervals. Thus j_λ quantities can be calculated as averages across the 10-nm intervals and the quantum yield for $O(^1D)$ atom production can be taken as unity. Summation of the j_λ values listed in Table 9.5 then yields the overall rate coefficient, conventionally denoted as J_{bO_3}, for the photodissociation of ozone to yield $O(^1D)$ atoms. Accordingly we deduce that J_{bO_3} is equal to $6.023 \times 10^{23} \times 6.81 \times 10^{19} = 1.1 \times 10^{-4}$ s^{-1}, with the overwhelming contribution coming from the two highest wavelength ranges given in Table 9.5.

The typical mid-latitude ozone concentration in the stratosphere at 25 km altitude is around 5×10^{-9} mol dm^{-3}. Hence the rate of production of $O(^1D)$ atoms for overhead sun would be given by the production of the above value of J_{bO_3} and the ozone concentration, which comes out as 5.7×10^{-13} in units of g-atoms dm^{-3} s^{-1} or as 3.4×10^{11} atoms dm^{-3} s^{-1}.

We now enter the final phase of our calculation of the steady-state concentration of $O(^1D)$ atoms at 25 km altitude, in which we must evaluate the rate of removal of this species. Table 9.7 summarizes absolute values for the rate constants applying to the removal of $O(^1D)$ atoms by interaction with other molecules (physical and/or chemical). Although these rate constants refer to 298 K, they are sufficiently large to allow the assumption that almost the same values will apply in the stratosphere.

Since none of the other rate constants is more than an order of magnitude larger than that for the interaction with molecular nitrogen, which is purely physical in nature, and none of the molecules other than nitrogen and oxygen have mixing ratios above 10^{-4}, it is evident that only the removal of $O(^1D)$ by interactions with N_2, and to a lesser extent O_2, need to be considered in the deduction of the rate of

Table 9.5. Calculation of J_{bO_3} for overhead Sun at 25 km altitude

Wavelength range/nm	Average ϵ/ dm^2 mol^{-1}	j_λ/photons s^{-1} mol^{-1}
305–313	280	3.37×10^{19}
295–305	919	3.20×10^{19}
285–295	3409	0.13×10^{19}
275–285	10,430	(5×10^{12})
215–225	4485	(2.2×10^{16})
205–215	1458	0.058×10^{19}
195–205	628	0.006×10^{19}
	Total	6.81×10^{19}

Table 9.6. Absolute rate constants (k) for the destruction of $O(^1D)$ atoms in bimolecular processes at 298 K

Coreactant molecule	$k/dm^3\ mol^{-1}\ s^{-1}$
N_2	3.3×10^{10}
O_2	4.5×10^{10}
O_3	3.2×10^{11}
CO_2	1.1×10^{11}
H_2O	2.1×10^{11}
N_2O	1.3×10^{11}
CH_4	2.4×10^{11}
NO	1.0×10^{11}
NO_2	1.7×10^{11}

Reproduced by permission of the National Research Council of Canada from R. J. Cvetanovic, 'Excited state chemistry in the stratosphere', *Canadian Journal of Chemistry*, **52**, 1452–1464 (1974).

destruction. The concentrations of these at 25 km altitude in the standard atmosphere are $[N_2] = 9.97 \times 10^{-4}\ mol\ dm^{-3}$ and $[O_2] = 2.65 \times 10^{-4}\ mol\ dm^{-3}$ and with the rate constants of Table 9.6 these yield a rate of removal of $O(^1D)$ of $4.48 \times 10^7\ [O(^1D)]$ g-atoms $dm^{-3}\ s^{-1}$. Then, since the rates of generation and removal of this species must be equal for a steady state, we generate the equality:

$$5.7 \times 10^{-13} = 4.5 \times 10^7\ [O(^1D)]$$

from which the concentration of $O(^1D)$ is 1.3×10^{-20} g-atoms dm^{-3} at 25 km altitude for overhead Sun.

Similar calculations can be made for different altitudes and for different solar angles. Figure 95 summarizes the results of such calculations for solar angles up to $80°$ over the altitude range 20 km to 60 km.

The profiles shown in the figure emphasize the compromise struck between decreasing ozone concentrations with increasing altitude above 25 km and the increasing flux of ultraviolet radiation capable of photodissociating ozone to produce $O(^1D)$ atoms. At the same time the removal rate decreases with increasing altitude in proportion to the decreasing concentrations of nitrogen and oxygen molecules. The net result is that the maximum concentrations of $O(^1D)$ atoms are found close to 50 km altitude.

Closely related to $O(^1D)$ is the excited oxygen molecule, $O_2(^1\Delta_g)$, in the sense that it is the coproduct of ozone photodissociation over much of the ultraviolet region. As pointed out earlier, the upper limiting wavelength for $O_2(^1\Delta_g)$ production is at 350 nm, as opposed to a threshold wavelength of 313 nm for $O(^1D)$ production. Therefore we should expect the rate of production of the excited oxygen molecule in the stratosphere to be quite significantly larger than that for the excited atom. However in the upper stratosphere and in the

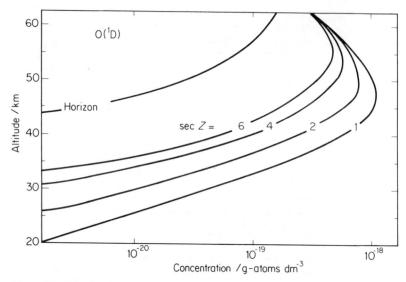

Figure 95. Calculated concentration profiles of $O(^1D)$ atoms as functions of altitude for various solar zenith angles, Z. From M. Nicolet, 'Aeronomic reactions of hydrogen and oxygen', *Aeronomica Acta*, A No. 79, 18 (1970). Reproduced by permission of Professor M. Nicolet.

mesosphere, where the bulk of ozone photodissociation results from the absorption of photons with wavelengths below 313 nm, we should predict that the rates of production of the two species would be much the same. The critical factor then determining the differences in the relative concentrations of $O(^1D)$ and $O_2(^1\Delta_g)$ will be the ratio of the rates of removal of the two species. Like $O(^1D)$ atoms, $O_2(^1\Delta_g)$ molecules are principally removed by physical quenching by nitrogen and oxygen molecules. In the latter case it is mainly molecular oxygen which is effective, with a rate constant generally accepted as being close to 1×10^3 dm^3 mol^{-1} s^{-1}, without much temperature dependence. The rate constant for the corresponding physical deactivation by nitrogen molecules is some three orders of magnitude less. Hence the total quenching rate for $O_2(^1\Delta_g)$ molecules will be around eight orders of magnitude less than that for $O(^1D)$ atoms and in fact the steady state concentrations of the two species are very close to the inverse of this ratio of the removal rates.

There are small concentrations of ozone above the stratosphere. These molecules will be exposed to virtually the extraterrestrial solar irradiance in the main ozone absorption band above 200 nm wavelength. We should therefore expect J_{O_3} to be fairly constant with increasing altitude between 60 and 100 km altitude, with $O_2(^1\Delta_g)$ the predominant molecular product. On this basis we can equate the rates of generation and removal, which leads to the proportionality:

$$[O_3] \propto [O_2(^1\Delta_g)] [O_2(^3\Sigma_g^-)]$$

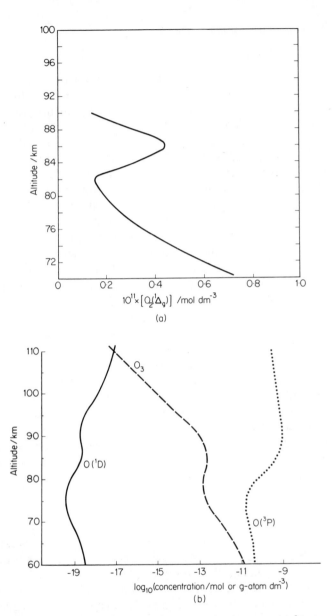

Figure 96(a). Concentration profile of $O_2(^1\Delta_g)$ versus altitude in the upper mesosphere, measured by rocket-borne spectrophotometric detection. From L. R. Megill, J. C. Haslett, H. I. Schiff and Copyrighted by the American Geophysical Union and reproduced with permission.

Figure 96(b). Midday O_3, $O(^3P)$ and $O(^1D)$ concentration profiles as functions of altitude in the upper mesosphere as computed by Hunt. From B. G. Hunt, 'Photochemical heating of the mesosphere and lower thermosphere', *Tellus*, **XXIV** 48 (1972). Reproduced by permission of the Swedish Geophysical Society.

and hence to the rearranged proportionality:

$$[O_2(^1\Delta_g)] \propto [O_3]/[O_2] \tag{9.3}$$

where the ground state of molecular oxygen is simply denoted by O_2 as opposed to the full term symbol, $O_2(^3\Sigma_g^-)$, used in the preceding proportionality for clarity.

Now, although the radiative transition from $O_2(^1\Delta_g)$ to the ground state is difficult, emission in the characterisitic (0,0) band of this transition at wavelengths around 1270 nm in the near infrared spectral region has been detected in the upper atmosphere. The intensity of such emission as a function of altitude, determined using rocket-borne spectrometers, is simply proportional to the concentration of $O_2(^1\Delta_g)$ and has led to the altitude/concentration profile shown in Figure 96(a). Figure 96(b) shows calculated concentration versus altitude profiles for ozone molecules, ground state $O(^3P)$ atoms, and $O(^1D)$ atoms for comparison. The general resemblance between the $O_2(^1\Delta_g)$ profile to that of ozone is striking and expected on the basis of the proportionality (9.3). Most remarkable is the mirroring of the kink in the ozone profile near 80 km altitude in the excited-molecule profile. The origin of this kink will be discussed in a later section, but it must be borne in mind that it is only at altitudes below 60 km that a near photochemical equilibrium model can be applied (Section 3.3) without consideration of significant perturbation by vertical diffusional transport. Hence the proportionalities developed above, such as (9.3), will not be strictly correct above 60 km altitude. However, we do not expect the diffusional perturbations to obliterate general features like the ozone kink and its counterpart in the $O_2(^1\Delta_g)$ concentration profile, but it will be observed that the profiles of the lighter mass species like $O(^3P)$ atoms will be more affected by diffusion so that the kink becomes less obvious. This situation is further complicated by the fact that in the range of altitude 60 to 100 km the atoms will have increasing rate coefficients for production from molecular oxygen by photodissociation due to the increasing availability of radiation of wavelengths less than 200 nm.

Case III – Nitrous Oxide

The last specific case study of primary photodissociation is nitrous oxide, N_2O. In Section 7.4 we discussed the microbiological production of this gas at the Earth's surface, and it was postulated that the tropospheric lifetime of this gas was long enough to allow a considerable upward flux into the stratosphere. Figure 97 shows the absorption cross-section of nitrous oxide as a function of wavelength together with the temperature dependence of the absorption cross-section.

The threshold for the onset of absorption lies in the vicinity of 260 nm wavelength and this is explicable on the basis that the absorption of a photon of lower wavelength produces photodissociation.

In principle the simplest mode of photodissociation for nitrous oxide would be represented by the equation:

$$N_2O(^1\Sigma) + h\nu \longrightarrow N_2(^1\Sigma_g^+) + O(^3P)$$

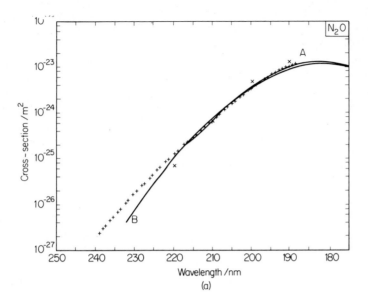

(a)

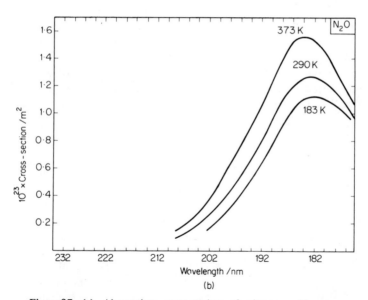

(b)

Figure 97. (a) Absorption cross-section of nitrous oxide as a function of wavelength at ambient temperature. Different points and letters refer to different determinations. The curves A and B are preferred at present. (b) Temperature dependence of nitrous oxide absorption cross-sections. Reproduced by permission of the National Research Council of Canada from R. D. Hudson, *Canadian Journal of Chemistry,* **52,** 1465–1478 (1974).

The minimum equivalent photon energy required to effect this dissociation would be $167.4 \text{ kJ mol}^{-1}$, corresponding to a wavelength of 715 nm, but nitrous oxide is a colourless gas, which precludes any absorption across the visible region of the spectrum. The fundamental reason forbidding the potential photodissociation process represented by the above equation is *spin conservation*, similar to that encountered in the spectroscopic selection rules. In our discussion of term symbols in Section 8.2, we saw that the left-hand superscript corresponded to $(2\Sigma + 1)$ for a molecule or $(2S + 1)$ for atoms, where Σ and S were net spin quantum numbers, with a conventional contribution of ½ from each unpaired electron. Let us consider how this applies to a general reaction situation represented by the equation:

$$\begin{array}{ccc} A + B & \longrightarrow & C + D \\ (S_1) \ (S_2) & & (S_3) \ (S_4) \end{array}$$

where S_1 and S_2 are the net spin quantum numbers of the reactants, and S_3 and S_4 are the net spin quantum numbers of the products. The *Wigner-Wittmer* spin conservation rule, which applies rigorously, states that for an elementary reaction there is an allowed reaction surface only if the vectorial sum of the net spins of the reactant species, i.e.

$$\overline{S_1 + S_2} = S_1 + S_2, S_1 + S_2 - 1, S_1 + S_2 - 2, \ldots \ldots, |S_1 - S_2|,$$

can be equal to the vectorial sum of the net spins of the product species, i.e. $\overline{S_3 + S_4}$. In the absence of such an allowed reaction surface a reaction cannot proceed at a significant rate.

The spin conservation rule applies equally to a photodissociation situation such as that represented above for nitrous oxide. Here we have a net spin of zero for the singlet ground state of the reactant, while the vectorial sum of the spins of the singlet and triplet states of the products can only be unity. Hence there is no potential reaction surface to allow this photodissociation to proceed at a significant rate.

The lowest energy electronically excited product from the photodissociation of nitrous oxide would be $O(^1D)$, with an excitation energy of $190.0 \text{ kJ mol}^{-1}$ compared to $O(^3P)$. Hence the new photodissociation route would be represented by the equation:

$$N_2O(^1\Sigma) + h\nu \longrightarrow N_2(^1\Sigma_g^+) + O(^1D)$$

On the basis of the Wigner–Wittmer rule this process is evidently allowed, with vectorial spins of zero on each side of the equation. The minimum energy requirement is $357.4 \text{ kJ mol}^{-1}$, equivalent in energy to a photon of wavelength 334.9 nm. However the Franck–Condon principle will also apply through the availability of potential surfaces. Since the absorption spectrum shown in Figure 97 is truly continuous, the electronic transition in the nitrous oxide molecule must be from the lowest vibrational level of the ground electronic state up to a part of the excited electronic state potential surface above its dissociation limit into N_2 and $O(^1D)$, where there is no quantization of energy. The disposition of potential energy curves in the two-dimensional representation of the molecule as N_2-O, which is likely to be a reasonable approximation for the photodissociation which breaks the N–O bond, shown in Figure 98, offers the explanation as to how the threshold of the main ultraviolet absorption lies at a wavelength corresponding to an energy substantially in excess of the apparent minimum requirement.

It is currently accepted that photodissociation of nitrous oxide inevitably follows the absorption of a photon with a wavelength shorter than 260 nm, to produce $O(^1D)$ atom with a quantum yield of unity.

It is of interest to us in our present context to consider how the rate of generation of $O(^1D)$ atoms from nitrous oxide is considerably lower than that from ozone in the lower regions of the stratosphere. The first point to be made is clearly the inability of nitrous oxide to absorb radiation above a wavelength of 260 nm, which in Table 9.5 is seen to be the main source of $O(^1D)$ atoms from ozone photodissociation. Secondly, on examination of the photon flux densities with wavelengths less than 260 nm at 25 km altitude in Table 9.4, we see that this is restricted to the window region between wavelengths of 230 and 190 nm for the penetration of any consequence; these photon flux densities are relatively low compared to those available above a wavelength of 290 nm. Moreover, even at a wavelength of 210 nm, the ozone absorption cross-section is some two orders of

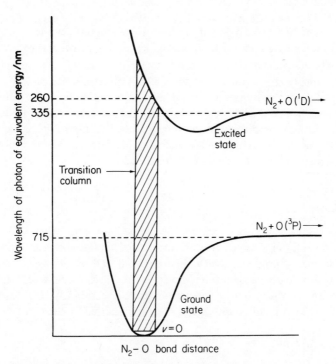

Figure 98. Diagramatic representation of the origin of the observed threshold for the photodissociation of nitrous oxide.

magnitude larger than that of nitrous oxide. On these bases it is hardly surprising that J_{N_2O} is only around 10^{-8} s^{-1} at 25 km altitude compared with the corresponding value of J_{bO_3} which is of the order of 10^{-4} s^{-1} (see Figure 100 to follow). The effectiveness of nitrous oxide as a source of $O(^1D)$ in the stratosphere around 25 km altitude is further reduced by the mixing ratios of around 2×10^{-7} for nitrous oxide as opposed to nearer 4×10^{-6} for ozone.

We may also consider at this stage whether photodissociation or chemical reaction with $O(^1D)$ atoms is the main fate of nitrous oxide in the middle stratosphere. There are two pathways for the chemical reaction represented by the equation:

$$O(^1D) + N_2O \longrightarrow N_2 + O_2 \tag{33}$$

$$O(^1D) + N_2O \longrightarrow 2 NO \tag{34}$$

These occur with approximately equal rates and the overall rate constant is 1.3×10^{11} dm^3 mol^{-1} s^{-1} as given in Table 9.6. We have calculated that the concentration of $O(^1D)$ atoms at 25 km altitude is 1.3×10^{-20} g-atoms dm^{-3}. Multiplication of the second order rate constant by this concentration yields a pseudo-first-order rate constant of 1.7×10^{-9} s^{-1}, which is an order of magnitude less than J_{N_2O}, so that photodissociation is the predominant fate of nitrous oxide in the middle stratosphere. Nevertheless, as we shall see, the reaction represented by equation (34) is the main natural source of nitric oxide in the stratosphere.

The lifetime of nitrous oxide in the stratosphere, determined in large measure by J_{N_2O}, is very long compared to those of the more reactive species such as ozone and exceeds the characteristic transport time of around 10^6 s discussed in Section 3.3. Accordingly it is incorrect to apply simple photochemical equilibrium considerations to the interpretation of nitrous oxide mixing-ratio profiles in the stratosphere; proper modelling demands that account be taken of eddy diffusive transport. Similar perturbations of purely chemical modelling calculations also come into play for other molecules with long lifetimes in the stratosphere, such as hydrogen, methane, and to some extent nitric acid.

General Cases

Carbon dioxide photodissociation is limited by the Wigner–Wittmer spin conservation rules in a similar way to nitrous oxide photodissociation. In principle the lowest energy products are $O(^3P)$ atoms and $CO(^2\Sigma)$ molecules, but we may now recognize that a singlet-state molecule like carbon dioxide cannot photo-dissociate to a triplet and a singlet product. Accordingly the lowest energy photodissociation of carbon dioxide must be represented by the equation:

$$CO_2(^1\Sigma) + h\nu \longrightarrow CO(^1\Sigma) + O(^1D)$$

The minimum energy required for this photodissociation is 721.4 kJ mol^{-1}, its equivalent corresponding to the photon of radiation of wavelength 166 nm. In fact,

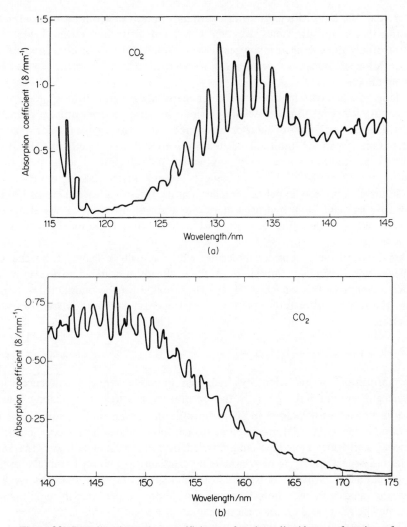

Figure 99. Decadic absorption coefficients of carbon dioxide as a function of wavelength in the vacuum ultraviolet spectral region, (above) from 115 to 145 nm, (below) from 140 to 175 nm. From E. C. Y. Inn, K. Watanabe and M. Zelikoff, *Journal of Chemical Physics,* **21**, (October) 1648, 1649, (1953). Reproduced with permission of the American Institute of Physics.

this wavelength marks the onset of strong absorption by carbon dioxide, as is shown in Figure 99.

The absorption spectrum consists of a banded structure superimposed upon an underlying continuum and the available evidence suggests a quantum yield of unity for $O(^1D)$ production below the threshold wavelength. However we can be sure that photodissociation of carbon dioxide cannot be significant in the stratosphere. In fact, the mixing ratio of carbon dioxide remains constant from ground level up to around 100 km altitude, at which altitude there is penetration of an effective photon flux of far ultraviolet radiation. One point of further note in Figure 99 is that not only does molecular oxygen have a window in its absorption spectrum for transmission of Lyman α radiation at a wavelength of 121.57 nm, but so also does carbon dioxide.

Many of the other natural trace component gases in the stratosphere are not subject to effective photodissociation therein because their absorption characteristics demand radiation of shorter wavelengths than are available. For example, water vapour can be photodissociated into ground state products, $H(^2S)$ and $OH(^2\Pi)$, but the minimum energy requirement is 498 kJ mol^{-1}, equivalent to the photon energy of radiation of wavelength 240 nm. In fact, strong absorption by water vapour only sets in below an upper limiting threshold wavelength of 190 nm. Similarly methane and hydrogen have no absorptions above a wavelength of 180 nm and can therefore play no part in the primary photochemistry of the stratosphere.

Finally we should consider molecules which actually originate from chemical cycles within the upper atmosphere. We have already considered nitric acid vapour which belongs to this category. Hydrogen peroxide can be considered as arising from the disproportionation reaction of hydroperoxy radicals represented by the equation:

$$HO_2 + HO_2 \longrightarrow H_2O_2 + O_2 \tag{35}$$

The absorption of radiation by hydrogen peroxide vapour commences at a wavelength around 310 nm with the absorption coefficient increasing rapidly towards shorter wavelengths. In general profile this absorption resembles that for the production of $O(^1D)$ atoms from ozone, but the absorption coefficients of hydrogen perioxide at corresponding wavelengths are considerably lower. Therefore we should expect $J_{H_2O_2}$ to be considerably less than J_{bO_3}, but of a simular profile as a function of altitude, as is borne out in Figure 100. The quantum yield of hydroxyl radicals in the photodissociation of hydrogen peroxide in its ultraviolet band may be taken as 2 on the basis of laboratory evidence.

Figure 100 summarizes computed values of J coefficients as functions of altitude in the atmosphere. J does not vary much with increasing altitude for molecules like nitrogen dioxide which have their principal absorption strength at wavelengths above 300 nm. This is also observed for J_{aO_3}, which refers to the photodissociation yielding $O(^3P)$ atoms.

The photodissociations which involve strong absorption in the wavelength range 200 nm to 300 nm such as ozone, nitric acid, and hydrogen peroxide, show rapidly increasing photodissociation rate coefficients with altitude in the stratosphere as the fraction of the ozone layer above decreases. Above 50 km these become almost invariant with further increase of altitude. Finally there is a group consisting of molecules which have absorptions restricted to wavelengths less than 200 nm , such as carbon dioxide, water vapour, and methane. The corresponding photodissociation rate coefficients are relatively insignificant in the stratosphere and rise rapidly into the mesosphere as the molecular oxygen column above decreases to allow some penetration of far ultraviolet radiation.

The most abundant constituent of all, molecular nitrogen, has a bond dissociation energy of $941.4 \text{ kJ mol}^{-1}$, corresponding to the equivalent energy of a photon of wavelength 127.2 nm. In fact nitrogen is effectively transparent above 100 nm wavelength and photodissociation of nitrogen has no significant rate below 100 km altitude.

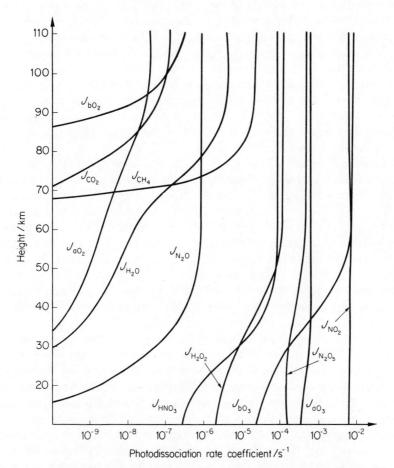

Figure 100. Photodissociation rate coefficients for overhead Sun as functions of height of atmospheric molecules. From I.S.A. Isaksen, 'Diurnal variations of atmospheric constituents in an oxygen–hydrogen–nitrogen–carbon atmospheric model and the role of minor neutral constituents in the chemistry of the lower ionosphere', *Geophysica Norvegica,* **30,** 4 (1973). Reproduced by permission of the Norwegian Academy of Science and Letters.

9.3 Secondary Chemistry of the Stratosphere

We may concede at the outset that a detailed account of stratospheric chemistry would be so complex as to be well beyond the scope of this book. The U.S. National Bureau of Standards has made a survey of rate-constant data for approximately 250 elementary reactions which enter into stratospheric chemistry, and at the same time considered that more than ten species underwent effective photolysis. Moreover comprehensive models of stratospheric chemistry must take into account both daily and seasonal variations in concentration profiles with altitude. One-dimensional models of the stratosphere have generally proved to be unsatisfactory because they cannot take into account properly the effects of atmospheric motions. The primary upwelling motion from the troposphere into the stratosphere takes place in equatorial regions where solar heating is most intense. Motions in the horizontal plane towards the poles of the Earth are generally much slower than movement around the planet. Atmospheric circulation is much faster in winter but the chemical reactions, driven primarily by photodissociation, are much faster in summer; this poses a problem for annual averaging approaches, exacerbated by the fact that eddy diffusional mixing is more significant in winter and spring than in summer or autumn.

Our reduced approach will be simply to identify elementary reactions which have rates sufficiently large under stratospheric conditions as to confer upon them a major role in the overall chemistry. Further, since it is apparent already from the variation of J values as a function of altitude that the relative importance of different reactions may also change with altitude, we shall address our attentions most specifically to the situation in the middle stratosphere at 25 km altitude. Such an altitude may be considered as the most critical since it corresponds closely with the peak concentration of ozone as a function of altitude. As we shall see in following sections, ozone is the stratospheric component of most concern at the present time in connection with potential anthropogenic perturbation of the stratosphere.

In the absence of any really successful chemical model of the stratosphere, we shall use as our framework of reference some of the direct measurements of the concentrations of various species in the vicinity of 25 km altitude. We must be cautious on several points in choosing the values to be adopted. In the first place we would look for measurements of concentrations of species involved in photo-chemical conversions performed under near overhead Sun conditions, since considerable changes in these would be expected under twilight conditions. Further, there is now strong evidence of considerable latitudinal variations in the concentrations of some species. This aspect may be illustrated in two ways. Figure 101 shows a contour diagram representing zonal average ozone concentrations for the southern hemisphere midsummer.

Figure 102 shows the variation in the nitric acid vapour column density above altitudes in the range 12 to 18 km, with the bulk of the data referring to the northern hemisphere Spring. The method of measurement was based upon the HNO_3 emissivity

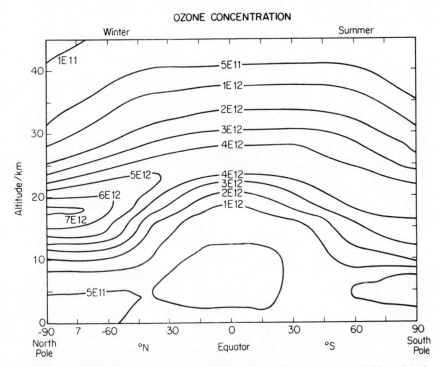

Figure 101. Standard (January 15) ozone concentrations, with contours marked in units of molecules cm^{-3} (divide by 6×10^{20} to convert to mol dm^{-3}), expressed as zonal average contour lines, using 7E12 to represent 7×10^{12}, etc. From H.S. Johnston and G. Whitten, *International Journal of Chemical Kinetics*, Symposium No. 1, p. 3 (1975). Reproduced by permission of John Wiley & Sons, Inc, New York.

in the infrared band at a central wavelength of 11,200 nm, which was directly proportional to the total column density of nitric acid vapour above the aircraft from which the observations were made. Associated balloon flight data yielded a mean altitude of 22 km for the nitric acid column density. The data obtained when the aircraft was flying at an altitude of 18 km were significantly lower than for flights at 12 to 16 km altitude, suggesting that the highest flight altitude was penetrating the nitric acid layer with, on average, about 75% of the layer lying above 18 km. Evidence of quite large seasonal variations in the nitric acid column density can be seen in the isolated points in Figure 102 corresponding to different months of 1970 at latitude 40° N.

Figures 103(a) and (b) show evidence of the latitudinal variations for the concentrations of nitric oxide and ozone measured from aircraft at altitudes of 18.3 km and 21.3 km respectively. These data were obtained in the northern hemisphere midsummer using *in situ* measuring techniques and generally correspond virtually to local solar zenith conditions. They show the expected poleward increase in ozone concentrations (compare with the southern hemisphere data of Figure 101). For the altitude of 21.3 km there is a steady poleward increase of

318

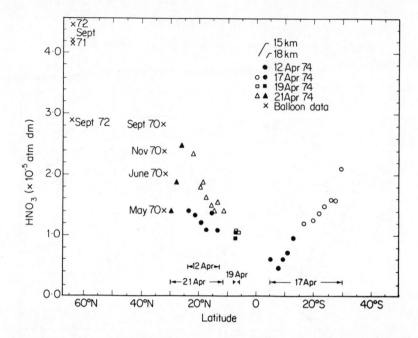

Figure 102 Measured variations of the stratospheric nitric acid vapour column with latitude and with an indication of the seasonal variation at 32° N. From D. G. Murcray, J. N. Brooks, A. Goldman and W. J. Williams, *Geophysical Research Letters*, **2**, 224 (1975). Copyrighted by the American Geophysical Union and reproduced with permission.

nitric oxide concentrations, but evidence of a minimum at around 45° N in the corresponding data for 18.3 km altitude.

The seasonal variations of ozone and nitric oxide concentrations at around 40° N for around midday conditions are shown in Figure 104, referring to measurements made at 21.3 km altitude.

There is a clear trend for the development of maximum nitric oxide concentrations in midsummer. In conjunction with the tendency for maximum column densities of nitric acid to develop during the winter months (see Figure 102), the likely interpretation of the minimum in nitric oxide concentrations in winter is that oxidation to nitric acid vapour is then favoured by the increased winter concentrations of ozone, evident in Figure 101 and 104, and the decreased rate of photolysis of nitric acid at higher solar angles.

Other measurements of the concentrations of species in the stratosphere have revealed the daily conversion cycles very clearly. Figure 105 shows measurements of the nitric oxide and nitrogen dioxide mixing ratios as functions of altitude made over Southern France in midsummer. These measurements were made using a balloon-borne radiometer which detected infrared emission from minor stratospheric constituents on the atmospheric limb. This figure shows clearly the maximum concentrations of nitrogen dioxide which exist at night and how these

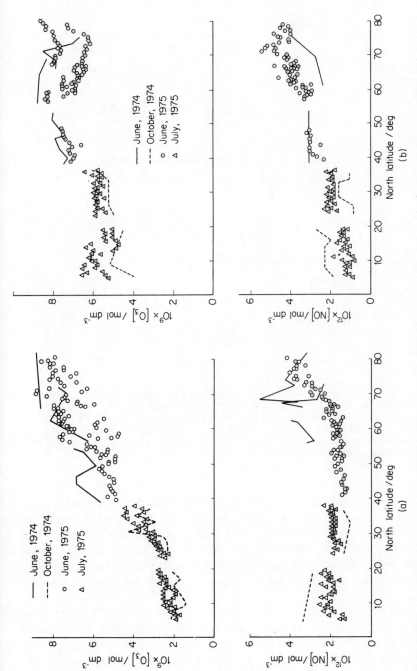

Figure 103. Measured nitric oxide and ozone concentrations at (a) 18.3 km altitude and (b) at 21.3 km altitude as functions of latitude at longitudes between 122° W and 158° W. From M. Loewenstein and H. Savage, *Geophysical Research Letters*, **2**, 448, 449 (1975). Copyrighted by the American Geophysical Union and reproduced with permission.

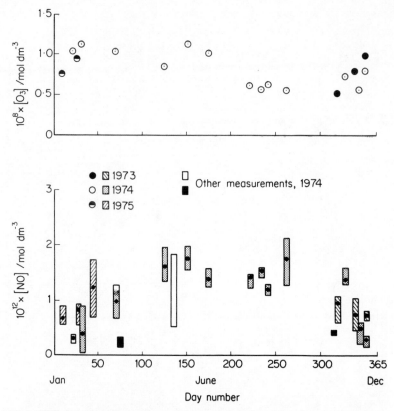

Figure 104. Average measured concentrations of nitric oxide and ozone at 21.3 km altitude at around 35° N, 120° W versus time of year, showing data obtained by various groups of investigators. From M. Loewenstein, H. F. Savage and R. C. Whitten, 'Seasonal Variations of NO and O_3 at altitudes of 18.3 and 21.3 km', *Journal of Atmospheric Sciences*, **32**, 2189 (1975). Reproduced by permission of the American Meteorological Society.

are rapidly reduced through sunrise to the much lower daytime steady-state levels. At the same time, nitric oxide concentrations were almost undetectably low overnight but rose rapidly as the nitrogen dioxide was photodissociated after dawn to reach a daytime level, which was still below that of nitrogen dioxide. Within the experimental errors of the analyses, the sum of the concentrations of nitric oxide and nitrogen dioxide remained constant at a particular altitude from night to day.

The disappearance of nitric oxide at sunset is shown in Figure 106, which shows concentrations at 28 km altitude measured over Texas using balloon-borne instrumentation. Despite the descent of the balloon towards 22 km altitude at the visible sunset, the rapid removal of nitric oxide is quite evident and the concentration has decreased by almost an order of magnitude in a 2-h period. This occurs by way of the bimolecular reaction (14) between nitric oxide and ozone, the latter species being present at concentrations about three orders of magnitude larger than those of the former. These observations then suggest a prominent role for reaction (14) in the ozone cycle in the middle stratosphere, a point which will be developed more fully later.

The slight divergences in the midday mixing ratios of nitric oxide in Figures 105 and 106 may simply reflect an inherent variability from one location to another and/or one day to another. There have been insufficient measurements as yet to confirm or deny this postulation.

Figure 107 shows the mixing ratio profiles as functions of altitude for some of the gases which originate within the troposphere and at the Earth's surface as measured over Texas. The mixing ratio of hydrogen hardly changes up to 30 km altitude, but those of methane and nitrous oxide decrease quite rapidly above the tropopause with increasing altitude. Carbon monoxide mixing ratios decrease across the tropopause (a point more clearly borne out in Figure 21) but there is evidence that the mixing ratio increases across the altitude range 20 to 30 km. There is likely to be a significant production associated with the oxidation of methane in the middle stratosphere. The oxidation of methane is also the major source of water vapour in the stratosphere: a number of measurements of the water vapour mixing ratios in the middle altitudes have yielded mixing ratios of a few parts per million, with a tendency for an increase in the range 25 km to 40 km altitude. This latter feature

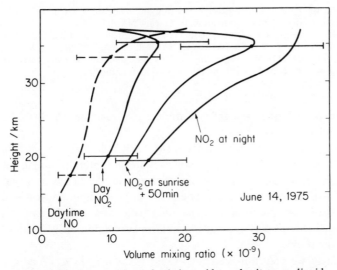

Figure 105 Mixing ratios of nitric oxide and nitrogen dioxide measured over southern France by balloon-borne radiometric monitoring of infrared thermal emission. Approximate error bars are indicated, including both systematic and random components. From C. P. Chaloner, J. R. Drummond, J. T. Houghton, R. F. Jarnot and H. K. Roscoe, 'Stratospheric measurements of H_2O and the diurnal change of NO and NO_2', *Nature*, **258**, 697, (1975). Reproduced by permission of Macmillan (Journals) Ltd.

322

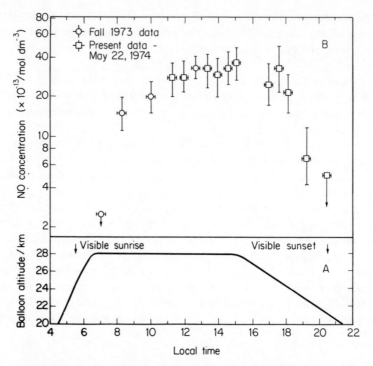

Figure 106 (A) Flight altitude as a function of time for the May, 22, 1974 balloon flight from Holloman A. F. B., New Mexico.
(B) Concentration of nitric oxide as a function of local time. From E. G. Burkhardt, C. A. Lambert and C. K. N. Patel, *Science*, **188** June, 1111–1113 (1975). Copyright 1975 by the American Association for the Advancement of Science and reproduced with permission.

might be expected since the same altitude range accounts for the main part of methane oxidation.

A recent report of measurements made with balloon-borne instrumentation (Lazrus and coworkers, 1976) has shown hydrogen chloride mixing ratios, on a volume basis, of 4×10^{-10} to 8×10^{-10} at altitude of 25 km over midlatitudes. The method of inhalation of stratospheric air through a filter impregnated with the alkaline reagent, tetrabutyl ammonium hydroxide, was used in this work, which also yielded the result that hydrogen bromide mixing ratios were about 8% of those of hydrogen chloride.

Accordingly we may now examine Table 9.7, which shows a reasonable set of mixing ratios and corresponding concentrations for trace components of the stratosphere under overhead Sun conditions at an altitude of 25 km. The table is merely devised to provide a framework upon which to hang our discussion of stratospheric chemistry; it is quite probable that actual mixing ratios at any one time and location could vary by a factor of 2 or 3 from those shown.

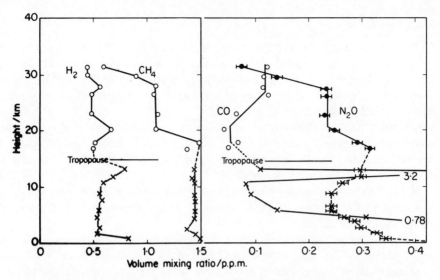

Figure 107 Vertical profiles of CH_4, N_2O, H_2 and CO mixing ratios as measured over eastern Texas in September, 1973. Except for N_2O the measurement error limits are about the size of the symbols. (Circles: balloon data; crosses: aircraft samples). From D. H. Ehhalt, L. E. Heidt, R. H. Lueb and W. Pollock, 'The vertical distribution of trace gases in the stratosphere', *Pure and Applied Geophysics,* **113**, 391, (1975). Reproduced by permission of Birkhäuser, Basle.

The concentrations of the more reactive atoms and radicals can only be obtained by calculation as yet, although there is some expectation that some of the radicals like hydroxyl may prove to be detectable by methods such as rocket-borne resonance fluorescence spectrometry. We may usefully perform approximate kinetic investigations of the reaction cycles involving such species in turn.

Table 9.7. A working set of volume mixing ratios for 25 km altitude in the middle latitude stratosphere for high Sun conditions

Component	Mixing ratio	Concentration mol dm^{-3}
Ozone	5×10^{-6}	6×10^{-9}
Water vapour	3×10^{-6}	4×10^{-9}
Methane	1×10^{-6}	1×10^{-9}
Hydrogen	5×10^{-7}	7×10^{-10}
Nitrous oxide	2×10^{-7}	3×10^{-10}
Carbon monoxide	1×10^{-7}	1×10^{-10}
Nitric acid vapour	3×10^{-8}	4×10^{-11}
Nitrogen dioxide	1×10^{-8}	1×10^{-11}
Nitric oxide	7×10^{-9}	9×10^{-12}
Hydrogen chloride	7×10^{-10}	9×10^{-13}

The O(^{3}P) Cycle

It is certain that O(^{3}P) atoms are generated with unit quantum efficiency by the photodissociation of ozone, either directly for wavelengths above 313 nm or indirectly through the predominant physical quenching of the O(^{1}D) atoms generated for wavelengths less than 313 nm. In Figure 100 we can read off the values at 25 km altitude of $J_{aO_3} = 5 \times 10^{-4}$ s^{-1} and $J_{bO_3} = 1 \times 10^{-4}$ s^{-1}, which means that photodissociation of ozone in the ultraviolet band below 313 nm wavelength is a minor source of O(^{3}P) atoms. Using the ozone concentration of 6×10^{-9} mol dm^{-3} from Table 9.7, the rate of production of O(^{3}P) atoms at 25 km altitude is 4×10^{-12} g-atoms dm^{-3} s^{-1}.

Table 9.8 shows calculated rates for the reactions which might be expected to be dominant in the removal of O(^{3}P) atoms at 25 km altitude, taking the average temperature to be 227 K.

Table 9.8 shows us immediately that to a good approximation, reaction (16) can be regarded as the unique removal route of O(^{3}P) atoms. As a result, the concentrations of oxygen atoms are related by a simply photoequilibrium in the sunlit middle stratosphere which may be written as:

$$O_3 + h\nu \underset{k_{16}[M]}{\overset{J_{O_3}}{\rightleftharpoons}} O(^3P) + O_2$$

Hence, using the parameter values developed above, we calculate that the concentration of O(^{3}P) atoms at 25 km altitude for near overhead Sun is $4 \times 10^{-12}/140 = 3 \times 10^{-14}$ g-atom dm^{-3}, which is five orders of magnitude less than the ozone concentration. At the same time this is over six orders of magnitude larger than the concentration of O(^{1}D) atoms calculated earlier.

The above photoequilibrium also leads us to the important concept of 'odd oxygen'. If we define O(^{3}P) and O$_3$ as 'odd oxygen' species for obvious reasons, it is clear that the photoequilibrium can shift in either direction without altering the 'odd oxygen' concentration. On the other hand, the slower reactions like (32) and (36) can alter the 'odd oxygen' concentration in conversions to O$_2$. The generation of 'odd oxygen' is accomplished by the photodissociation processes represented by the equation:

$$O_2 + h\nu(\lambda \leqslant 242 \text{ nm}) \longrightarrow O + O$$

Table 9.8. Estimated rates of reactions removing O(^{3}P) atoms at 25 km altitude

Reaction		Estimated rate constant for $T = 227$ K	Instantaneous rate/s^{-1}
O + O$_2$ + M → O$_3$ + M	(16)	4×10^8 dm^6 mol^{-2} s^{-1}	1.4×10^2 [O]
O + O$_3$ → 2 O$_2$	(32)	5×10^5 dm^3 mol^{-1} s^{-1}	3×10^{-3} [O]
O + NO$_2$ → NO + O$_2$	(36)	5×10^9 dm^3 mol^{-1} s^{-1}	5×10^{-2} [O]

Table 9.9. Estimated rates of reactions removing ozone at 25 km altitude

Reaction	Estimated rate constant for $T = 227$ K	Instantaneous rate/s^{-1}
$O_3 + h\nu \rightarrow O + O_2$	J_{O_3}	6×10^{-4} [O_3]
$NO + O_3 \rightarrow NO_2 + O_2$ (14)	3×10^6 dm^3 mol^{-1} s^{-1}	3×10^{-5} [O_3]
$O + O_3 \rightarrow 2 O_2$ (32)	5×10^5 dm^3 mol^{-1} s^{-1}	2×10^{-8} [O_3]
$OH + O_3 \rightarrow HO_2 + O_2$ (37)	1×10^7 dm^3 mol^{-1} s^{-1}	1×10^{-8} [O_3]^a

[a]Based upon a concentration of OH radicals at 25 km altitude in the sunlit stratosphere of 1×10^{-15} mol dm^{-3}, indicated by various modelling studies.

with the photodissociation rate coefficient J_{O_2}. From the point of view of 'odd oxygen' removal rates, Table 9.8 is only half the story; Table 9.9 summarizes the rates of reactions removing ozone at 25 km.

Despite considerable scope for variations in the above rates, on account of some uncertainties in the values of rate constants and the likelihood of variations in concentrations of species from day to day and between one locality to another, the data of Table 9.9 emphasize the significance of reaction (14) in the removal of ozone in the middle stratosphere. Together reactions (14) and (36) constitute a catalytic chain mechanism for the removal of ozone, a fact made clear by the equations written as:

$$NO + O_3 \longrightarrow NO_2 + O_2 \qquad (14)$$

$$O + NO_2 \longrightarrow NO + O_2 \qquad (36)$$

Net reaction $\quad O + O_3 \longrightarrow 2 O_2$

Evidently this cycle could have a large role to play in determining the concentration of 'odd oxygen' in the main part of the ozone layer, the quantitative extent of which can be estimated as follows.

Now, in addition to oxygen atoms and ozone, nitrogen dioxide is also to be considered as an 'odd oxygen' species; it has a labile oxygen atom and is equilibrated with atomic oxygen and ozone in the sunlit stratosphere through its photodissociation and reactions (14) and (36). On the other hand nitric oxide is not an 'odd oxygen' species since the oxygen atom therein is not labile. Under steady-state conditions, the rate of generation of 'odd oxygen' can be equated to its rate of removal. On this basis we can make the equality:

$$2 J_{O_2} [O_2] = 2 k_{32} [O] [O_3] + 2 k_{36} [O] [NO_2] \qquad (9.4)$$

The coefficients in equation (9.4) appear because two 'odd oxygen' particles are generated by the photodissociation of molecular oxygen and two 'odd oxygen' particles are destroyed in each of reactions (32) and (36). Evidently the photodissociation of nitrogen dioxide and reaction (14) merely exchange one 'odd oxygen' species for another.

At the same time, the rates of production of nitric oxide and its rates of removal must also be equal, generating a second equality:

$$J_{NO_2}[NO_2] + k_{36}[O][NO_2] = k_{14}[NO][O_3] \qquad (9.5)$$

Rearrangement of equation (9.4) yields the alternative form:

$$[O] = J_{O_2}[O_2]/(k_{32}[O_3] + k_{36}[NO_2]) \qquad (9.6)$$

On the basis of the photoequilibrium between the oxygen atoms and ozone mentioned above, there will be a direct proportionality between the concentrations for a defined solar photon-flux density. Accordingly equation (9.6) leads to the proportionality at a specific altitude:

$$[O_3] \propto (k_{32}[O_3] + k_{36}[NO_2])^{-1} \qquad (9.7)$$

This equation then makes it clear that it is the apparently minor rates of reactions (32) and (36) (see Tables 9.8 and 9.9) which actually exert the major control over the overhead Sun concentrations of ozone around the maximum in the ozone layer. This seeming anomaly arises because the photodissociation of ozone produces no effect upon the 'odd oxygen' concentration, producing an oxygen atom for which the overwhelming fate is association with molecular oxygen to reform ozone.

On the basis of the concentrations of ozone and nitrogen dioxide given in Table 9.7, the relative magnitudes of the denominator terms are as given in the right hand column of Table 9.8. This suggests that the reaction of atomic oxygen with nitrogen dioxide (36) is an order of magnitude faster than the reaction of atomic oxygen with ozone (32). On this basis we conclude that the ozone concentration in the middle stratosphere is controlled predominantly by the total concentration of NO_x (representing NO and NO_2 together), a defined fraction of which corresponds to the midday nitrogen dioxide concentration being considered, according to equation (9.5).

We may now test the ability of equation (9.6) to predict the steady state concentration of $O(^3P)$ atoms calculated solely from the ozone photoequilibrium above. The value of J_{O_2} at 25 km altitude for near overhead Sun is, not surprisingly, very small, being around $3 \times 10^{-12} \, s^{-1}$; the concentration of molecular oxygen at this altitude is some $2.6 \times 10^{-4} \, mol \, dm^{-3}$ in the standard atmosphere model. Hence equation (9.6) leads to the equalities:

$$[O] = J_{O_2}[O_2]/k_{36}[NO_2] = 2 \times 10^{-14} \text{ g-atom dm}^{-3}$$

The agreement with the result of 3×10^{-14} g-atom dm^{-3} calculated before is quite acceptable considering that there are no measurements of contemporary ozone and nitrogen dioxide concentrations available.

The direct observation of the rate of decay of the nitric oxide in the middle stratosphere at sunset, shown in Figure 106, confirms the importance of the reactions of NO_x in the ozone layer. At sunset we expect the photodissociation of ozone to cease fairly sharply so that oxygen atom concentrations decay. Then there is no route for the regeneration of nitric oxide from nitrogen dioxide, since the rates of reaction (36) and direct photodissociation of nitrogen dioxide will be severly inhibited in the absence of a significant photon flux density. Therefore reaction (14) will become an effective net reaction and, as is seen in Figure 106, it reduces the nitric oxide concentration by a factor of approximately 7 in 2 h. Since the ozone concentration is so much larger than that of nitric oxide, we can consider reaction (14) to occur under pseudo-first-order conditions, with an effective rate constant of $k_{14}[O_3] = 3 \times 10^6 \times 6 \times 10^{-9} = 2 \times 10^{-2}$ s^{-1}. Letting $[NO]_0$ represent the sunlit stratosphere concentration of nitric oxide and $[NO]_t$ be the concentration after time t of comparative darkness, the integrated pseudo-first-order rate law is given by the equation:

$$\ln([NO]_0/[NO]_t) = \ln 7 = 1.945 = 2 \times 10^{-2} \cdot t.$$

This, somewhat artificial, equation then leads to a time of around 100 s for the defined conversion of nitric oxide. The extension of the actual time scale no doubt reflects the imprecision in the term 'sunset', since there will be sky-diffused radiation available for some time thereafter to photodissociate nitrogen dioxide; moreover the oxygen atom concentration will take some time to decay below the level at which it can effect significant conversion of nitrogen dioxide back to nitric oxide.

The Hydroxyl Radical Cycle

At the outset we must anticipate the reactions likely to be involved in the production and removal of hydroxyl radicals and attempt to assess their relative importance in terms of instantaneous rates. The major difficulty in this procedure is that there have been no measurements of the concentrations of some of the species, like hydrogen peroxide and nitric acid vapours, which are potentially of high importance. Uncertainties in the values of the rate constants for some of the possible reactions further increase the degree of imprecision. Table 9.10 sets out what are likely to be the major production reactions with some indications of the rates, even if some of these can only be quoted as upper limits based upon modelled estimates for concentrations.

The data in this table strongly suggest that the photolysis of nitric acid vapour is a major source of hydroxyl radicals in the stratosphere. However this constitutes

Table 9.10. Estimated rates of reactions producing hydroxyl radicals at 25 km altitude under near-overhead Sun conditions

Reaction		Estimated rate constant for $T = 227$ K	Instantaneous rate/ mol dm^{-3} s^{-1}
$HNO_3 + h\nu \rightarrow OH + NO_2$		5×10^{-6} s^{-1}	2×10^{-16}
$H_2O_2 + h\nu \rightarrow 2\,OH$		6×10^{-6} s^{-1}	$\leqslant 2 \times 10^{-17}$
$O(^1D) + H_2O \rightarrow 2\,OH$	(21)	2×10^{11} dm^3 mol^{-1} s^{-1}	1×10^{-17}
$O(^1D) + CH_4 \rightarrow OH + CH_3$	(38)	2×10^{11} dm^3 mol^{-1} s^{-1}	3×10^{-18}
$O(^3P) + HO_2 \rightarrow OH + O_2$	(39)	$\leqslant 3 \times 10^{10}$ dm^3 mol^{-1} s^{-1}	$\leqslant 3 \times 10^{-17}$
$HO_2 + NO \rightarrow OH + NO_2$	(18)	$\leqslant 1 \times 10^8$ dm^3 mol^{-1} s^{-1}	$\leqslant 4 \times 10^{-17}$

part of a photoequilibrium represented by the equation:

$$HNO_3 + h\nu \rightleftharpoons OH + NO_2$$

At the same time it is doubtful if the effects of diffusive transport of nitric acid vapour can be properly ignored. The computer modelling approaches have suggested that the rates of other reactions of nitric acid vapour in the stratosphere, for example with $O(^3P)$ atoms or OH radicals, are much slower than the rate of photodissociation. The half-life of nitric acid molecules under these circumstances is then of the order of 2×10^5 s or something over 2 days, which is close to the characteristic transport time of 10^6 s mentioned in Section 3.3. Therefore in this connection the photostationary state approximation will be of, at best, only marginal validity. On this basis also, nitric acid is stable enough in the lower stratosphere to serve as the major sink of NO_x; this will ultimately be transported downwards to be precipitated to the Earth's surface, accounting for such phenomena as the nitrate ions detected in the Antarctic ice-sheets.

This brings us to the consideration of the relative importance of the reactions removing hydroxyl radicals in the middle stratosphere. The combination with nitrogen dioxide to reform nitric acid will be close to its third order regime at the low total concentrations of molecules, [M], which are only about 3% of those at ground level for an altitude of 25 km. Hence we should represent the reaction more meaningfully by the equation:

$$OH + NO_2 + M \longrightarrow HNO_3 + M \tag{40}$$

Table 9.11 then shows rate estimates for the reactions removing hydroxyl radicals in the middle stratosphere.

It is clear that reaction with ozone is a major fate of hydroxyl radicals in the middle stratosphere. However this may not represent an effective consumption of these radicals since there is good reason to believe that the dominant reaction of hydroperoxy radicals in the middle stratosphere is also with ozone, according to the equation:

$$HO_2 + O_3 \longrightarrow OH + 2 O_2 \tag{42}$$

The chain cycle consumption of ozone represented by equations (37) and (42) is

Table 9.11. Estimated rates of reactions removing hydroxyl radicals at 25 km altitude under near-overhead Sun conditions

Reaction		Estimated rate constant for T = 227 K	Instantaneous rates/s^{-1}
$OH + NO_2 + M \rightarrow HNO_3 + M$		2×10^{12} dm^6 mol^{-2} s^{-1}	0.02 [OH]
$OH + NO + M \rightarrow HNO_2 + M$		4×10^{11} dm^6 mol^{-2} s^{-1}	0.004 [OH]
$OH + O_3 \rightarrow HO_2 + O_2$	(37)	1×10^7 dm^3 mol^{-1} s^{-1}	0.07 [OH]
$OH + HO_2 \rightarrow H_2O + O_2$	(41)	$\leqslant 1 \times 10^{11}$ dm^3 mol^{-1} s^{-1}	$\leqslant 0.02$ [OH]
$CO + OH \rightarrow CO_2 + OH$	(13)	9×10^7 dm^3 mol^{-1} s^{-1}	0.009 [OH]
$O(^3P) + OH \rightarrow O_2 + H$	(−1)	2×10^{10} dm^3 mol^{-1} s^{-1}	0.0006 [OH]

many orders of magnitude less significant than the nitric oxide catalytic cycle represented by equations (14) and (36); the rate constants k_{37} and k_{14} are of the same order of magnitude but the nitric oxide concentration is expected to be about four orders of magnitude larger than that of hydroxyl radicals.

The number of reactions capable of contributing significantly to the hydroxyl radical cycle in the middle stratosphere, and the uncertainties evidenced by the upper limit signs in Table 9.10 and 9.11 point out clearly the limitations in attempting to model the chemistry of the stratosphere, and thence to obtaining a profile of concentration versus altitude for the hydroxyl radical. The actual measurement of hydroxyl radical concentrations in the middle stratosphere may be awaited with interest.

Carbon Cycles

We have remarked earlier (Section 7.2) upon the large upward flux of the methane generated at the Earth's surface. Figure 107 shows that this is completely mixed in the troposphere but the decrease of the mixing ratio with altitude in the stratosphere points to an efficient consumption mechanism operating therein. The principal species responsible for the primary attack on methane are hydroxyl radicals and $O(^1D)$ atoms. Rate constants are well established for these elementary reactions and Table 9.12 compares the rates in the middle stratosphere.

The present uncertainty about the hydroxyl radical concentration does not allow a conclusion any firmer than a statement that it is likely that both reactions of $O(^1D)$ atoms and OH radicals are significant in initiating methane destruction. However, it is clear that the reaction of $O(^3P)$ atoms with methane is insignificant; this stems from the relatively high temperature coefficient of the rate constant, corresponding to an Arrhenius activation energy of the order of 40 kJ mol^{-1}, which makes the Arrhenius factor, $\exp(-E/RT)$, very small for a temperature of 227 K (see Table 9.2).

A very complex chemistry follows the initial production of methyl radicals in the middle stratosphere. It seems likely that the first step will be the formation of methylperoxy (CH_3OO) radicals by direct association with molecular oxygen, and formaldehyde could be formed in either parallel or secondary steps. Subsequently the inevitable course of reaction appears to be through formyl radicals (HCO) to carbon monoxide. There is a strong case for considering that the bulk of the

Table 9.12. Estimated rate parameters for methane consumption reactions at 25 km altitude under near-overhead Sun conditions

Reaction	Estimated rate constant for $T = 227$ K	Instantaneous rate/s^{-1}
$O(^1D) + CH_4 \rightarrow CH_3 + OH$	2×10^{11} dm^3 mol^{-1} s^{-1}	3×10^{-9} [CH$_4$]
$OH + CH_4 \rightarrow CH_3 + H_2O$	8×10^5 dm^3 mol^{-1} s^{-1}	8×10^{-10} [CH$_4$]
$O(^3P) + CH_4 \rightarrow CH_3 + OH$	42 dm^3 mol^{-1} s^{-1}	1×10^{-12} [CH$_4$]

methane molecules oxidized in the stratosphere are converted to carbon monoxide and water vapour. Accordingly the flux of methane from the troposphere into the stratosphere can be thought of as equivalent to parallel fluxes of carbon monoxide and water vapour. The balancing removal processes for these are reaction with hydroxyl radicals [reaction (13)] to form carbon dioxide in the former case, and downward mass transport of carbon dioxide and water vapour back into the troposphere. Estimates have been made of the integrated flux of methane into the stratosphere and it is believed that no more than one-quarter of the total annual production rate at the Earth's surface (see Section 7.1) does penetrate through the tropopause, providing further evidence of the need for tropospheric sinks for methane. It is interesting to note that the total anthropogenic production of carbon monoxide (Section 7.2) is of the same order as the natural production from methane oxidation in the stratosphere.

There is a possibility that some methane is transported into the mesosphere above 50 km altitude. But since it seems certain that the mixing ratio of methane decreases by at least an order of magnitude through the stratosphere, the fate of the gas in the mesosphere is comparatively unimportant and most of the consumption is effected below 40 km altitude. Since methane can only be photodissociated by radiation of wavelengths less than 165 nm, this process will be of no significance in the chemistry of methane in the upper atmosphere.

Nitrogen Cycles

In Section 9.2 we have already discussed the generation of nitric oxide from the nitrous oxide which is transported upwards from the troposphere. As with methane it is only a fraction, believed to be between a quarter and a third, of the total production of the gas at the Earth's surface which enters the stratosphere. We have already deduced that most of the nitrous oxide will be photodissociated to molecular nitrogen and $O(^1D)$ atoms rather than react with $O(^1D)$ atoms to yield nitric oxide according to the equation:

$$O(^1D) + N_2O \longrightarrow 2\,NO \tag{34}$$

As a result the production rate of nitric oxide, integrated throughout the stratosphere, corresponds to only around 2% of the production rate of nitrous oxide at the Earth's surface according to current stratospheric modelling estimates.

Ammonia must also be a trace component of the stratosphere, since the ammonium ion detected in the sulphate aerosols (see next subsection) cannot be easily explained otherwise. This gas is presumably carried up from the troposphere, although it must be recognized that the overwhelming bulk of the ammonia evolved at the Earth's surface will be removed from the troposphere by rain-out, chemical attack by species such as hydroxyl radicals, and aerosol formation, etc., to the extent that even the tropospheric mixing ratio is highly variable with time and location. However, even if as little as 0.1% of the total ammonia production at the surface penetrated into the stratosphere and the nitrogen content were converted to nitric oxide, this could produce an effective nitric oxide flux comparable in magnitude to that derived from nitrous oxide. The chemistry of ammonia in the stratosphere is apparently initiated by the reaction with hydroxyl radicals represented by the equation:

$$OH + NH_3 \longrightarrow H_2O + NH_2 \tag{43}$$

for which the rate constant is some two orders of magnitude larger than that for the corresponding reaction of methane with hydroxyl radicals at stratospheric temperatures. The subsequent reactions of the NH_2 radical are not well enough understood as yet to predict the stoichiometry of the yields of nitric oxide.

From what has been said above, and the mixing ratio profile of nitrous oxide shown in Figure 107, it may be deduced that most of the production of nitric oxide will take place in the lower and middle stratosphere. Therefore we should expect the nitric oxide mixing ratios to decrease with increasing altitude in the upper stratosphere above 35 km altitude on the basis of the profile shown in Figure 105. The definition of the sign of such concentration gradients is important since mass transport of a particular species will always be preferentially down rather than up concentration gradients. Hence the overwhelming direction of transport of NO_x species in the lower and middle stratosphere will be downwards. Under these circumstances the net downward flux through the tropopause of nitric acid molecules, the overwhelming sink species for NO_x, should be on average equal to the effective equivalent flux of nitric oxide into the stratosphere through nitrous oxide and ammonia. More will be said on the quantitative aspects of this transport in the following section in connection with the potential for anthropogenic perturbation of stratospheric chemistry.

Sulphur Cycles

The main repository of sulphur in the stratosphere consists of the so-called 'sulphate layers', composed of aerosol particles.

Twilight light-scattering effects ascribed to the presence of particulate matter in the stratosphere have been known for approaching 90 years. The most spectacular instances of such effects have occurred subsequent to large volcanic explosions which have thrown inorganic dust through the tropopause. The Mount Agung explosion on the island of Bali in 1963 generated outstandingly coloured twilight skies, for example, but in the intervening periods between relatively infrequent volcanic injections of lava dust, there is a background aerosol component within the stratosphere. Remote-sensing techniques like *LIDAR* (laser radar) have provided clear evidence of a worldwide stratospheric aerosol layer, located at around 20 km altitude in middle latitudes. The *LIDAR* method relies upon the ability of particulate material to backscatter a vertically directed laser beam from the Earth's surface. Measurement of the intensity distribution of the backscattered light as a function of angle from the vertical at various detection stations allows estimation of the aerosol density as a function of altitude. Other methods may be used such as the balloon-borne photoelectric-particle counter and some typical results derived in this way are shown in Figure 108.

332

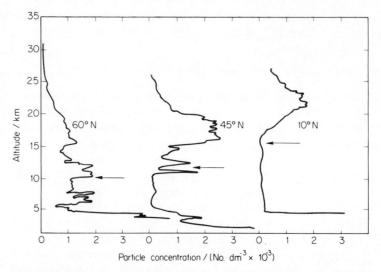

Figure 108 The vertical distribution of dust at three latitudes as determined by a photoelectric particle counter. Arrows mark the local position of the tropopause. From J. M. Rosen, *Space Science Reviews*, **9**, 76 (1969). Reproduced by permission of D. Reidel Publishing Company, Dordrecht, Holland.

More specific evidence has come from particle collection procedures applied from high-flying aircraft or balloons. Impactor surfaces (e.g. platinum foil with or without coating materials) have been used to collect samples of the larger aerosol particles, generally above the Aitken size range, i.e. greater than 10^{-7} m diameter. On the other hand, various filters composed of submicron diameter polystyrene fibres are very efficient in the collection of even Aitken particles; these filters have the added virtue of not absorbing gaseous species like nitric acid, in contrast to earlier types of filter, the use of which often indicated unusually high nitrate ion levels on this account. Table 9.13 gives some typical analysis data for samples of aerosol particles collected in the mid-latitude stratosphere.

The sulphate ion is the dominant component, hence the name 'sulphate layer'. In fact even higher concentrations of this ion have been measured in samples collected at altitudes around 37 km, giving an indication of the large scale of this phenomenon throughout the stratosphere. It should be emphasized that the sulphate layer is highly variable, both in the total aerosol density per unit volume of air and in its composition from one sampling time to the next, as is borne out to some extent by the data of Table 9.13 which were all obtained on December 4, 1969, at latitudes in the range 34° N to 48° N with 30-min sampling times. No potassium ion or nitrate ion was detected in any of these samples and the ratios of chloride to bromide are in the range 12 to 19, in contrast to the ratio of around 300 found in seawater. There is, therefore, no possibility that any component of the stratospheric aerosol represents dried sea spray. The high degree of variability was emphasized by a further set of measurements made in October 1970 by the

Table 9.13. Analyses of particles collected on polystyrene filters in the middle latitude stratosphere at around 18 km altitude (Concentrations expressed as μg of ion per m^3 air.)

SO_4^{2-}	Si	Na^+	Cl^-	NO_3^-	NH_4^+	Mn	Br^-
0.32	0.18	0.004	0.042	0.0012	0.0034	0.0036	0.0021
0.24	0.19	0.003	0.041	0	0.017	0.0025	0.0026
0.22	0.17	0.002	0.030	0	0.012	0.0012	0.0020
0.36	0.17	0.003	0.051	0.0036	0.043	0.0049	0.0024

No potassium or nitrite ion was detected in any of the samples.
From R. D. Cadle, *Transactions of the American Geophysical Union*, **53**, 812–820 (1972). Copyright by the American Geophysical Union and reproduced with permission.

same workers, when the concentrations of sulphate ion in particular were between one and two orders of magnitude lower than those given in Table 9.13.

Two persistent features are found in the analysis of sulphate aerosols from the stratosphere. The sulphate ion concentration is almost always greater than would correspond to ammonium sulphate as the only sulphate component. Moreover the ammonium ion content of the air as determined by impactor techniques (larger particles only) is closely in agreement with the content indicated by the filter techniques (all particles) at the same time. This is not the case with the sulphate ion content, which is always greater in the filter technique. This strongly suggests that the bulk of the ammonium ion content resides in the larger particles and that the smaller particles are predominantly a sulphuric acid aerosol. There is also some evidence that the sulphuric acid aerosol predominates in the lower stratosphere with a progressive increase in ammonium sulphate content with increasing altitude. This trend is illustrated by the analyses of samples collected on impregnated polystyrene filters at various altitudes at around $10°$ N, $80°$ W in the February-to-April period of 1971. At altitudes around 17 km it was quite common for the sample to contain no detectable ammonium ion. Table 9.14 shows a selection of the analyses obtained.

These data are compatible with a process of mass transport of sulphur dioxide

Table 9.14. Composition of sulphate aerosol collected by filters at various altitudes around $10°$N, $80°$W in February to April, 1971

Altitude/km	$SO_4^{2-}/\mu g.m^{-3}$	$NH_4^+/\mu g.m^{-3}$	Maximum % of SO_4^{2-} bound to NH_4^+
17	0.054	0.0035	15
21	0.113	0.0096	27
24	0.072	0.0050	25
27	0.019	0.0088	100

From A. L. Lazrus, B. W. Gandrud, and R. D. Cadle, *Journal of Geophysical Research*, **76**, 8085 (1971). Copyrighted by the American Geophysical Union and reproduced with permission.

from the troposphere into the stratosphere. In passage upwards through the stratosphere the gas is progressively oxidized to sulphur trioxide and hence a sulphuric acid aerosol. As the aerosol mixes upwards it may be envisaged as growing by absorption of further sulphur dioxide from the air and by its photooxidation within the droplet. Now if such a sulphuric acid droplet comes into contact with any traces of ammonia in the stratosphere, not only will these be absorbed but the resultant ammonium ion will enhance the ability of the liquid medium to oxidize sulphur dioxide, as was detailed in Section 8.6 in connection with the growth of tropospheric aerosol particles. Therefore it is to be expected that the main ammonium content will be found in the larger particles, which have been enabled to grow through its catalytic action, and in the presumably older aerosol at the higher altitudes. On the basis of the catalytic action of the ammonium ion, it will be the availability of ammonia in the stratosphere which will control the growth of the sulphate aerosol particles in general. The high degree of variability of the density of the sulphate aerosol is then to be expected in view of the variability of the ammonia content of the stratosphere, which must reflect the very patchy situation in the troposphere.

On penetrating the tropopause into the stratosphere, sulphur dioxide will be increasingly likely to be subject to attack by the increasingly abundant reactive species. Reactions of importance in this connection are represented by the equations:

$$O(^3P) + SO_2 + M \longrightarrow SO_3 + M \tag{30}$$

$$OH + SO_2 + M \longrightarrow HSO_3 + M \tag{31}$$

$$HO_2 + SO_2 \longrightarrow OH + SO_3 \tag{44}$$

A model of the stratospheric sulphate aerosol system has been developed (Harrison and Larson) which has matched the observed densities of sulphate aerosols in the vicinity of 18km altitude to an average upward flux of sulphur dioxide through the tropopause of 3×10^{10} molecules $dm^{-2} s^{-1}$, in balance with a downward flux of sulphate particles. This injection rate corresponds to around 3×10^9 kg year^{-1} of sulphur. Comparison with the sulphur turnover terms in the main tropospheric budget in Figures 66 and 67 show this to be, not unexpectedly, a rather small term, of the same order of magnitude as the sulphur release to the atmosphere by volcanic activity. It is then obvious that the stratospheric contribution to sulphate deposition of ground is virtually negligible.

9.4 Potential Anthropogenic Perturbations of Stratospheric Chemistry

Two principal areas of anthropogenic activity have been recognized at present as posing potential perturbations of the natural chemical cycles of the stratosphere. These are the exhaust emissions of Supersonic Transport (SST) aircraft and the release of chlorofluorocarbons, entirely man-made in origin. The former will be emitted directly to the stratosphere since the cruise altitudes of the SST aircraft such as Concorde are in the range 15 to 20 km. The latter are released within the troposphere but are sufficiently resistant to destruction therein for the chloro-fluorocarbon gases to mix inevitably into the stratosphere by upward diffusion. We shall examine the general chemical bases upon which these give cause for concern. However we must recognize at the outset that there is considerable dispute as to the scale of the threats and we cannot expect, therefore, to come to any definite conclusion on this score.

The SST Problem

It was in 1970 that warnings of a potential threat to the stratospheric ozone layer, principally from the nitric oxide component of the SST exhaust gases, gathered force. Concern with regard to other exhaust gases, such as carbon monoxide, water vapour, and sulphur dioxide, and the particulate materials also injected through the cruise operation of these aircraft has now abated somewhat. We are aware of the shielding properties of the ozone layer towards ultraviolet solar radiation of wavelength below 320 nm. In Table 2.1 it was shown that the 3 nm equivalent thickness of the presently existing ozone layer was quite sufficient to attenuate the solar ultraviolet irradiances of wavelengths below 290 nm by factors approaching one million or more. In the next section we shall see that the radiations in a comparatively narrow range of wavelengths from 290 nm to 320 nm, which do penetrate to the Earth's surface to some extent currently, represent a deadly threat to life on this planet due to their ability to inflict photodamage on DNA, the stuff of life itself. We must ask what would be the enhancement factor for these radiations if it did come to pass that the ozone layer was partially destroyed.

The problems induced by SST aircraft exhaust emissions into the stratosphere will depend in the first place upon the total number of aircraft flying, that is the total amount of nitric oxide added to the natural cycle terms on say an annual basis. Initially one would wish to assess how the anthropogenic injection term compared in magnitude with the natural rate of generation of nitric oxide from nitrous oxide, as discussed in the last section. If the anthropogenic term could be of the same magnitude or greater, then obviously the perturbation of the ozone layer would be expected to be severe due to the operation of the chain mechanism for the destruction of ozone represented by the equations:

$$NO + O_3 \longrightarrow NO_2 + O_2 \tag{14}$$

$$O + NO_2 \longrightarrow NO + O_2 \tag{36}$$

As was seen in the last section, the 'odd oxygen' concentration, mainly ozone, in the middle stratosphere was close to being inversely proportional to the nitrogen dioxide (and hence total NO_x) concentration. On the simplest basis, therefore, an SST nitric oxide exhaust emission term on an annual basis equal to the natural term for generation of nitric oxide should lead to a halved equivalent thickness of the ozone layer.

How then may we obtain an estimate of the natural source strength of nitric oxide in the stratosphere? One approach could be by computer modelling of the stratospheric chemistry and air motions, but this approach has hardly been totally successful on account of uncertainties in various of the input parameters such as rate constants and eddy diffusion coefficients. At best, one would need an independent measurement of some sort for confirmation. This appears to have been produced by assessment of the annual downward flux of nitric acid vapour from the stratosphere to the troposphere, a term which must be equal to the rate of generation of nitric oxide within the stratosphere. Over a period of 2 years, the concentration profiles of nitric acid vapour as a function of altitude were measured along a path from northern Alaska to southern South America (Lazrus and Gandrud, 1974). The observed concentration gradients were incorporated into a model of air motions, which was calibrated in terms of downward flux on the basis of the established rates of removal of radioactive debris injected into the stratosphere by the atmospheric nuclear weapons testing of the early 1960s. The indicated downward flux of nitric acid vapour through the tropopause was the equivalent of between 0.3×10^9 kg and 1×10^9 kg of nitric oxide on an annual basis. This turns out to be in good agreement with the estimates averaging around 0.5×10^9 kg of nitric oxide per annum computed for the natural injection rate deriving from nitrous oxide.

Due to the cancellation of the American SST development project and the present sparsity of sales of Concorde and the rival Russian Tupolev, early estimates that a fleet of some 500 SST aircraft could be in service by 1985 can be considered unrealistic. A better assessment of the likely perturbation effects might come from consideration of a fleet of 100 SST aircraft with some contribution to the nitric oxide injection in to the stratosphere coming from some 2000 subsonic aircraft cruising largely in the upper troposphere. Nitric oxide emission indices for the current Concorde engines under simulated supersonic cruise conditions are approximately 12 g per kg of fuel, and estimates of the stratospheric injection rate from the above fleets of aircraft come out at around 5×10^7 kg of nitric oxide per annum. It is therefore clear that there is no question of anthropogenic inputs of nitric oxide into the stratosphere becoming comparable to the natural generation term within the remainder of this century.

Even if the SST injection rate of nitric oxide did ultimately become significant compared to the natural generation rate, there is considerable doubt as to whether this would produce a *pro rata* increase in the NO_x levels throughout the stratosphere. A large fraction of the nitric oxide emitted by the SST aircraft would be injected in northern latitudes, within rather localized flightpaths or corridors and at altitudes below 20 km. On the other hand, most of the natural generation of nitric oxide takes place above 20 km altitude. In any attempt to model the effects of the anthropogenic emission it is then clear that the horizontal and vertical transport motions of the stratosphere must be taken into the computational procedures. As has been remarked earlier, dispersal of the emission latitudinally around the globe will be more rapid than longtitudinal dispersal, while there will be higher resistance to vertical dispersal against the temperature gradient. At the same time, seasonal variations in the diffusion coefficients are significant and in the reverse direction to the photochemical rates. The type of model required is therefore a three-dimensional one with seasonally varying parameters, and few of the computations so far made have approached this degree of sophistication. Once again it would be useful if some actual situation could be found with similar nitric oxide injection characteristics to what is likely to be the case with the operating SST fleets.

The atmospheric testing programmes for nuclear weapons, which reached a peak in 1961 to 1962, turn out in retrospect to have been wider tests than was first imagined. The nuclear fireball produced by the bomb heated a substantial amount of the surrounding air to temperatures in the region of 6000 K within a few seconds of the explosion. From our discussion of the chemical changes in heated air of Chapter 5 we may realize that considerable synthesis of nitric oxide must have resulted from the inmixing of air into the rising fireball. Bombs of above 10 Mton explosive force are considered to have injected considerable amounts of NO_x into the stratosphere, amounting to something of the order of 80% of the total nitric oxide synthesized. Moreover, with particular regard to the extensive Russian programme, the latitudinal points of injection were very similar to those proposed for the main SST flightpaths, and of course the vertical mode of injection is also a direct analogy requiring upward eddy diffusion to penetrate into the main ozone layer. Restricting examination only to the total yields, the 1961 to 1962 tests could have injected as much as one-third of the total natural NO_x content of the stratosphere. This stiuation then provides an excellent test of the ability of the stratospheric ozone layer to withstand the type of nitric oxide injection which could result from the SST aircraft. If there is a serious problem in prospect then the ozone records in the following years to 1963 should show a marked recovery from an anticipated minimum equivalent thickness. In fact, the examination of the ozone records from many observation stations around the world shows no noticeable increases in the ozone column as measured by the attenuation of the shorter wavelengths of incident solar radiation; at most a few of these might indicate a 5% increase over a 7 year period but even these could reflect the sunspot cycle rather than any NO_x effect. In any event the seasonal variation in the thickness of the ozone column can be a factor of up to 2, with a maximum in early spring and a minimum in autumn, so that it will be difficult to pick out any small average variation.

The evidence so far available, therefore, tends to suggest that SST aircraft pose less of a potential hazard to the ozone layer than has been considered over recent years.

The SST aircraft will also emit particulate materials at exhaust and thereby could increase the stratospheric aerosol denstiy. Such an increase could lead to an alteration of the delicate heat balance at the surface of the Earth. Two oppositely acting effects could come into play. The most obvious is the possible change in the average albedo for solar radiation of the atmosphere, whereby increased particulate density in the stratosphere would reduce the flux penetrating to the Earth's surface, producing an effective cooling. Concurrently the particulate material would reduce the radiative cooling term by trapping infrared radiation in backscattering and reflection, which would tend to increase the temperature. Hence the net surface temperature change could be in either direction. Recent estimates have suggested that an increase of 0.2% in the hemispherically averaged aerosol optical thickness of the stratosphere could result in a 0.07 K decrease in mean global temperature. However, it is likely that particulates originating in SST aircraft exhaust emissions would not be evenly spread and if they were concentrated closer to the northern latitude flightpaths, where the solar rays have a longer stratospheric pathlength; localized temperatures could be decreased by as much as 0.3 K by the potential rises in stratospheric aerosol densities on this account. Two types of SST-generated aerososl can be envisaged. Direct emissions, such as soot, will mostly be particles well above Aitken nuclei size; accordingly these should settle down from the stratosphere fairly rapidly and hence be of little consequence. Of more significance could be the emission of sulphur dioxide, converted into sulphate aerosols of fairly long residence times. An abating factor in this connection could be the unaltered availability of ammonia, so that the growth rate of the particles would be limited in view of the catalytic role of the ammonium ion (Section 8.6). The sulphur content of the fuel of the SST aircraft will be emitted as sulphur dioxide with an estimated index of around 1 g of sulphur per kg of fuel. The total annual injection of sulphur to the stratosphere would then probably be less than 10^8 kg of sulphur. At the end of Section 9.3 the estimated natural injection rate from the troposphere to the stratosphere of 3×10^9 kg of sulphur was given. On this basis the anthropogenic term in prospect is unlikely to lead to any significant increase in the density of the stratospheric sulphate aerosols.

The Chlorofluorocarbon Problem

At first sight the propellant gases of aerosol spray cans and the working substances of refrigerating systems would appear to be totally unlikely threats to the stratospheric ozone layer. The molecules in question, mainly $CFCl_3$ in the first application and CF_2Cl_2 in the second, are however totally man-made with no natural sources known. Moreover they have few if any sink reactions in the troposphere and their residence times are likely to be more than 30 years, perhaps much longer. Therefore, like nitrous oxide and methane, these molecules must be transported into the stratosphere. In fact carbon tetrachloride might be regarded similarly, except that its normal melting point of 250 K suggests that the upward flux may, like that of water vapour, be severely inhibited in its ability to cross the low temperature barrier at the tropopause. However, the chlorofluorocarbons are sufficiently volatile to cross the barrier with ease.

The general usage of the chlorofluorocarbons (commonly known as freons) began around 1950 with dramatic growth in annual production rates thereafter. For example in 1960 the British production of aerosol spray cans was approaching 50 million; by 1974 the figure was an order of magnitude larger. Worldwide, the annual production rates of $CFCl_3$ and CF_2Cl_2 are approximately equal and of the order of half a million tons for each. These species are now omnipresent but variable components of tropospheric air with average mixing ratios of the order of

10^{-10} by volume. As might be expected, the mixing ratios can range up to two orders of magnitude larger than these average values in urban environments or when the sampled air is downwind of such, even at distances of many miles.

The total worldwide production of $CFCl_3$ from 1950 to 1976 is estimated to be of the order of 3×10^9 kg, corresponding to of the order of 2×10^{10} mol or 10^{34} molecules. In Table 1.4 we see that the estimated mass of the atmosphere is 5×10^{18} kg, which may be taken as the mass of the troposphere in good approximation. On the basis of an average molecular weight of air of 28.96, the total number of molecules in the troposphere is close to 10^{44}. It is then evident that currently measured mixing ratios of $CFCl_3$ of the order of 10^{-10} concur with postulates that there is no significant destruction mechanism for $CFCl_3$ in the troposphere, so that the bulk of the molecules released to air over the past 20 or so years are still present as such. Similar considerations apply to the other species, CF_2Cl_2.

Eventually the freon molecules must be transported into the stratosphere and there they will be subject to shorter wavelength radiation effects absent in the troposphere. Figure 109 shows the relevant ultraviolet absorption cross-sections for $CFCl_3$ and CF_2Cl_2. The important feature is that these molecules have significant

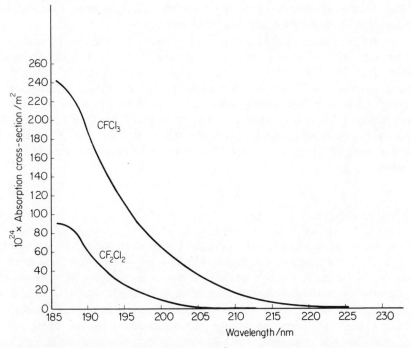

Figure 109. Profile of absorption cross-sections of chlorofluoromethanes versus wavelength in the middle ultraviolet atmospheric window spectral region. Based upon measurements tabulated by F. S. Rowland and M. J. Molina, *Reviews of Geophysics and Space Physics*, **13**, 7 (1975).

absorption characteristics across the wavelength region 190 to 220 nm. Table 9.4 illustrated the 'window' which exists in the transmission of solar radiation through the atmosphere in this wavelength region, and Figure 93 shows the significance of such radiation in photodissociating nitric acid vapour in the stratosphere above 20 km altitude. In fact the freon gases have absorption characteristics which may be likened to those of nitrous oxide in the sense that it is solar radiation which penetrates through the window between ozone and molecular oxygen absorptions which provides the main part of the photodissociation rate coefficient for each gas. In each case we may expect the variation of J to ressemble curve A in Figure 94, which represents the altitude variation of the component of J_{HNO_3} corresponding to the absorption of radiation of wavelengths in the range 185 to 250 nm. At the same time we might expect the mixing-ratio profiles for the freons as functions of altitude to tend towards that given for nitrous oxide in Figure 107. The reason why the nitrous oxide mixing-ratio profile may only suggest a limiting situation for the freons is that the latter are effectively newcomers to the stratosphere, continuously adjusting on a delayed time scale to the increasing mixing ratios in the troposphere; nitrous oxide, as a long-established natural component, must be in a steady state balance. Nevertheless the above analogies allow us to infer that, like nitrous oxide, the main photodissociation of freon gases will occur in the middle stratosphere and we may expect the mixing ratios to fall quite sharply with increasing altitude above 20 km. This postulation is borne out by recent measurements of stratospheric mixing ratios in combination with modelling techniques.

The transport rate of $CFCl_3$ into the stratopshere may be gauged by comparison with nitrous oxide. The tropospheric background mixing ratio of nitrous oxide is around 2.5×10^{-7} (Table 7.1). In the earlier part of this section it was estimated that the annual production of nitric oxide in the stratosphere (presumed to be substantially derived from nitrous oxide) was around 5×10^8 kg. In Section 9.2, it was concluded that in the middle stratosphere the photodissociation rate of nitrous oxide was around an order of magnitude larger than the rate of the chemical reaction with $O(^1D)$ atoms to yield nitric oxide, processes represented by the equations:

$$N_2O + h\nu \longrightarrow N_2 + O(^1D)$$

$$O(^1D) + N_2O \longrightarrow 2 NO \tag{34}$$

Taking this situation to represent an average for the stratosphere, this suggests that the annual rate of transport of nitrous oxide into the stratosphere must be of the order of 5×10^9 kg or 7×10^{34} molecules. Now it seems that this will be carried through the tropopause by the upwelling of air, particularly within the tropics, and it would not be unreasonable to presume that the relative rates of transport of nitrous oxide and freon gases into the stratosphere would simply be in proportion to their tropospheric mixing ratios. Since nitrous oxide is some three orders of magnitude more abundant than $CFCl_3$ in the troposphere, a transport rate of between 10^{31} and 10^{32} molecules per annum for $CFCl_3$ would seem appropriate. The current annual release rate into the troposphere is of the order of

2×10^{33} molecules of $CFCl_3$, so that something less than 5% of that is transported into the stratosphere at present. This illustrates the time-delayed effect of the transport into the stratosphere in the sense that the current rate corresponds closely to the total annual production rate of 20 years ago. Because of the dilution effect of the tropospheric air and the fact that only around 3% of that is exchanged with the stratosphere in a year, it will be at least 10 years on the basis of current production trends before the present annual production rate corresponds to the annual rate of transport into the stratosphere.

We have yet to look at the chemical nature of the problem posed by the presence of freons in the stratosphere. As implied above, the result of the absorption of a photon by these molecules is photodissociation, represented by the equations:

$$CFCl_3 + h\nu \longrightarrow CFCl_2 + Cl$$

$$CF_2Cl_2 + h\nu \longrightarrow CF_2Cl + Cl$$

The crux of the problem is the intervention of the chlorine atom into stratospheric chemistry, in particular the ozone reaction cycles. Further chlorine atoms are generated from the reaction of the carbon radical products with molecular oxygen and perhaps from subsequent steps also. Like nitric oxide, these chlorine atoms can induce a chain cycle of reactions for the destruction of ozone, represented by the equations:

$$Cl + O_3 \longrightarrow ClO + O_2 \tag{45}$$

$$O(^3P) + ClO \longrightarrow Cl + O_2 \tag{46}$$

$$NO + ClO \longrightarrow Cl + NO_2 \tag{47}$$

Net reaction $\quad O(^3P) + O_3 \longrightarrow 2 O_2 \quad$ (taking into account reaction (36) which regenerates NO)

The critical parameter is the rate constant k_{45}, which has a value of the order of 6×10^9 dm^3 mol^{-1} s^{-1} at the temperatures of the middle stratosphere. This rate constant is then some three orders of magnitude larger than k_{14} for the reaction between nitric oxide and ozone, which means that a very small concentration of chlorine atoms could control the equivalent thickness of the stratospheric ozone layer. At the same time the values of the rate constants for the subsequent reactions, k_{46} and k_{47}, are of the same order of magnitude as k_{36} for the reaction of $O(^3P)$ atoms with nitrogen dioxide. It is likely that reaction (46) is more important at higher altitudes and reaction (47) more important at lower altitudes within the stratosphere.

A steady-state analysis is required for a quantitative appreciation of the potential effects of chlorine atom concentrations in the middle stratosphere. Once again we shall invoke the concept of 'odd oxygen' in this. To the previously designated 'odd oxygen' species, $O(^3P)$ atoms, ozone and nitrogen dioxide molecules, we must now add ClO radicals since these evidently have a labile oxygen atom, as instanced in reactions (46) and (47). Thus the set of reactions which are effective in changing the 'odd oxygen' concentration are represented by the equations:

$$O_2 + h\nu \, (\lambda \leqslant 242 \text{ nm}) \longrightarrow O + O$$

$$O + NO_2 \longrightarrow NO + O_2 \tag{36}$$

$$O + ClO \longrightarrow Cl + O_2 \tag{46}$$

Reactions such as (45) and (47) and the photodissociation of ozone merely convert one 'odd oxygen' species into another. Also it is unnecessary to consider reaction (32) between $O(^3P)$ atoms and ozone since that has already been shown to be considerably slower than reaction (36) in the existing middle stratosphere.

For the steady state of 'odd oxygen' the equation is now:

$$2 J_{O_2} [O_2] = 2 k_{36} [O] [NO_2] + 2 k_{46} [O] [ClO] \tag{9.8}$$

At the same time the steady state for nitrogen dioxide is represented by the equation:

$$J_{NO_2} [NO_2] + k_{36} [O] [NO_2] = k_{14} [NO] [O_3] + k_{47} [NO] [ClO] \tag{9.9}$$

Also the steady state for the chlorine atom is represented by the equation:

$$k_{45} [Cl] [O_3] = k_{46} [O] [ClO] + k_{47} [NO] [ClO] \tag{9.10}$$

Equation (9.8) then rearranges to the equation:

$$[O] = J_{O_2} [O_2] / (k_{36} [NO_2] + k_{46} [ClO]) \tag{9.11}$$

and this will in turn be proportional to the ozone concentration because of the maintainance of the ozone photodissociation equilibrium.

Presently available estimates of the ratio of rate constants, k_{46}/k_{36}, for stratospheric temperatures average around 6. For the daytime mixing ratio of NO_2 of 10^{-8} indicated for altitudes around 25 km in Figure 105, the required mixing ratio for ClO radicals to produce a local 10% decrease in the ozone mixing ratio is around 2×10^{-10}. No measurements of existing ClO mixing ratios in the stratosphere have been made as yet but in Table 9.7 a measured mixing ratio for hydrogen chloride, at 25 km altitude, of 7×10^{-10} was quoted. Since hydrogen chloride is a relatively stable molecular species while ClO is evidently a rather reactive radical species, it would be surprising if the mixing ratio of ClO were greater than that of hydrogen chloride, especially as it is shown below that the two species are involved in an interconversion reaction scheme. Therefore there are reasonable grounds for believing that the chlorine-based reduction in the equivalent thickness of the ozone layer can only be fairly minor at present. However stratospheric modelling schemes, such as that in the paper by Rowland and Molina cited at the end of this chapter, have indicated that the ratio of the concentrations of hydrogen chloride to ClO is less than 10 at 25 km latitude and moves close to unity near 35 km altitude in the present-day stratosphere. Hence the numbers considered above are close enough to significance to give rise to serious concern for the future equivalent thickness of the ozone layer when the mixing ratios of freons (and consequently the rates of generation of chlorine containing species) must inevitably increase.

If chlorine is not to accumulate in the stratosphere, there must be an effective sink. Current thinking casts hydrogen chloride in this role. The principal routes to the formation of this molecule are attack by chlorine atoms upon methane and on molecular hydrogen, according to the equations:

$$Cl + CH_4 \longrightarrow HCl + CH_3 \tag{48}$$

$$Cl + H_2 \longrightarrow HCl + H \tag{49}$$

The rate constant, k_{48}, has a temperature coefficient characterized by an Arrhenius activation energy of around 10 kJ mol^{-1} while that attaching to k_{49} is somewhat larger. Since the frequency factors of the two rate constants are of similar magnitude, and we see in Table 9.7 that methane is around twice as abundant as molecular hydrogen in the middle stratosphere, it is clear that reaction (48) must be dominant in that region. However, if we examine Figure 107, we see that the methane mixing ratio decreases sharply with increasing altitude above 30 km, while that of molecular hydrogen remains substantially constant. Therefore reaction (49) is expected to play a more important part in the conversion of chlorine atoms in the upper regions of the stratosphere. Other reactions of chlorine atoms with hydrogen-containing molecules or radicals could also enter into the reaction mechanism, but the rates are uncertain due to either or both of lack of information on the values of rate constants and on the mixing ratios of the coreactant. Instances are the reactions of chlorine atoms with hydrogen peroxide vapour, nitric acid vapour, or hydroperoxy radicals. One reaction which can be ruled out is that of chlorine atom attack on water vapour since that is endothermic by some 42 kJ mol^{-1}, which value then corresponds to the minimum Arrhenius activation energy for the reaction and is too high for the reaction to have any significance at the low temperatures of the stratosphere.

It is instructive to compare the relative values of the rate constants for the main chain propagation reaction (45) and for the main chain termination reaction (48) in the middle stratosphere. For a temperature of 227 K, the ratio k_{45}/k_{48} is of the order of 1000. Moreover the mixing ratio of ozone to methane, the coreactants with chlorine atoms in these reactions, is given as 5 in Table 9.7. Therefore in principle the chain length of the chlorine atom chain cycle consuming ozone approaches 5000, so that each chlorine atom produced by the photodissociation of a chlorofluorocarbon in the middle stratosphere has the potential to destroy 5000 ozone molecules. This may be compared with a chain length of nearer to 200 for NO_x cycles in the same region (the ratio of the rates of reactions (36) and (40)). Further the chlorofluorocarbons carry fluorine atoms as well as chlorine atoms into the stratosphere, so that there might be some concern with regard to a fluorine chain cycle analogous to the chlorine atom cycle represented by equations (45), (46), and (47). However, the chain termination reaction of fluorine atoms with methane and molecular hydrogen is much faster than that for chlorine atoms. Accordingly the potential chain length is less than 100 and this cycle is of far less concern.

Once formed, hydrogen chloride is still subject to attack, principally by hydroxyl radicals according to the equation:

$$OH + HCl \longrightarrow H_2O + Cl \tag{50}$$

The rate constant for this reaction at middle stratospheric temperatures is of the order of 3×10^8 dm^3 mol^{-1} s^{-1}, about three-times larger than the corresponding reaction for nitric acid vapour, the principal sink of the NO$_x$ cycles. On the basis of an estimated concentration of hydroxyl radicals of 1×10^{-15} mol dm^{-3} at 25 km altitude, this predicts a half-life of some 2×10^6 s for hydrogen chloride, evidently large enough for atmospheric motions (with a characteristic transport time of around 10^6 s) to have a considerable effect upon the hydrogen chloride mixing-ratio profile. By comparison with the better-established situation for nitric acid vapour, we could expect effective downward transport of hydrogen chloride into the troposphere with subsequent rain-out to be the effective sink process.

What we have achieved, therefore, with regard to the chlorofluorocarbon problem is to demonstrate in a semiquantitative manner that there is a potentially serious interaction with the stratospheric ozone layer operating on the basic set of elementary reactions given above in this subsection. With the present level of uncertainty in the values of critical parameters and the high degree of complexity of the actual situation, we cannot hope to go further within the scope of this book. However so much effort is being expended in research work in this area that one might consider that the understanding and assessment of crucial trends cannot be delayed by many years. It may well turn out that our usage of chlorofluorocarbons will be restricted before long, if only on the basis that 'guilty until proven innocent' might be the best advised attitude for mankind; time could hardly be said to be on our side.

9.5 Biological Effects of Ultraviolet Radiation

The principal aim of this subsection is to outline the biological basis for concern with regard to the additional solar ultraviolet flux which would penetrate to the Earth's surface if the stratospheric ozone layer were significantly depleted.

The commonest effect upon man of the present solar ultraviolet flux density is sunburn, properly termed *erythema*. In this the minute blood vessels located just below the surface of the skin are dilated and manifest their condition in the typical reddening and pain. It is generally considered that erythema is the least serious of a series of photoinduced skin complaints, culminating in skin cancer, which have a sufficiently common origin to be related. Accordingly erythemal sensitivity is considered to give a basis for insight into the less well established mechanism of skin cancer induction.

When the erythemal effectiveness of monochromatic radiation has been studied as a function of wavelength, it has been found that there is a marked decrease in the effectiveness with increasing wavelength across the important range of 280 nm to 320 nm. In other words, the higher the wavelength, the larger the dose of radiation required to produce a specified degree of erythema; radiation of wavelength greater than 320 nm is considered not to induce erythema for practical purposes. Now of course solar photon flux densities increase in the opposite direction. Therefore it is to be expected that the absolute rate of erythema induction in sunlight as a function of wavelength would be proportional to the product of these two parameters and accordingly would go through a maximum somewhere close to 305 nm.

We may explore the origin of the erythemal effectiveness profile as a function of wavelength. A prime factor is the extent of penetration of the ultraviolet radiation through the various skin layers. From the surface downwards there are three principal layers. The outermost, the *stratum corneum*, consists of dead horny cells without nuclei and is 20 to 80 μm in thickness. It is a coherent membrane and, although a basically inhomogeneous medium, transmittance as a function of wavelength is meaningful since it is the *Melanin* content which gives rise to the principal attenuation. Below the stratum corneum is the *rete Malpighii*, or Malpighian layer, which is about 20 to 30 μm in thickness. This consists of living cells which continually migrate upwards to renew the horny layer with the cell relics. There are protein and nucleoprotein molecules in this layer which can absorb ultraviolet radiation of wavelengths below 300 nm. Together the stratum corneum and the Malpighian layer form the *epidermis*.

Below the epidermis is the *dermis*, about 2 mm thick.

Figure 110 shows the penetration of the skin of a forearm by ultraviolet radiation. It is apparent that radiation of wavelength less than 300 nm is mostly absorbed in the epidermis but in the critical region between 300 and 320 nm, from the point of view of erythema induction, there is significant penetration into the dermis. Many of the experimental investigations have taken advantage of the high intensity output of a mercury arc. This emits several lines within what may be termed the epidermal transmission region, for example, at wavelengths in the vicinity of 297 nm, 302 nm, and 313 nm. Lines of lower wavelengths, for example, 253.7 nm

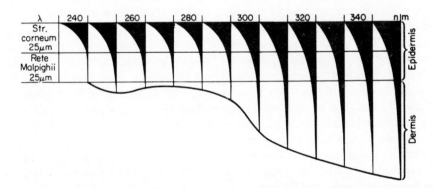

Figure 110 Penetration of the skin by ultraviolet radiation inside a forearm. (Adapted from Tronnier) From 'Effects of ultraviolet light on the skin', in *An Introduction to Photobiology* (ed. C. P. Swanson), A. Wiskemann, Prentice-Hall Inc., New Jersey, 1969, p. 83. Reproduced by permission of Prentice-Hall Inc. New Jersey.

and 265.4 nm will be absorbed in the epidermis. Radiation from the mercury arc has been shown to produce both cancers of the epidermis (carcinomas) and of the dermal tissues (sarcomas). Using optical filters the effects of radiation of wavelengths above and below 290 nm have been investigated. The shorter wavelength radiation produces only epidermal carcinomas while the longer wavelength radiation produces sarcomas in the dermis, more or less what would be anticipated. These findings then suggest that radiation absorbed in particular layers gives rise to cancers localized close to the site of photon absorption. It therefore becomes apparent that with erythema and presumably cancer induction by sunlight we are concerned with the dermal tissues to a large extent.

We can say a little on the mechanism of sarcoma induction. Currently held ideas suggest that carcinogenesis by ultraviolet light involves a component of the action which results in genetic change in tissue cells. The source of the genetic specification within a cell is encoded in the nucleotide sequences of the deoxyribonucleic acid (DNA) of the chromosomes. There may be only a few copies of this cell-building blueprint in a cell so that damage to the DNA chain is serious and can lead to mutation or death. It might therefore be considered that DNA represents a highly sensitive target for ultraviolet radiation penetrating the epidermis. Figure 111 shows the absorption or biological sensitiveness spectrum of DNA in conjunction with the erythemal-effectiveness spectrum, the absorption cross-section of ozone and the solar photon flux density spectrum. It is clear that the biological sensitivity of DNA increases markedly with decreasing wavelength across the shortest wavelengths of the solar flux, with the region around a wavelength of 305 nm being particularly important because of the maximum values of the product of the photon flux density and the biological sensitivity (or the erythemal effectiveness).

It is known that the absorption of ultraviolet radiation results in damage to the DNA chain. This damage may take such forms as localized denaturation or induced protein cross-linking across the two strands of the famous double helix structure; alternatively there may be dimerization of pyrimidine components of the chains with consequent destruction of hydrogen-bonding links between the two strands. All of these would be expected to interfere with the DNA replication mechanism in which the two complementary strands of the double helix uncoil as daughter complementary chains are synthesized. Repair processes must be operative, otherwise life could not exist in hereditary forms even with present levels of the solar ultraviolet flux. One process is what is called *enzymic photoreactivation*, induced by radiation of wavelengths longer than 320 nm. The enzyme involved in this photoreactivation binds to ultraviolet-da..naged DNA in the dark and upon absorption of a suitable photon by the complex, processes like the scission of pyrimidine dimers are induced *in situ* with the release of the enzyme. The radiation photon hence provides the energy for effecting the repair and the photoreactivation mechanism is thought to be rather specific for this dimer splitting. There also appears to be a dark repair process, where the complementary intact strand of the damaged DNA double helix acts to recover the information destroyed in its partner, utilizing the highly specific base-pairing relationship between complementary

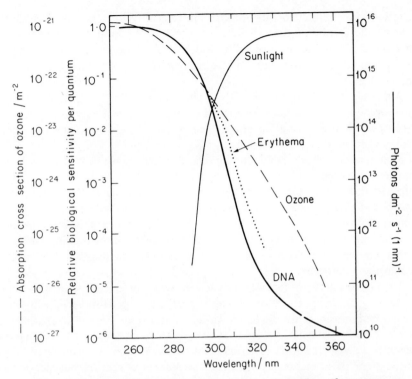

Figure 111 Combination diagram of the optical cross-section of ozone as a function of wavelength, the relative biological activity spectrum for DNA and for erythema and the transmitted radiation (sunlight) as functions of wavelength for an equivalent thickness of ozone layer of 2.3 mm and solar angle of 25°. Reprinted with permission from H. S. Johnston, 'Ground-level effects of supersonic transports in the stratosphere', *Accounts of Chemical Research* **8**, 293 (1975). Copyright by the American Chemical Society.

strands. The mechanism acts in what is termed a 'cut and patch' mode, wherein the damaged section is somehow excised and a newly synthesized segment is incorporated in its place. Enzymes must play a prominent role in this dark repair–replication process. There can be little doubt that further repair processes are also operative.

Photosensitization of DNA, as opposed to direct absorption of photons by DNA itself, may also have a role in extending the damaging effects of ultraviolet radiation out to longer wavelengths. In this another molecule present in the cell absorbs the photon and subsequently either transfers its energy to, or reacts with, a neighbouring part of a DNA strand, inducing damage.

An obvious deduction is that a balance must become established in living cells between the rate of radiation damage and the rate of repair. Just as nature has evolved life around DNA with its high sensitivity to solar ultraviolet radiation, so it has at the same time evolved efficient repair mechanisms for renewal of most or all of the damaged sections of the DNA molecule. Indeed the generation of the latter facilities was essential for life to have the capability of hereditary reproduction through the agency of genes. In the prehistory of the Earth's living cycles, when photosynthesis began in plant life which was of necessity exposed to solar irradiation, the development of the cell repair mechanisms was of crucial importance.

Although the link between DNA photodamage in living tissue cells and carcinogenesis by radiation is somewhat empirical, the relationship between these two aspects and the mutation of cells and genetic change would seem to point in the right direction in general terms.

Skin cancer is clearly related to the solar flux density which is incident. Its most frequent occurrence is on parts of the body exposed to highest incident and reflected fluxes, for example, the head, neck, arms and hands. Pigmented races are less subject to erythema than are the white races and accordingly are less prone to skin cancer. As might be expected also, the highest incidence of the condition occurs in white-skinned peoples living in areas of the highest insolation on a daily basis within the year, for example, the desert areas of the southwestern United States and northern Australia.

To put matters in further perspective, we may remark that skin cancer, accounting for at most 2% of all cancer deaths among white races, is by far the most common cancer of all. In the United States it has been estimated that skin cancer causes some 5000 deaths annually. There is some evidence that total exposure to solar radiation of wavelengths below 320 nm at locations at different latitudes correlates reasonably well with the number of skin tumours per person at that location.

Hence we are led to focus our attention on solar radiation of wavelengths in the vicinity of 305 nm as being the most critical in connection with any depletion of the ozone layer. At this wavelength the absorption coefficient of ozone may be taken as $\delta = 0.23 \text{ mm}^{-1}$, to apply in the form of equation (2.5). The present attenuation factor for a 3 mm equivalent thickness of the ozone layer is then given as:

$$\frac{I_{3.0}(305 \text{ nm})}{I_0(305 \text{ nm})} = 10^{-0.23 \times 3.0} = 0.204$$

where $I_{3.0}(305 \text{ nm})$ is the photon flux density of wavelength 305 nm transmitted by an equivalent thickness of 3.0 mm of ozone. The corresponding attenuation factors at wavelengths of 300 nm and 310 nm are 0.059 and 0.44, respectively, on the basis of the absorption coefficients given in Table 2.1. The Rayleigh optical thicknesses of the atmosphere, given in Table 2.5, give additional molecular-scattering attenuation factors of 0.304, 0.326, and 0.353 at the wavelengths of 300 nm, 305 nm, and 310 nm respectively. With the approximation that half of the light scattered eventually reaches the Earth's surface, the effective attenuation factors become 0.65, 0.66, and 0.68 at the three respective wavelengths. It must be remembered that the vast bulk of the atmosphere lies below the tropopause so that this molecular scattering occurs after the attenuation by the stratospheric ozone layer. In fact, for the radiation of wavelength 300 nm only about 13% of the photons are scattered above 15 km altitude after passage through about 12% of the total atmospheric column. So at a wavelength of 305 nm, at present only some 13% of the extraterrestrial solar photon flux density penetrates to the surface for a vertical path through the atmosphere, with the ozone column being the main source of the attenuation.

Consider now the effect of a 10% decrease in the equivalent thickness of the ozone layer. The ozone attenuation factors at the wavelengths of 300 nm, 305 nm, and 310 nm will now be 0.078, 0.239, and 0.474 respectively, which are to be multiplied by the same molecular scattering attenuation factors as before. The penetration of the extraterrestrial solar photon flux density at 305 nm wavelength is now almost 16%, so that the 10% decrease in the ozone layer has produced a disproportionate 17% rise in the terrestrial solar photon flux density of wavelength 305 nm. At 300 nm wavelength the effect is even more disproportionate with the rise being close to 32%. Figure 112 shows plots of the product of the erythemal effectiveness and solar photon flux density at the surface for ozone layer equivalent thicknesses of 3.0 and 2.7 mm as a function of wavelength.

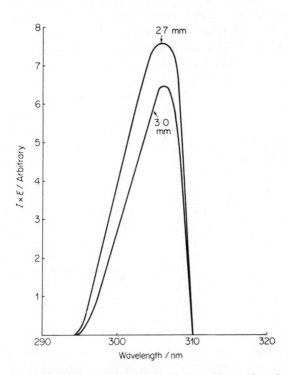

Figure 112 Plots of the products of relative erythemal effectiveness (ϵ) and solar photon flux density (I) at ground-level for overhead Sun and equivalent thickness ozone columns of 3.0 mm and 2.7 mm.

The area within the envelope in each case can be expected to be related to the probability for the incidence of skin cancer. It turns out that the above 10% decrease in the effective thickness of the ozone layer results in a near 25% increase in the area contained by the profile of the product. The disproportionate increase is partly the result of the evident broadening of the envelope on the short wavelength side, where the ozone layer attenuation effect is most severe. More or less in line with this, recent estimates for the United States have suggested a minimum increase of 8000 cases of skin cancer per year for every 1% decrease in the average thickness of the ozone layer. This then emphasizes the potential seriousness of the problem posed by the continuing use of chlorofluorocarbons.

9.6 Chemistry of the Atmosphere above the Stratopause

The concepts of full photochemical equilibrium cannot be applied in the mesosphere and beyond since diffusion processes occur on a time scale comparable with that for quite a few of the chemical conversions. Accordingly the detailed computational procedures which are used to obtain concentration profiles of species as functions of altitude are beyond our present scope.

An approach whereby we accept the results of a typical computation and examine the inferred rates of the various chemical reactions can prove interesting and instructive. We shall adopt as our framework the concentration profiles for hydrogen—oxygen species developed by Hunt and reported in the paper cited in connection with Figure 113 to follow. Some of these have already appeared in Figure 96(b); the effects of diffusional processes upon the purely photochemical equilibrium concentration profiles is emphasized in Figure 26. The nitrogen chemistry is largely independent of the hydrogen—oxygen cycles and will be considered separately for clarity. Moreover, because carbon dioxide is virtually the only carbon-containing species with any significant mixing ratio in the mesosphere, and is only subject to photodissociation by very short wavelength radiation (section 9.2), we may ignore carbon chemistry to a large extent.

The computed profiles of the concentrations of species are shown in Figure 113. Confidence in these concentration profiles is generated by the general agreement with isolated measurements of some of the concentrations from rocket-borne instrumentation. For example, hydroxyl-radical concentrations in the altitude range 45 km to 70 km have been measured by a resonance fluorescence technique near ground sunset (Anderson, 1971). The profile showed little variation and at 70 km the concentration was in the range of $(3 - 12) \times 10^{-15}$, with a mean value of 7×10^{-15} mol dm^{-3}, in good agreement with the concentration for 70 km altitude shown in Figure 113. Atomic oxygen concentration profiles for nighttime conditions above 80 km altitude have also been measured using a resonance fluorescence technique combined with vacuum ultraviolet absorption spectrometry calibration. These are shown in Figure 114 and compared favourably with the midnight concentration profile for atomic oxygen shown in Figure 113. In particular there is good agreement between the two profiles on the maximum in the concentration profile in the vicinity of 90 km altitude.

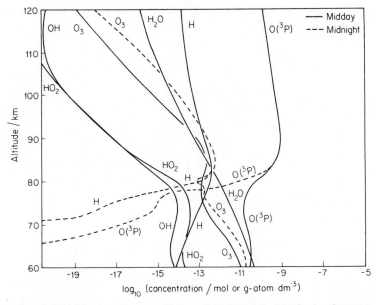

Figure 113 Concentration profiles as functions of altitude in the upper atmosphere as computed by Hunt. (Full lines: midday; dashed lines: midnight). From B. G. Hunt, 'Photochemical heating of the mesosphere and lower thermosphere', *Tellus,* **XXIV**, 48 (1972). Reproduced by permission of the Swedish Geophysical Society.

Eddy diffusion is the predominant mode of vertical mass transport below the turbopause near 105 km altitude (preamble to Chapter 3) so that it will be the eddy diffusion coefficient, K_v, which will determine the characteristic transport time defined in equation (3.13), or more specifically for the mesosphere as:

$$\text{Characteristic Transport Time } (\tau_v) = H^2/K_v \qquad (9.12)$$

H is the scale height defined by equation (3.12) and will vary slightly from one species to another because of the appearance of the molecular weight. The gravitational constant, g, does not vary much as a function of altitude, being 9.80 m s^{-2} at ground level and 9.50 m s^{-2} at 100 km altitude. Temperature is therefore the main parameter determining the variation of scale height with geometrical altitude. In the mesosphere at altitude 80 km the Standard Atmosphere temperature is 181 K (Table 9.1) so that for the average molecule the scale height, H, is about 5.5 km. Values of the eddy diffusion coefficient, K_v, in this region are taken generally to be of the order of $10^2 \text{ m}^2 \text{ s}^{-1}$, so that the characteristic transport time, τ_v, is of the order of 3×10^5 s. The photochemical characteristic time for a species, τ_c, may be defined as the concentration of that species at a particular altitude divided by the rate of generation or rate of removal of that species assuming photochemical equilibrium. The question of which process, mass transport or chemical conversion, is more important in determining the concentration profile of a species depends upon the relative magnitudes of the two

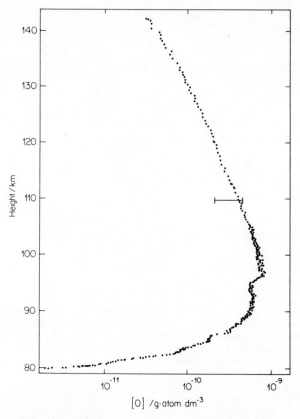

Figure 114 O(^{3}P) atom concentration profile versus height measured near midnight above South Uist, Scotland, on April 1, 1974. The error bar indicates the maximum range of uncertainty in the absolute scale. From P. H. G. Dickinson, R. C. Bolden and R. A. Young, 'Measurement of atomic oxygen in the lower ionosphere using a rocket-borne resonance lamp', *Nature*, **252**, 290 (1974). Reproduced with permission of Macmillan (Journals) Ltd.

characteristic times. For a photochemical steady state to operate, the photochemical charactistic time must be considerably shorter than the characteristic transport time. If these are of comparable magnitude then there will be a photochemical steady state modified substantially by diffusion.

Let us examine the feasibility of a photochemical steady state for 'odd oxygen' on the above basis. The generation of 'odd oxygen' is accomplished by the photodissociation of molecular oxygen and, on examining Figure 100, we find a photodissociation rate coefficient, J_{O_2}, of around 10^{-8} s^{-1} for an altitude of 80 km. The corresponding concentration of molecular oxygen is derived from Table 9.1 by multiplying [M] by 0.2095 to give 1.4×10^{-7} mol dm^{-3}. Hence the rate of generation of 'odd oxygen' at this altitude for overhead Sun conditions is

$2 J_{O_2} [O_2] = 2.8 \times 10^{-15}$ g-atoms dm^{-3} s^{-1}. The midday concentration of 'odd oxygen' is close to being that of atomic oxygen, approximately 5×10^{-11} g-atoms dm^{-3}. Hence the characteristic photochemical time, τ_c, as defined above, is close to 2×10^4 s at 80 km altitude for 'odd oxygen', about an order of magnitude less than the characteristic transport time. Therefore one would tend to consider that the modification of the 'odd oxygen' photochemical equilibrium would be slight at altitudes of 80 km or less. However, on ascending only 10 km in altitude, there is a dramatic change in the status of the photoequilibrium. Since neither the scale height nor the eddy diffusion coefficient alter significantly from 80 to 90 km, τ_v is 3×10^5 s. However it may be observed in Figure 113 that the oxygen atom concentration rises sharply from 80 to 90 km altitude by a factor of close to 30. At the same time J_{O_2} increases by a factor of around 5, reflecting mainly the rapid increase with altitude in the photon flux density within the Schumann–Runge continuum, but this, in part offset by the decrease in the molecular oxygen concentration, is not enough to prevent τ_c rising sharply to around 7×10^5 s. It is therefore clear that diffusion takes over from photochemical equilibrium in controlling the concentrations of 'odd oxygen' species in this fairly narrow altitude range including the mesopause.

The source of 'odd hydrogen' at 80 km altitude is the photodissociation of water vapour into a hydrogen atom and a hydroxyl radical, both contributing to 'odd hydrogen'. In Figure 100 we may read off a value of J_{H_2O} of 2×10^{-6} s^{-1}, largely due to Lyman α radiation at wavelength 121.5 nm. The corresponding concentration of water vapour is probably somewhat variable but might be taken as 10^{-12} mol dm^{-3} for a reasonable average. Hence the production rate of 'odd hydrogen' is $2 J_{H_2O} [H_2O] = 4 \times 10^{-18}$ g-atom dm^{-3} s^{-1}. In Figure 113 we can see that several species contribute significantly to the 'odd hydrogen' concentration, notably H with lesser contributions from OH and HO$_2$. A concentration of 3×10^{-13} g-atoms dm^{-3} s^{-1} may be derived for 'odd hydrogen'. On this basis the photochemical characteristic time, τ_c, comes out as 8×10^4 s, not much less than the characteristic diffusion time of 3×10^5 s. Hence even at 80 km altitude, the 'odd hydrogen' photochemical equilibrium will be subject to some modification due to diffusion. In the same way as was argued for 'odd oxygen' above, we may expect diffusive processes to be dominant for 'odd hydrogen' above the mesopause.

Despite the above interventions of physical perturbations to a purely chemical approach to the mesosphere as a whole, we can see that a photochemical equilibrium model will serve as a good enough approximation for the purpose of gaining an insight into the factors which produce the concentration profiles with altitude in the lower part of Figure 113, and some of the associated airglow phenomena. Obviously this approximation will not serve in the thermosphere above the mesopause.

The Day and Night Concentration Profiles of the Atoms

The main reactions which will remove hydrogen atoms and oxygen atoms overnight are the three-body recombinations represented by the equations:

$$O + O_2 + M \longrightarrow O_3 + M \tag{16}$$

$$H + O_2 + M \longrightarrow HO_2 + M \tag{17}$$

As is indicated by the oxygen atom concentration profile in Figure 114, and by the two sets of lines in Figure 113, oxygen atoms almost disappear overnight below 80 km altitude but persist at almost unchanged levels above 90 km. It may be presumed that hydrogen atoms behave similarly in view of the parallel nature of reactions (17) and (16). We should explore why this difference should set in so suddenly with increasing altitude.

This rate constant k_{16} has a value of the order of 6×10^8 dm^6 mol^{-2} s^{-1} at 80 km altitude. From Table 9.1 we see that the total species concentration, $[M]$, drops by a factor of 6.33 from 80 to 90 km altitude; the coreactant concentration product, $[O_2][M]$, will therefore decrease by a factor of 40 across this range. The integrated rate law for reaction (16) is expressed by the equation:

$$\ln([O]_0/[O]) = k_{16}[O_2][M] \cdot t \tag{9.13}$$

where $[O]_0$ is the oxygen atom concentration at, say, sunset for 80 km altitude and $[O]$ is the residual concentration after time t has elapsed. Insertion of the above values of the rate constant and concentrations shows that 90% decay of the oxygen atom concentration requires around 10 h. It is hence apparent that at this altitude the oxygen atoms should not be able to disappear completely overnight. On the other hand, at 70 km altitude the total species concentration is over four-times larger so that the time scale for 90% decay is contracted by a factor of almost 20 to around 30 min; oxygen atoms at this altitude should virtually disappear overnight. However, at 90 km the time scale for 90% removal has expanded to nearly 18 days so that the oxygen atoms here should persist overnight largely undiminished.

Reaction (16) is not a sink for 'odd oxygen', but it is followed by reactions which accomplish the formation of molecular oxygen as represented by the equations:

$$H + O_3 \longrightarrow OH + O_2 \tag{51}$$

$$O + OH \longrightarrow O_2 + H \tag{-1}$$

Both of these reactions have rate constants approaching 10% of the collision frequency. At the same time reaction (17) generates the hydroperoxy radical and subsequent reactions of this serve as sinks for both 'odd oxygen' and 'odd hydrogen'. The main reactions are represented by the equations:

$$O + HO_2 \longrightarrow OH + O_2 \tag{39}$$

$$H + HO_2 \longrightarrow H_2O + O \tag{52}$$

$$H + HO_2 \longrightarrow 2\,OH \tag{53}$$

$$H + HO_2 \longrightarrow H_2 + O_2 \tag{54}$$

$$OH + HO_2 \longrightarrow H_2O + O_2 \tag{41}$$

Table 9.15. Estimated rate parameters for 'odd oxygen' and 'odd hydrogen' conversion in the daytime mesosphere at 80.km altitude

Reaction			Rate constant/ $dm^3\ mol^{-1}\ s^{-1}$	Rate/ $mol\ dm^{-3}\ s^{-1}$
$H + O_3$	$\longrightarrow$	$OH + O_2$ (51)	2×10^{10}	2×10^{-15}
$O + OH$	$\longrightarrow$	$O_2 + H$ (−1)	2×10^{10}	2×10^{-15}
$O + HO_2$	$\longrightarrow$	$OH + O_2$ (39)	3×10^{9}	3×10^{-16}
$H + HO_2$	$\longrightarrow$	$H_2O + O$ (52)	3×10^{9}	2×10^{-18}
$H + HO_2$	$\longrightarrow$	$2\ OH$ (53)	1×10^{9}	5×10^{-19}
$H + HO_2$	$\longrightarrow$	$H_2 + O_2$ (54)	4×10^{9}	2×10^{-18}
$OH + HO_2$	$\longrightarrow$	$H_2O + O_2$ (41)	$\leqslant 1 \times 10^{11}$	$\leqslant 5 \times 10^{-19}$

Of these reactions, (52), (54), and (41) act as sinks for 'odd hydrogen', while (39) in conjunction with (−1) acts as a sink for 'odd oxygen'. Reaction (53) is effectively neutral if followed by reaction (−1) but serves as a sink for 'odd hydrogen' if followed by reaction (41). It is evident that we must consider the relative rates of this array of reactions in order to gain some idea of the likely overall effect under mesospheric conditions. Table 9.15 sets out estimates of the rate parameters.

The data of Table 9.15 clarify the dominant roles of reactions (51) and (−1) as bimolecular reactions in the upper mesosphere. Moreover, it becomes clear that reaction (39) accounts for most of the destruction of the hydroperoxy radical formed by reaction (17). This latter point is important since it means that reaction (17), followed overwhelmingly by reactions (39) and (−1), is more of a sink for 'odd oxygen' than for 'odd hydrogen'. The relative rates of reactions (16) and (17) at an altitude of 80 km at the end of the daylight hours are approximately as 2:1. Therefore we should expect atomic oxygen concentrations to decrease somewhat faster at sunset than in our above estimate, and considerably faster than do atomic hydrogen concentrations. Nevertheless since at least 2% of the hydroperoxy radicals are destroyed by reactions (52), (54), and (41) at the outset, and the fraction will obviously increase as the oxygen atom concentration falls, it is clear that hydrogen atom concentrations below 80 km altitude will tend to follow the oxygen atom concentrations but on a significantly longer time scale.

The Kink in the Ozone Concentration Profile near 75 km Altitude

It has become apparent that the concept of photochemical equilibrium can be applied to the mesosphere at altitudes around 75 km, without the introduction of any great error arising from the ignoring of atmospheric motions.

The steady state approximation of 'odd oxygen' during the midday period will generate the equation:

$$J_{O_2} [O_2] = k_{-1} [O] [OH] \tag{9.14}$$

This equation is restricted to the major rates, as indicated by the data in Table 9.15, and evidently the hydroxyl radical is to be regarded as a component of 'odd oxygen'. Now the left hand side of equation (9.14) decreases with increasing altitude since the molecular oxygen concentration decreases more rapidly with increasing altitude than J_{O_2} increases, a point which has been made earlier in this section (the product $J_{O_2}[O_2]$ goes through a maximum with increasing altitude somewhere in the vicinity of 40 km). The oxygen atom concentration would not be expected to do other than increase with altitude in this region; the rate of removal in reaction (16) falls far more sharply with altitude increase than could the production rates from the photodissociation of ozone and molecular oxygen. Therefore equation (9.14) explains the sharp fall in the hydroxyl radical concentration as the oxygen atom concentration rises sharply with increasing altitude above 80 km, as shown in Figure 113.

In Figure 100 it can be seen that the photodissociation coefficient for water vapour rises strongly with increasing altitude in the range 70 to 85 km, which largely reflects the increasing Lyman α photon flux density (see Figure 92). In fact, the value of J_{H_2O} increases by a factor of 6.4 across this range which almost offsets the factor by which the water vapour concentration falls. Hence $J_{H_2O}[H_2O]$ can be regarded as being almost constant in the upper mesosphere. At the same time the rate of the main removal reaction for hydrogen atoms, reaction (17), will decrease by a factor of 110, solely on the basis of the decrease of molecular oxygen and total species concentrations from 70 to 85 km altitude. Therefore the hydrogen atom concentrations must rise sharply across this range, as is seen to be the case in Figure 113.

Ozone as such is destroyed in two ways in the mesosphere: by photodissociation and by the attack of hydrogen atoms in reaction (51). Figure 100 shows that the photodissociation rate coefficient. J_{O_3}, is constant throughout the mesosphere. However, because of the rising hydrogen atom concentration, the effective competing first-order rate constant, $k_{51}[H]$, will be increasing sharply with altitude. The steady-state concentration of ozone is expressed by the equation:

$$[O_3] = k_{16}[O][O_2][M]/(J_{O_3} + k_{51}[H]) \qquad (9.15)$$

In Figure 100 we may read off $J_{O_3} = 10^{-2}$ s^{-1} and k_{51} has a value of the order of 2×10^{10} dm^3 mol^{-1} s^{-1}. Hence the two terms on the denominator of equation (9.15) will tend toward comparable magnitudes for hydrogen atom concentrations exceeding 5×10^{-14} g-atom dm^{-3}, as is the case near 75 km altitude since the concentrations in Figure 113 are in excess of 10^{-13} g-atom dm^{-3}. On the other hand, the hydrogen atom concentration at 60 km altitude is less than 10^{-14} g-atom dm^{-3} so that there the only effective term in the denominator of equation (9.15) is the constant J_{O_3}. Hence the pattern which emerges is that up to 70 km altitude the ozone concentration profile is largely determined by the numerator product $[O_2][M]$. Then, with further increase in altitude, reaction (51) in effect 'takes a bite out of' the ozone profile, and the gradient of the ozone concentration versus altitude profile becomes more negative. This trend is halted at 80 km or so by the rising oxygen atom concentrations with increasing altitude, resulting in a reversal of the sign of the gradient of the ozone concentration profile in the layer between 80 and 85 km. Above the mesopause, the numerator product $[O_2][M]$ again becomes the dominant variable with altitude, while the denominator terms remain of comparable magnitude. At the same time, diffusion also becomes a dominant influence so that equation (9.15) progressively deviates from the real situation with increasing altitude.

Photoequilibrium Concentration Ratios

The rate parameters shown in Table 9.15 make it evident that the steady-state concentration of hydroxyl radicals in the upper mesosphere is primarily determined by reactions (51) and (−1). On this basis we may equate the rates:

$$k_{51}[H][O_3] = k_{-1}[O][OH]$$

This equation may be recast to yield the proportionality:

$$[H]/[OH] \propto [O]/[O_3] \tag{9.16}$$

The right hand side of this relationship increases sharply with increasing altitude above 75 km on the basis of the profiles shown in Figure 113. Since hydrogen atom concentrations vary relatively little with altitude above this altitude, the dramatic fall-off of the hydroxyl radical concentrations above 80 km is inferred.

The rate parameters of Table 9.15 also show that the primary determinants of the hydroperoxy radical concentration are reactions (17) and (39). Hence we may write down another approximate equality:

$$k_{17}[H][O_2][M] = k_{39}[O][HO_2]$$

This may be recast into the proportionality:

$$[HO_2]/[H] \propto [O_2][M]/[O] \tag{9.17}$$

From what has been said before, it is clear that the numerator on the right hand side of (9.17) will be the dominant variable and will decrease rapidly with increasing altitude. Hence the rapid decrease of hydroperoxy radical concentrations above about 80 km altitude can be anticipated. Combination of the two proportionalities (9.16) and (9.17) yields the further relationship:

$$[HO_2]/[OH] \propto [O_2][M]/[O_3] \tag{9.18}$$

Now equation (9.15) reveals that the product $[O_2][M]$ is a dominant variable in determining the ozone concentration profile. Hence it is to be expected that the concentration profiles with altitude of hydroxyl and hydroperoxy radicals would be similar, as is shown in Figure 113.

The Airglow

The night sky is not dark even if the light from the moon and stars is excluded. There is a weak background emission originating from the mesosphere, and the lower parts of the thermosphere, which is a permanent feature and not to be confused with occasional phenomena like aurorae. The background emission spectrum extends from the ultraviolet, through the visible, to the infrared region and is composed of bands and lines. The emissions in the infrared are the most prominent and spectral analysis has shown that the emitting species is predominantly a vibrationally excited hydroxyl radical, in levels up to and including $v = 9$. This emission has been termed the *Meinel bands* and its principal components have been identified as the $\Delta v = 2$ sequence [i.e. (9,7), (8,6) etc.] of the hydroxyl radical appearing in the wavelength range up to 2200 nm, and the $\Delta v = 1$ sequence extending from 2650 nm to longer wavelengths. Figure 115 illustrates the night airglow spectrum from 1200 nm to 2300 nm, showing bands of the $\Delta v = 2$ sequence clearly.

Also notable in this spectrum is the $O_2(^1\Delta_g \rightarrow {}^3\Sigma_g^-)$ emission in the vicinity of 1270 nm wavelength from which the concentration profile of $O_2(^1\Delta_g)$ as a function of altitude shown in Figure 96(a) was obtained.

Further progressions, $\Delta v = 3,4,5$, of the hydroxyl radical emission are also detected at wavelengths closer to the visible part of the spectrum, for example the (9,4) band is detected near 770 nm while the (9,5) band is found near 990 nm wavelength.

The chemical origin of the Meinel band emission has been demonstrated conclusively in laboratory studies in discharge flow systems. When flowing hydrogen atoms are mixed with flowing ozone at very low pressures and temperatures around 200 K, almost exactly the same set of bands are found in the spectrum of the resultant chemiluminescent emission. This leaves no doubt that the overwhelming source of vibrationally excited hydroxyl radicals is reaction (51), which may be written more explicitly from this point of view as:

$$H + O_3 \longrightarrow OH^* + O_2 \tag{51}$$

The asterisk denotes the vibrationally excited nature of the product. The exothermicity of reaction (51) to produce ground state hydroxyl radicals in the lowest vibrational level is 322 kJ mol^{-1}. The excitation energy of the ninth vibrational level (the highest observed) with respect to the lowest level of the hydroxyl radical is almost exactly this available amount of energy. This explains the absence of any emission from $v = 10$ or larger. The laboratory experiments, moreover, indicate that some 90% of the initially produced OH* radicals are almost equally divided between the two highest possible levels, $v = 9$ and $v = 8$, emphasizing the efficiency of reaction (51) in the conversion of chemical energy into vibrational energy.

The mean radiative lifetime of OH($v = 9$) is of the order of 0.06 s and the vibrational deactivation rate by colliding air molecules corresponds to an efficiency of about one collision in 10^5 resulting in vibrational deactivation or a rate constant of 3×10^6 dm^3 mol^{-1} s^{-1}. At an altitude of 80 km, the total concentration of molecules is approximately 7×10^{-7} mol dm^{-3}, so that the effective rate constant on a first order basis for deactivation by collision is about 2 s^{-1}, quite a bit less than the radiative rate constant of 17 s^{-1}, the reciprocal of the mean radiative lifetime. Hence the bulk of the vibrationally excited hydroxyl radicals formed in the

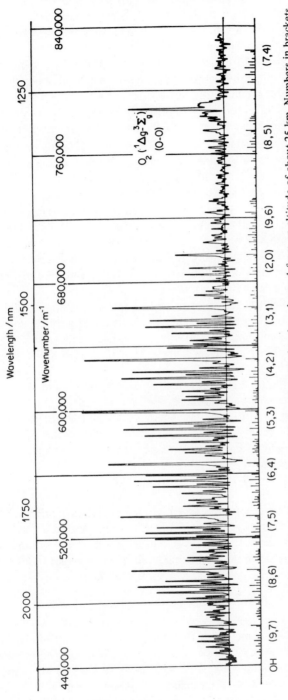

Figure 115. Night-time airglow spectrum in the near infrared spectral region observed from an altitude of about 25 km. Numbers in brackets below identify Meinel bands of OH. Reproduced by permission of the National Research Council of Canada from H. P. Gush and H. L. Buijs, *Canadian Journal of Physics*, **42**, 1037–1045 (1964).

upper mesosphere will cascade down the vibrational levels emitting infrared photons in the course of each stage. This is the origin of the emission from all levels seen in Figure 115, even though reaction (51) populates only the two highest levels to a large extent in the first place.

As would be expected from our earlier conclusion that hydrogen atoms persist through the night above 80 km altitude, the Meinel bands only show a fall in overnight intensity of around 50%. However it is interesting that this intensity falls sharply by a factor of around 3 at sunrise, which cannot be a reflection of any change in hydrogen atom concentrations. Since the intensity of Meinel band emission is simply proportional to the rate of reaction (51) and in turn, therefore, to the ozone concentration in the mesosphere, the origin of the dawn decrease is more obvious. Overnight we would expect ozone concentrations to increase since the molecule is no longer subject to photodissociation. This is seen to be the case in Figure 113. However, at dawn there is a sudden recommencement of photodissociation of ozone so that the concentration will drop quite quickly, taking the Meinel band intensity down with it. Subsequently, over the first few hours of daylight, the Meinel band intensity recovers to near its nocturnal level, an effect due to the recovery of atomic hydrogen concentrations in the mesosphere below 80 km and the renewed onset of reaction (51).

Rocket-borne spectrometers have measured the integrated intensity of Meinel-band emission in the vertical direction as a function of altitude, thus producing profiles of the volume emission rate. Recent results indicate that the hydroxyl emission is concentrated in a layer some 10 km thick centred on an altitude of around 86 km. This is what would be expected from the hydroxyl radical concentration profile of Figure 113, related strongly to the rate of reaction (51), and hence to the atomic hydrogen and ozone concentration profiles.

In principle the measured volume emission rate profile for the Meinel bands should provide information on the concentration profiles of the reactant species as well as hydroxyl radicals themselves. We may express the emission rate per unit volume (I/V) by the equation:

$$I/V = X \cdot Y \cdot k_{51} [H] [O_3]$$

X is a factor which takes account of the multiple generation of photons by cascading following the initial act of reaction (51); recent estimates suggest that X has a value close to 5. The other factor, Y, is introduced to correct for the fact that as the OH* species are cascading down the vibrational levels, they are under attack from ground state oxygen atoms in the predominant hydroxyl radical removal reaction (-1):

$$O + OH^* \longrightarrow O_2 + H \tag{-1}$$

On the basis that k_{-1} does not vary with the degree of vibrational excitation of OH* and preserves the value of 2×10^{10} dm^3 mol^{-1} s^{-1} for the lowest vibrational level radical, we can estimate the significance of Y. The oxygen atom concentration at 80 km altitude is around 5×10^{-11} mol dm^{-3}, so that the effective first-order rate constant represented by the product $k_{-1}[O]$ has a value of 1 s^{-1}, acting in competition with the radiative decay with rate constant 17 s^{-1}. It then follows that for the average five-step radiative decay to the lowest vibrational level, the depletion effected by reaction (−1) is of the order of 25% so that emission from the lowest vibrational levels should show a significant effect. This has been detected in a recent rocket-borne study of the altitude variation of the intensity of emission from higher vibrational levels in the wavelength region 1850 nm to 2120 nm concurrently with that from the lower vibrational levels in the region 1400 nm to 1650 nm (Rogers and coworkers, 1973). The intensity of the latter emission fell away with increasing altitude above 80 km compared with that of the former, in line with the sharp rise in oxygen atom concentrations at this altitude which are evident in Figure 113. The atomic oxygen concentration profile which was found to be consistent with the measured volume emission rates of the Meinel bands in the altitude range 82 km to 88 km was very similar to the comparable part of Figure 114 under comparable nighttime conditions.

Another strong feature of the airglow is the so-called Auroral Green Line of atomic oxygen, located at a wavelength of 557.7 nm and corresponding to the transition $O(^1S \rightarrow {}^1D)$ between the first two electronically excited states of atomic oxygen. This transition is 'difficult' in the sense that it contravenes the $\Delta L = 0,1$ selection rule discussed in Section 8.2 and the other potential transition to the ground state, ^3P, will also be 'difficult' because it contravenes the spin conservation selection rule. Therefore the long radiative lifetime of $O(^1S)$ of 0.74 s^{-1} is to be expected. The corresponding radiative decay rate constant is then small enough to allow quenching processes by other species to be significant fates of $O(^1S)$ atoms. The principal quenching processes are represented by the equations:

$$O(^1S) + O(^3P) \longrightarrow 2\,O(^3P) \tag{55}$$

$$O(^1S) + O_2(^3\Sigma_g^-) \longrightarrow O(^3P) + O_2 \tag{56}$$

Figure 116 shows the altitude variation of the Auroral Green Line emission intensity as measured at night over Manitoba, Canada using a rocket-borne photometric detector.

In the layer near 86 km altitude, from which the maximum volume emission rate comes, the atomic oxygen concentration is about two orders of magnitude less than that of molecular oxygen, as may be seen in Figure 113. Currently available measurements of the rate constants for the above quenching processes suggest average values of $k_{55} = 9 \times 10^8 \text{ dm}^3 \text{ mol}^{-1} \text{ s}^{-1}$ and $k_{56} = 4 \times 10^7 \text{ dm}^3 \text{ mol}^{-1} \text{ s}^{-1}$ for the temperature of 181 K. On this basis, the rate of reaction (56) is around 4.5-times larger than that of reaction (55) at this altitude of maximum emission. At the same time it is to be noted that the two concentrations move towards one another with further increase of altitude, so that we may expect reaction (55) to become the predominant removal route for $O(^1S)$ atoms in the lower thermosphere. The effective first order rate constant representing the combined effects of reactions (55) and (56) in the vicinity of 86 km altitude has a value of the order of 3 s^{-1}, to be compared with the radiative rate constant of 1.35 s^{-1}.

The Chapman mechanism is the classical theory for the excitation of the Auroral green line near the mesopause. This is an elementary reaction represented by the

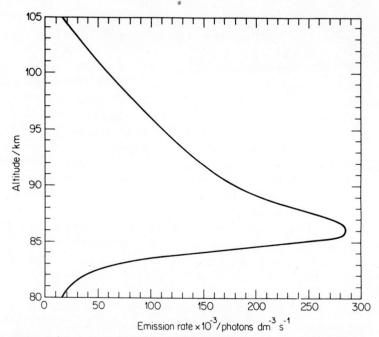

Figure 116. Altitude dependence of the emission rate of the auroral green line at wavelength 557.7 nm overnight above Manitoba, Canada, on August 20, 1968. From B. S. Dandekar, *Planetary Space Science*, **20**, 1782 (1972). Reproduced by permission of Pergamon Press Ltd.

equation:

$$O + O + O \longrightarrow O(^1S) + O_2 \tag{57}$$

In this, it is considered that three $O(^3P)$ ground state atoms interact so that the energy released by the combination of two finds its way, in part, into electronic excitation of the third into the $O(^1S)$ state. The excitation energy of $O(^1S)$ from $O(^3P)$ is 402.8 kJ mol^{-1}, which corresponds to a large fraction of the recombination energy of 493.6 kJ mol^{-1} of molecular oxygen formation.

We may consider how the Chapman mechanism accords with the observed emission profile as a function of altitude. The steady state approximation will apply to $O(^1S)$ atoms and leads to an equation:

$$k_{57}[O(^3P)]^3 = (k_R + k_{55}[O(^3P)] + k_{56}[O_2])[O(^1S)]$$

from which is derived the expression for the state concentration of $O(^1S)$ atoms:

$$[O(^1S)] = k_{57}[O(^3P)]^3/(k_R + k_{55}[O(^3P)] + k_{56}[O_2])$$

The volume emission rate of the auroral green line is then simply given by the right hand side of this equation multiplied by k_R, the radiative rate constant. From the above assessment of the relative sizes of the denominator terms, it is clear that the volume emission rate should show a more than square power dependence upon the concentration of ground-state oxygen atoms. This then explains why the maximum volume emission rate is detected at roughly the same altitude as that where the ground-state oxygen atom concentration profile shows its maximum. Moreover the peak in the volume emission rate profile with altitude would be expected to be much sharper than that in the oxygen atom concentration profile; this is evidently the case when Figures 113 and 116 are compared.

We should therefore be able to calculate the volume emission rate at around the maximum for comparison with the measured value of Figure 116. Unfortunately k_{57} has only been measured at 300 K as around 2×10^9 dm^6 mol^{-2} s^{-1} and it is uncertain as to how much this value will be altered at 181 K. However, application of this value of this rate constant for the conditions appropriate to 86 km altitude leads to a predicted volume emission rate for the auroral green line of the order of 2×10^5 photons dm^{-3} s^{-1}, compared very favourably with the observed value of 2.9×10^5 photons dm^{-3} s^{-1} of Figure 116. The agreement should be improved by the likelihood that k_{57} will have a negative temperature coefficient so that the value used in our calculation is too small, but the success of the prediction gives strong support to the mechanism proposed above. However, the position above the maximum is less secure. In Figure 114 it may be seen that the oxygen atom concentration changes relatively little within the altitude range 86 to 100 km, whereas in Figure 116 there is a substantial fall in the volume emission rate across this range. Despite the fact that diffusion will be important in determining concentration (and hence emission rate) profiles in this region, it may well be necessary to consider other possible components of the mechanism, the precise nature of which is uncertain at present.

Nitric Oxide in the Mesosphere

In Section 9.3 it was established that nitric oxide was generated in the stratosphere, mainly through the attack of O(^{1}D) atoms upon nitrous oxide. Now Figure 107 shows that the nitrous oxide mixing ratio falls sharply towards the stratopause and it is therefore evident that nitric oxide in the mesosphere does not originate from nitrous oxide. From the measurements which have been made from rocket-borne instruments, the nitric oxide concentration profile appears to decrease with increasing altitude through the stratosphere into the lower mesosphere to reach a minimum in the vicinity of 70 km altitude. The concentrations then increase with further increase in altitude to go through a maximum somewhere near 110 km altitude. This makes it clear that there must be a source of nitric oxide in the mesosphere or lower thermosphere, and a sink in the vicinity of 70 km altitude. We shall now investigate the origins of these phenomena.

The first step is to consider the application of photochemical equilibrium to the reactions involved in the nitric oxide cycles. Figure 117 shows the concentration profile of nitric oxide for daytime conditions as generated by modelling computations. This is a situation where there is good reason to believe that the concentrations actually measured in the mesosphere are generally too high because of interference by other species in the method of measurement. Justification of the low concentrations shown for altitudes around 70 km compared with the 'measured values' will be discussed in Chapter 10 in connection with the electron densities measured in this region.

At the same time, the calculational approach allows the generation of the concentration profiles of species such as $N(^4S)$ (ground state) and $N(^2D)$ atoms in the mesosphere, which have not been measured because of the absence of a suitable

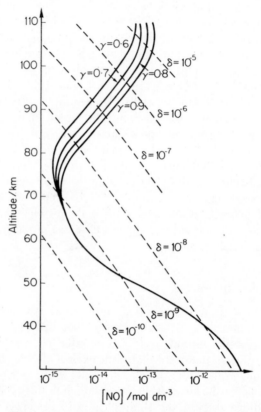

Figure 117. A realistic modelled nitric oxide concentration profile versus altitude for the upper atmosphere, found to be largely invariant with time. The uncertainty above 70 km is the fraction (γ) of dissociative recombinations of NO^+ which yield $N(^2D)$ as opposed to $N(^4S)$; results for a likely range of values of γ appear. Dashed lines correspond to concentrations for a range of constant mixing ratios with altitude for nitric oxide. From I. S. A. Isaksen, 'Diurnal variations of atmospheric constituents in an oxygen–hydrogen–nitrogen–carbon atmospheric model and the role of minor neutral constituents in the chemistry of lower ionosphere', *Geophysica Norvegica*, **30**, 31 (1973). Reproduced by permission of the Norwegian Academy of Science and Letters.

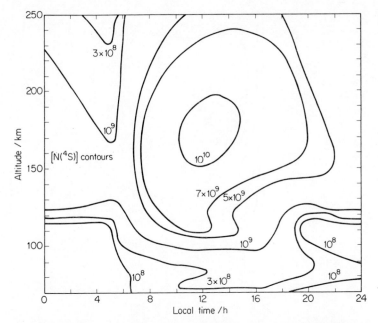

Figure 118. The diurnal variation of $N(^4S)$ atom concentrations in a modelled upper atmosphere, calibrated against sunrise and twilight observations of nitric oxide concentration profiles. The contour lines correspond to concentrations expressed in atoms dm^{-3}; divide by 6×10^{23} to express in g-atoms dm^{-3}. From E. S. Oran, P. S. Julienne, and D. F. Strobel, *Journal of Geophysical Research*, **80**, 3073 (1975). Copyrighted by the American Geophysical Union and reproduced with permission.

technique. Contour maps from a recent computer-modelling approach and representing the diurnal variations of concentrations as functions of altitude are shown in Figures 118 and 119.

Near to 70 km altitude, two reactions (−5) and (14), are mainly responsible for nitric oxide consumption:

$$N(^4S) + NO \longrightarrow N_2 + O \qquad\qquad (-5)$$

$$NO + O_3 \longrightarrow NO_2 + O_2 \qquad\qquad (14)$$

As will be shown shortly, photodissociation of nitric oxide is a comparatively small term. The photochemical characteristic time of nitric oxide is defined as the concentration divided by the rates of the reactions removing it, which turns out as the reciprocal of the sum of $k_{-5}[N(^4S)]$ and $k_{14}[O_3]$. On the basis of a midday concentration of 5×10^{-16} g-atom dm^{-3} of $N(^4S)$ and an ozone concentration of 2×10^{-12} mol dm^3, the photochemical characteristic time at 70 km altitude is of the order of 7×10^4 s. The characteristic transport time is around 3×10^5 s, hence the concept of a photochemical equilibrium is of reasonable validity for nitric oxide at 70 km altitude. Since the concentration of $N(^4S)$ atoms increases roughly by a

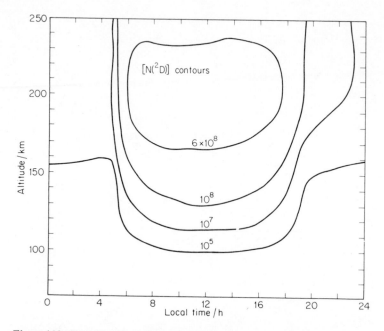

Figure 119. The diurnal variation of $N(^2D)$ atom concentrations in the same modelled atmosphere as Figure 118. The contour lines correspond to concentrations expressed in atoms dm^{-3}; divide by 6×10^{23} to express in g-atoms dm^{-3}. From E. S. Oran, P. S. Julienne, and D. F. Strobel, *Journal of Geophysical Research*, **80**, 3073 (1975). Copyrighted by the American Geophysical Union and reproduced with permission.

factor of 3 from 70 to 100 km, reaction (-5) is the main route for removal of nitric oxide by virtue of the high value of $k_{-5} = 3 \times 10^{10}$ dm^3 mol^{-1} s^{-1} compared with k_{14} being of the order of 2×10^6 dm^3 mol^{-1} s^{-1} at the temperatures in question. Hence it can be considered that a reasonable condition of photochemical equilibrium applies to nitric oxide up to the lower thermosphere.

It is now generally accepted that the main production of nitric oxide above the stratopause takes place above 100 km altitude in the thermosphere. Nitric oxide then enters the mesosphere as a result of downward diffusive transport. It has also become clear that the main production reactions involve ionized species or electrons as reactants, so that to some extent this subsection will slightly overlap Chapter 10. Molecular nitrogen photodissociation probably makes a small contribution, yielding $N(^4S)$ atoms by absorption in the banded Lyman–Birge–Hopfield absorption system in the wavelength range below 127 nm, the longest wavelength where the photon possesses sufficient energy to effect the dissociation of molecular nitrogen. However none of the well defined absorption bands overlap strongly with the Lyman α solar irradiance at wavelength 121.5 nm, precluding an important contribution to 'odd nitrogen' generation. The main 'odd nitrogen' generating reactions effective above about 90 km altitude are represented by the following set of equations:

$$e^- \text{ (fast)} + N_2 \longrightarrow 2 N \,(^4S \text{ or } ^2D) \tag{58}$$

$$N_2^+ + e^- \longrightarrow 2 N \,(^4S \text{ or } ^2D) \tag{59}$$

$$O(^3P) + N_2^+ \longrightarrow NO^+ + N \,(^4S \text{ or } ^2D) \tag{60}$$

$$O^+ \,(^4S) + N_2 \longrightarrow NO^+ + N \,(^4S) \tag{61}$$

Reaction (58) is the dissociative transfer of the kinetic energy of a fast electron to molecular nitrogen while reaction (59) is referred to as dissociative recombination. Of these processes, (58) and (60) are considered to have the highest rates, reaching maximum values by 120 km as functions of increasing altitude. As would be anticipated on energetic grounds alone, the formation of $N(^2D)$ atoms tends to occur at higher altitudes than does the formation of $N(^4S)$ atoms.

None of these 'odd nitrogen' generating reactions give rise to nitric oxide as such, that being formed by subsequent exchange reactions in terms of 'odd nitrogen'. The role of the ground-state nitrogen atoms, $N(^4S)$, in this is severely restricted by the slowness of its potential reactions. The reaction with ground-state molecular oxygen appears to be favourable in terms of exothermicity, which is 134 kJ mol^{-1} for the reaction represented by the equation:

$$N(^4S) + O_2(^3\Sigma_g^-) \longrightarrow NO(^2\Pi) + O(^3P) \tag{6}$$

Moreover reaction is not precluded by considerations of spin conservation $(3/2 - 1 = 1 - 1/2)$(Section 9.2). However laboratory studies of the rate constant for this reaction have shown that its temperature dependence is governed by an Arrhenius activation energy of around 27 kJ mol^{-1}. Temperatures in the altitude range 80 km to 110 km rise from 181 K to 290 K; reference to Table 9.2 then suggests that reaction (6) will be rather slow, especially when the rapidly decreasing concentration of molecular oxygen with altitude above 90 km is taken into account (Figure 113).

Another potential route for the conversion of $N(^4S)$ atoms to nitric oxide is reaction with carbon dioxide, which is exothermic by 96 kJ mol^{-1} as represented by the equation:

$$N(^4S) + CO_2(^1\Sigma) \longrightarrow NO(^2\Pi) + CO(^1\Sigma) \tag{62}$$

The mixing ratio of carbon dioxide at 100 km altitude is around 2×10^{-4}, not much reduced from its tropospheric value by any photodissociation caused by the short-wavelength ultraviolet solar irradiance penetrating to this altitude. However, reaction (62) is spin 'forbidden' on the basis of the Wigner–Wittmer rules $(3/2 \pm 0 \neq 1/2 \pm 0)$ and its occurrence has never been detected even in laboratory studies at elevated temperatures (Section 5.5).

The electronically excited atom $N(^2D)$ is much more reactive than $N(^4S)$. A significant reaction is that with molecular oxygen represented by the equation:

$$N(^2D) + O_2 \longrightarrow NO + O \tag{63}$$

The rate constant k_{63} is about seven orders of magnitude larger than k_6 for the comparable reaction of $N(^4S)$ atoms at 200 K. In Figures 118 and 119 we can see that the electronically excited atom is about four orders of magnitude less abundant than the ground-state atom at 100 km altitude, which must mean that reaction (63) is the dominant source of the nitric oxide diffusing down into the mesosphere.

The principal sink reaction of 'odd nitrogen' in the mesosphere must be the reaction of ground-state nitrogen atoms with nitric oxide, (-5). As already mentioned, this has a large rate constant, which corresponds to reaction occurring at about 1 collision in 10. Such a sink mechanism demands that there be a source of $N(^4S)$ atoms effective down to altitudes of 70 km or so.

The predissociation which follows the absorption of a photon in the $\delta(0,0)$ band of nitric oxide at a wavelength in the vicinity of 191 nm has been discussed in Section 8.2. The photon energy excites nitric oxide from the ground $X^2\Pi$ state up to the electronically excited $C^2\Pi$ state; the molecule then transfers onto the $a^4\Pi$ potential energy curve and thereby dissociates into ground state atoms, $N(^4S)$ and $O(^3P)$. The apposition of the potential energy curves is shown in Figure 72. Also in the wavelength range 175 nm and 190 nm, above the onset of the Schumann–Runge continuum of molecular oxygen (Figure 70) which attenuates the solar irradiance of lower wavelengths to insignificant levels below 100 km altitude (Table 9.3), are several other absorption bands of nitric oxide which lead to dissociation. This part of the nitric oxide absorption spectrum is shown in Figure 120.

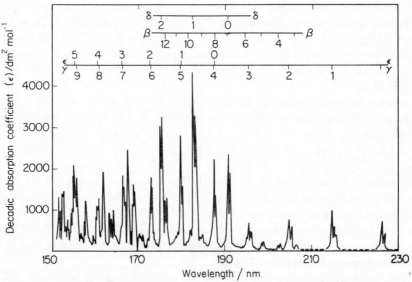

Figure 120. The absorption spectrum of nitric oxide in the wavelength range 150 to 230 nm, showing bands of the designated series. From J. R. McNesby and H. Okabe, 'Vacuum ultraviolet photochemistry', in *Advances in Photochemistry* (Eds. W. A. Noyes, Jr., G. S. Hammond and J. N. Pitts, Jr.), Vol. 3, Interscience, New York, 1964. p. 179. Reproduced with permission of Interscience, New York.

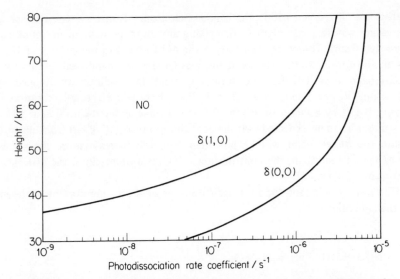

Figure 121. The vertical variations of the photodissociation rate coefficients of nitric oxide for overhead Sun, through predissociation in the δ (1,0) and δ (0,0) bands. From J. H. Park, 'The equivalent mean absorption cross-sections for the O_2 Schumann–Runge bands: Application to the H_2O and NO photodissociation rates', *Journal of Atmospheric Sciences*, **31**, 1896 (1974). Reproduced by permission of the American Meteorological Society.

The absorption coefficients are large within the bands but the bands are comparatively narrow with respect to the widths of the gaps between. The attenuation of solar irradiance in this spectral region is very difficult to estimate with the precision required to obtain a good approximation for the photodissociation rate of nitric oxide in the mesosphere. The main source of this difficulty is the banded nature of the Schumann–Runge absorption spectrum of molecular oxygen which produces the attenuation. It will not suffice in this instance to use the familar averages over 10-nm wavelength intervals for attenuations and absorption coefficients, since it is critical that an accurate assessment is made of the extent to which any molecular oxygen band overlaps a particular nitric oxide absorption band. It is therefore essential to use exceedingly small wavelength intervals (as small as 3×10^{-4} nm) in any realistic calculation. Figure 121 shows the resultant photodissociation coefficient, $J_{NO}(N(^4S))$, as a function of altitude from 30 km to 80 km showing that it is absorption within the $\delta(0,0)$ band which provides the overwhelming contribution to this.

It is seen that the attenuation by molecular oxygen of the effective radiation only becomes significant below altitudes of 60 km or so. Thus $J_{NO}(N(^4S))$ is only reduced by an order of magnitude from its limiting high-atmosphere value at about 40 km altitude. It can be considered that above this altitude the photodissociative sink for nitric oxide is a substantial process. The basic reason why bands other than the $\delta(0,0)$ band are ineffective in inducing photodissociation is that the solar irradiance decreases with decreasing wavelength (Table 3.4) while the absorption coefficients of oxygen within the banded Schumann–Runge system increase with decreasing wavelengths (Figure 70), making any overlaps with nitric oxide bands more significant. Therefore even though the $\delta(1,0)$ band in the vicinity of 183 nm has an absorption coefficient about twice as large at the band peak as that for the $\delta(0,0)$ band near 191 nm, the former is much more closely overlapped by a Schumann–Runge band, (9,0), than is the latter, while the solar photon flux density is less by a factor of around 2.5 at 183 nm compared with that at 191 nm. It is therefore to be expected that the contribution to $J_{NO}(N(^4S))$ from absorption within the $\delta(1,0)$ band will be less than, but will increase more sharply with increasing altitude than, the contribution from absorption within the $\delta(0,0)$ band, as shown in Figure 121.

The nitric oxide (and 'odd nitrogen') sink process can therefore be represented by the equation:

$$NO + h\hat{\nu} \; (\delta \text{ bands}) \longrightarrow N(^4S) + O$$

$$N(^4S) + NO \longrightarrow N_2 + O \tag{-5}$$

Hence each photon absorbed can lead to the destruction of two nitric oxide molecules. However this depends upon the $N(^4S)$ product of the initial photodissociation reacting exclusively with nitric oxide. Now, despite its low rate constant, the reaction of $N(^4S)$ with molecular oxygen may be able to compete with reaction (-5) because of the sheer abundance of the latter. This reaction is represented by the equation:

$$N(^4S) + O_2 \longrightarrow NO + O \tag{6}$$

It is clear that if reaction (6) follows the initial photodissociation of nitric oxide, the net effect in terms of nitric oxide destruction is nil. Accordingly we must gain some insight into the relative rates of reactions (-5) and (6) at different altitudes in the upper atmosphere. The ratio of these two rates as a function of altitude is shown in Table 9.16. The calculated nitric oxide concentration profile as a function of altitude of Figure 117 is applied in this assessment, along with a rate constant expression of $k_6 = 3 \times 10^9 \exp(-26800/RT) \, dm^3 \, mol^{-1} \, s^{-1}$ and a temperature independent value of $k_{-5} = 3 \times 10^{10} \, dm^3 \, mol^{-1} \, s^{-1}$. This table illustrates that efficient photodestruction of nitric oxide, associated with the dominance of reaction (-5) in the consumption of the $N(^4S)$ atoms produced by predissociation of nitric

Table 9.16. Estimated ratio of rates (R_{-5}/R_6) as a function of altitude.

Altitude/km	T/K	$[NO]/mol \, dm^{-3}$	$[O_2]/mol \, dm^{-3}$	R_{-5}/R_6
30	227	8×10^{-12}	1.2×10^{-4}	0.97
40	250	2×10^{-12}	2.8×10^{-5}	0.24
50	271	7×10^{-14}	8.0×10^{-6}	0.013
60	256	4×10^{-15}	2.5×10^{-6}	0.005
70	220	2×10^{-15}	7.0×10^{-7}	0.066
80	181	3×10^{-15}	1.2×10^{-7}	12.7

oxide, only sets in above 70 km altitude. At the same time the rate of photo-destruction of nitric oxide will be determined by the value of $J_{NO}(N(^4S))$ at a particular altitude; the low value of this at an altitude of 30 km means that photo-destruction of nitric oxide is insignificant despite the apparently favourable ratio of R_{-5}/R_6 shown in Table 9.16.

The minimum in the altitude concentration profile of nitric oxide with altitude near to 70 km is evidently a reflection more of the variation of production rate with altitude than of the variation of the photodestruction rate, which in terms of the value of R_{-5}/R_6, moves in the wrong direction from 70 km to 80 km in Table 9.16. Since 'odd nitrogen' production depends upon the reaction of charged species (58) to (61), which only exist in substantial concentrations in the thermosphere, it is hardly surprising that the very small rates in the upper mesosphere increase very rapidly with increase in altitude.

Below an altitude of 90 km, photodissociation of nitric oxide is the dominant source of atomic nitrogen. Since Table 9.16 shows that the reaction with nitric oxide (-5), is the dominant removal process at 80 km altitude, then, on the assumption of photoequilibrium, we may generate the equation:

$$J_{NO}(N(^4S))[NO] = k_{-5}[N(^4S)][NO]$$

The steady-state concentration of ground-state nitrogen atoms is hence simply given by $J_{NO}(N(^4S))/k_{-5}$. The photodissociation rate coefficient can be read from Figure 121 as approximately 10^{-5} s^{-1} while k_{-5} is 3×10^{10} dm^3 mol^{-1} s^{-1} for 80 km altitude, so that the nitrogen atom concentration is predicted as being of the order of 3×10^{-16} g-atom dm^{-3}. This is in reasonable agreement with the midday concentration shown in Figure 118; the difference reflects the effects of diffusion to a large extent since it is clear that the photochemical characteristic time represented by the reciprocal of $J_{NO}(N(^4S))$ is only a factor of three less than the characteristic transport time.

It was noted in Section 9.1 that, because of the 'window' in the molecular oxygen absorption spectrum near the wavelength of the Lyman α line, there was a considerable penetration of the solar radiation at 121.5 nm below 100 km altitude, as is shown in Figure 92. Nitric oxide has a considerable photoionization coefficient, which is of the order of 5000 dm^2 mol^{-1} on a decadic basis, at the wavelength of 121.5 nm. Accordingly we expect significant ionization of nitric oxide down to about 70 km altitude. We must investigate whether this can contribute to the production or removal of nitric oxide in the mesosphere.

The initial photoionization step proceeds with approaching unit quantum efficiency according to the equation:

$$NO + h\nu(\text{Lyman } \alpha) \longrightarrow NO^+ + e^-$$

Above an altitude of 85 km, NO^+ is a terminal ion (i.e. it is not converted into other ions) and the only loss processes are the dissociative recombinations represented by the equations:

$$NO^+ + e^- \longrightarrow N(^4S) + O(^3P) \tag{64}$$

$$NO^+ + e^- \longrightarrow N(^2D) + O(^3P) \tag{65}$$

The ratio of the rate constants of these two processes is uncertain at present, but it seems likely that reaction (65) accounts for between 50% and 90% of the total rate of dissociative recombination of NO^+. The ground-state atomic nitrogen will, at altitudes above 80 km, induce reaction (−5) predominantly and thus serve as a sink reaction for 'odd nitrogen'. On the other hand, at altitudes below 80 km, the $N(^4S)$ atom will react mainly with molecular oxygen on the basis of the data in Table 9.16 and thus cancels out the effect of the photoionization cycle in this context. Similarly, reaction (65) will involve a nil effect since the reaction of $N(^2D)$ with molecular oxygen is the overwhelming fate, replacing the nitric oxide molecule destroyed by dissociative recombination.

In Table 3.4 we see that the extraterrestrial solar irradiance within a range of 10 nm about wavelengths of 190 nm and 120 nm are of the same order of magnitude. As functions of wavelength, the photon flux density in the former band can be assumed to be fairly evenly spread across the wavelengths but, of course, the latter band will have the photon flux density almost completely concentrated in the Lyman α line at 121.5 nm wavelength. Now the $\delta(0,0)$ band of nitric oxide spreads across the wavelength range 190.2 nm to 191.5 nm, so that it is this approximately one-seventh part of the 10-nm range which we must consider in connection with the photodissociation of nitric oxide. The average absorption coefficient of nitric oxide across the $\delta(0,0)$ band is about one-sixth of that at the wavelength of the Lyman α line. It is therefore clear that, in the absence of attenuation of the solar photon flux densities, the rate of absorption of Lyman α line photons by nitric oxide would be about an order of magnitude larger than the rate of absorption of photons within the δ bands. However Figure 92 shows the Lyman α line photon flux densities are attenuated by a factor of about 4 at 80 km altitude and by a factor of over 100 at 70 km. Since the absorption coefficients of molecular oxygen are extremely small around 190 nm wavelength, the solar photon flux density in that spectral region will penetrate to 70 km altitude virtually undiminished, as is illustrated in Figure 121. Hence the nitric oxide photoionization cycle is only significant in the upper reaches of the mesosphere in comparison with the photodissociation cycle from the point of view of the sink processes for 'odd nitrogen'.

Finally we may assess the relative contributions of nitric oxide and nitrogen dioxide to the 'odd nitrogen' concentration in the mesosphere. The interconversion of these will be governed by the familiar cycle composed of the equation:

$$NO \quad + O_3 \quad \longrightarrow \quad NO_2 + O_2 \tag{14}$$

$$O(^3P) + NO_2 \quad \longrightarrow \quad NO \quad + O_2 \tag{36}$$

The photodissociation of nitrogen dioxide can be ignored since at every altitude in the mesosphere it has a rate at least an order of magnitude less than the rate of reaction (36). On this basis the ratio of the two nitrogen oxide concentrations is given by the equation:

$$[NO]/[NO_2] = (k_{36}/k_{14})([O]/[O_3])$$

The ratio of the two rate constants is of the order of 4000 in the middle mesosphere, and in Figure 96(b) it is clear that the oxygen atom concentration is always greater than that of ozone. Therefore nitrogen dioxide concentrations in the mesosphere will be totally insignificant.

9.7 Concluding Remarks

In this chapter we have investigated the chemistry of the stratosphere and mesosphere on the basis of a photoequilibrium model, which has been shown to be a good enough approximation for our fairly general purposes.

It has been demonstrated that the ozone content of the present stratosphere is largely determined by the natural nitrogen oxide mixing ratios, which originate in the chemical interaction of $O(^1D)$ atoms produced by ozone photodissociation with nitrous oxide which has diffused up from the Earth's surface. The balancing removal of the nitrogen oxides is accomplished through the downward diffusion of the sink species, nitric acid vapour, back through the tropopause. Reassurance with regard to the potential output of nitric oxide from the exhausts of SST aircraft, and the resultant possible anthropogenic perturbation of the stratospheric ozone layer, can be derived from the realization that atmospheric nuclear tests have already injected comparable amounts of nitrogen oxides without any noticable effect on the ozone layer thickness. On the other hand there is no reassurance that the chlorine atoms resulting from the photodissociation of chlorofluoromethane molecules in the stratosphere will not significantly reduce the thickness of the ozone layer through a chain destruction mechanism. The potential risks of increased incidence of skin cancer have been shown to depend critically on the thickness of the Earth's protective ozone layer.

The hydrogen–oxygen chemistry of the mesosphere is largely determined by the relative rates of photodissociation of molecular oxygen and water vapour, the sources of 'odd oxygen' and 'odd hydrogen' respectively. The concurrent nitrogen chemical conversions proceed almost independently and 'odd nitrogen' production is largely a spin-off from the ionospheric chemistry.

Bibliography

1. A Collection of Papers presented at the Ozone Symposium in Arosa, Switzerland, 1972. *Pure and Applied Geophysics*, **106–108**, 1139–1519 (1973).
2. A Collection of Papers on the Chemistry of the Upper Atmosphere. *Canadian Journal of Chemistry*, **52**, 1381–1634 (1974).
3. 'Chemical Reactions in the Atmosphere as Studied by the Method of Instantaneous Rates', by H. S. Johnston and G. Whitten, *International Journal of Chemical Kinetics*, **7**, 1–26 (1975).
4. 'Ground-level Effects of Supersonic Transports in the Stratosphere', by H. S. Johnston, *Accounts of Chemical Research*, **8**, 289–294 (1975).
5. 'Chlorofluoromethanes in the Environment', by F. S. Rowland and M. J. Molina, *Reviews of Geophysics and Space Physics*, **13**, 1–35 (1975).

374

6. 'Diurnal Variations of Atmospheric Constituents in an Oxygen–Hydrogen–Nitrogen–Carbon Atmospheric Model, and the Role of Minor Neutral Constituents in the Chemistry of the Lower Ionosphere', by I. S. A. Isaksen, *Geophysica Norvegica,* **30**, 1–63 (1973). (In English)

7. 'Ozone Photochemistry and Radiative Heating of the Middle Atmosphere', by J. H. Park and J. London, *Journal of Atmospheric Sciences,* **31**, 1898–1916 (1974).

8. 'The Vertical Distribution of Trace Gases in the Stratosphere', by D. H. Ehhalt, L. E. Heidt, R. H. Lueb, and W. Pollock, *Pure and Applied Geophysics,* **113**, 389–402 (1975).

9. *An Introduction to Photobiology,* Ed. C. P. Swanson, Prentice-Hall, Inc., New Jersey, 1969.

10. *The Biologic Effects of Ultraviolet Radiation,* Ed. F. Urbach, Pergamon Press, London, 1969.

11. 'Nitrogen Oxides, Nuclear Weapon Testing, Concorde and Stratospheric Ozone', by P. Goldsmith, A. F. Tuck, J. S. Foot, E. L. Simmons, and R. L. Newson, *Nature,* **244**, 545–551 (1973).

12. *Particles in the Atmosphere and Space,* by R. D. Cadle, Reinhold Publishing Corporation, New York, 1966.

13. 'A Generalised Aeronomic Model of the Mesosphere and Lower Thermosphere including Ionospheric Processes', by B. G. Hunt, *Journal of Atmospheric and Terrestrial Physics,* **35**, 1755–1798 (1973).

14. 'The Aeronomy of Odd Nitrogen in the Thermosphere', by E. S. Oran, P. S. Julienne, and D. F. Strobel, *Journal of Geophysical Research,* **80**, 3068–3076 (1975).

15. 'The Infrared Spectrum of the Airglow', by A. Vallance Jones, *Space Science Reviews,* **15**, 355–400 (1973).

16. 'Eddy Diffusion and Oxygen Transport in the Lower Thermosphere', by F. D. Colegrove, W. B. Hanson, and F. S. Johnson, *Journal of Geophysical Research,* **70**, 4931–4941 (1965).

Chapter 10

Ionic Species in the Mesosphere and Lower Thermosphere

For the sake of completeness in our consideration of the upper atmosphere, the ionic components of the lower ionosphere will be dealt with briefly in this short chapter. As far as is known there is little possibility of the ionospheric chemistry of the atmosphere being perturbed by anthropogenic activity, so that this is therefore peripheral to our main theme. Our approach will therefore be simply to pick out a few aspects of outstanding interest rather than to attempt an overall view.

10.1 Primary Photoionization in the Lower Ionosphere

Our discussion in the preceding chapter of 'odd nitrogen' chemistry in the upper atmosphere and the penetration of Lyman α line solar photon fluxes of significance into the mesosphere has already made contact with ionic species. We can appreciate that very short-wavelength solar radiation will be the primary source of positive ions and electrons. The ionization potentials and energies of relevant species present in the upper atmosphere are listed in Table 10.1.

From the limiting upper wavelength data (at which the photon has an energy equivalent to the ionization potential) in this table, it is clear that only nitric oxide can be ionized by Lyman α radiation at wavelength 121.5 nm. Species other than those listed in the table need not be considered as their concentrations in the lower ionosphere (extending down to around 70 km altitude) are too small to give rise to significant rates of ion production.

Since Lyman α radiation, by virtue of the highly localized 'window' in the absorption spectrum of molecular oxygen, is the only short wavelength radiation to penetrate on any scale into the mesosphere (see Table 9.3), it follows that photoionization of nitric oxide must be the overwhelming source of the electron density in the mesosphere. However, above 90 km altitude the rates of formation of N_2^+ and O_2^+ become appreciable because of the availability of shorter wavelength radiation in conjunction with the major concentrations of molecular nitrogen and oxygen. In fact, the N_2^+ formation rate at around 95 km altitude is enhanced by the strong absorption of penetrating X-radiation of wavelengths below 10 nm, which is much less attenuated at this altitude than is the radiation of wavelength around 80 nm which would produce the more obvious type of photoionization. It

Table 10.1. Ionization parameters of upper atmosphere atoms and molecules

Species	Ionization potential/eV	Ionization energy/kJ mol^{-1}	Equivalent wavelength/nm
$N(^4S)$	14.54	1403.4	85.3
$O(^3P)$	13.61	1314.0	91.1
$H(^2S)$	13.54	1307.2	91.6
$N_2(X^1\Sigma_g^+)$	15.58	1503.9	79.6
$O_2(X^3\Sigma_g^-)$	12.09	1166.7	102.6
$O_2(a^1\Delta_g)$	11.10	1070.7	111.8
$NO(X^2\Pi)$	9.26	893.3	134.0

Data are generally available and are collected from several sources.

turns out that NO^+ formation above the mesopause is dominated by the secondary conversion of N_2^+ represented by the equations:

$$N_2^+ + O \longrightarrow NO^+ + N \tag{60}$$

$$N_2^+ + NO \longrightarrow NO^+ + N_2$$

These reactions are highly exothermic, essentially because the ionization potential of nitric oxide is the lowest of all the species present. Towards the mesopause the X-ray source of N_2^+ is rapidly attenuated with decreasing altitude so that Lyman α radiation becomes the predominant source of ionization. Nitric oxide does not contribute significantly to the attenuation of Lyman α radiation as a function of altitude. The absorption coefficients of molecular oxygen and nitric oxide at 121.57 nm are approximately as 1 to 250 respectively but the preponderance of about 10^7 of the former compared with the latter makes molecular oxygen the primary source of the attenuation. We may apply the weak absorption approximation form of equation (8.13) to obtain the rate of photoionization of nitric oxide by Lyman α radiation at a particular altitude. For a 1-dm vertical path with overhead Sun, the resultant equation takes the form:

$$\text{Rate of NO photoionization} = 2.303 I_\alpha \epsilon [NO] \tag{10.1}$$

I_α represents the Lyman α photon flux per square decimetre for the altitude in question and ϵ is the decadic photoionization coefficient (almost identical with the decadic absorption coefficient) at the wavelength of 121.57 nm. Hence we expect the photoionization rate in the mesosphere to be directly proportional to the product of the Lyman α solar photon flux density and the nitric oxide concentration at a particular altitude. From the profiles of these factors as functions of altitude in Figures 92 and 117 it is apparent that the rate of NO^+ production falls off sharply below the mesopause to be largely insignificant below 70 km altitude.

The Lyman β line of atomic hydrogen lies at a wavelength of 102.6 nm and like the Lyman α line is enhanced well above the solar irradiance at neighbouring wavelengths. Although there is a partial 'window' in the molecular oxygen absorption spectrum, in the vicinity of the β line, carbon dioxide has a high absorption coefficient in this spectral region with the result that potential photoionization on molecular oxygen itself (see Table 10.1) in the mesosphere turns out to be insignificant. The O_2^+ ions which have been detected in the mesosphere are likely to have originated through photoionization of $O_2(^1\Delta_g)$ by radiation of wavelengths up to 111.8 nm (Table 10.1), even though this excited state is much less abundant than the ground state.

10.2 The Major Positive Ionic Species in the Mesosphere

Following our discussion of the primary photoionization processes above it may come as some surprise that NO^+ is not the dominant positive ion component of the mesosphere. Figure 122 shows the ionic concentration profiles as functions of altitude as detected by a rocket-borne mass spectrometer. The major ions identified below 80 km altitude have mass-to-charge ratios of 19^+ and 37^+, while there is a considerable presence of ions with mass to charge ratios in excess of 45^+. The 30^+ profile will correspond to NO^+ while the less abundant 32^+ profile will correspond to O_2^+. It is only above 80 km altitude that the last two come into real prominence. The identities of the major ions below 80 km altitude are hydrated proton species $H^+(H_2O)(19^+)$ and $H^+(H_2O)_2(37^+)$, with the likelihood

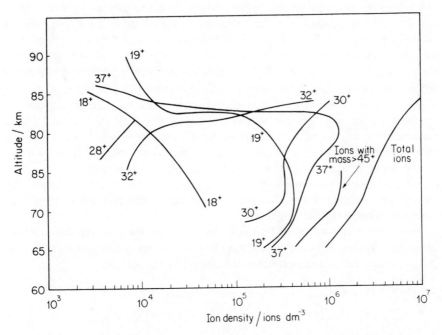

Figure 122. Concentration profiles ions detected by a rocket-borne mass spectrometer in the altitude range 64 to 83 km. From R. S. Narcisi and A. D. Bailey, *Journal of Geophysical Research*, **70**, (1965) 3693. Copyrighted by the American Geophysical Union and reproduced with permission.

that a large fraction of the ions with mass-to-charge ratios above 45^+ are further hydrated species such as $H^+(H_2O)_3(55^+)$.

Much of the current difficulty in modelling the chemistry of the lower ionosphere stems from explaining how the primary NO^+ ion transfers its charge to the hydrogen atom, and how the so-called cluster ions $(H^+(H_2O)_n)$ are formed. Before the carbon dioxide attenuation of the Lyman β solar radiation was considered, it appeared that the mechanism could be based upon the existence of relatively high concentrations of O_2^+ below 80 km altitude. Good agreement could be obtained on this basis between computed models and the actual detected ion concentration profiles using a mechanism commencing with O_2^+ rather than NO^+ according to the equations:

$$
\begin{aligned}
O_2^+ + O_2 + O_2 &\longrightarrow O_4^+ + O_2 \\
O_4^+ + H_2O &\longrightarrow O_2^+ \cdot H_2O + O_2 \\
O(^3P) + O_4^+ &\longrightarrow O_2^+ + O_3 \qquad\qquad (66) \\
O_2^+ \cdot H_2O + H_2O &\longrightarrow H_3O^+ \cdot OH + O_2 \\
O_2^+ \cdot H_2O + H_2O &\longrightarrow H_3O^+ + OH + O_2 \\
H_3O^+ \cdot OH + H_2O &\longrightarrow H^+(H_2O)_2 + OH
\end{aligned}
$$

This scheme of reactions was made particularly attractive by the fact that reaction (66) appeared to produce the sharp decrease in the concentrations of the cluster ions evident above 80 km in Figure 122. Reference back to Figure 113 shows that the concentration profile of $O(^3P)$ atoms with altitude rises sharply in this region; the passage of O_4^+ on to $O_2^+ \cdot H_2O$ is therefore a much less effective step above 80 km altitude and thereby cluster ion formation is inhibited. A further attraction of the above reaction scheme is that all the reactions bar the first are bimolecular reactions, so that the rates of these will be less restricted by the low total-species concentrations in the upper mesosphere than would be the case for three-body processes. However, in view of the very low Lyman β photon flux densities which will be available at 80 km altitude on account of the attenuation by the carbon dioxide column density above this altitude, there is doubt as to whether this scheme can generate the production rate of cluster ions required to match the observed profiles below 80 km altitude.

There are reactions of NO^+ which can produce the hydrated cluster ions. Laboratory studies of potential reactions have led to the conclusion that the most rapid reaction mechanism available is represented by the equations:

$$
\begin{aligned}
NO^+ + CO_2 + M &\longrightarrow NO^+ \cdot CO_2 + M \\
NO^+ \cdot CO_2 + H_2O &\longrightarrow NO^+ \cdot H_2O + CO_2 \\
NO^+ \cdot H_2O + H_2O + M &\longrightarrow NO^+(H_2O)_2 + M \\
NO^+ \cdot (H_2O)_2 + H_2O + M &\longrightarrow NO^+(H_2O)_3 + M \\
NO^+(H_2O)_3 + H_2O &\longrightarrow H^+(H_2O)_3 + HNO_2 \qquad (67)
\end{aligned}
$$

The initial formation of the monohydrated NO^+ ion represented by the first two equations above is therefore catalysed by carbon dioxide. This indirect mechanism has been shown to be much faster than the direct process represented by the equation:

$$NO^+ + H_2O + M \longrightarrow NO^+ \cdot H_2O + M$$

under the typical mesospheric conditions where carbon dioxide mixing ratios are some two orders of magnitude larger than those of water vapour.

The most obvious point to be made with regard to the NO^+ hydration mechanism is that it involves at least three three-body reactions which would be expected to be relatively slow on account of the very low total species concentrations, $[M]$. Hence it should come as no great surprise that computer-modelled rates of production of hydrated clusters by this route are too slow to satisfy the measured parameters. Further, the computations on this basis suggest that the concentration of NO^+ itself should be approaching 80% of the total positive ion content at around 80 km altitude, which allows no compromise with the detected profiles of Figure 122.

A further difficulty with the NO^+ mechanism is that, as represented, the triply hydrated species, $H^+(H_2O)_3(55^+)$, is the major product rather than the 19^+ monohydrate and 37^+ dihydrate. The laboratory studies of the hydration of NO^+ have demonstrated clearly that the ion must be triply hydrated before any switching reactions like (67) can occur to release the hydrated proton species. It appears that $n = 3$ is the first of the hydrated species $NO^+(H_2O)_n$ with increasing degrees of hydration where such a switching process can be exothermic. At temperatures below 200 K in the upper mesosphere, any endothermicity of a reaction will be likely to create an Arrhenius activation barrier to reaction, which cannot be easily overcome.

Moreover the NO^+ mechanism does not appear to have an efficient chain breaking step analogous to reaction (66) in the O_2^+ mechanism. Such a step is essential to explain the rapid fall-off in the concentrations of hydrated ions above 80 km altitude.

Electron densities in the upper mesosphere increase sharply above 80 km altitude which is consistent with the fact that laboratory studies have established that the rate coefficient for the dissociative recombination of $H^+(H_2O)_n$ species with electrons is larger than that for the dissociative recombination of NO^+. Hence the conversion from dominance of $H^+(H_2O)_n$ ions below 82 km to dominance of the NO^+ ion above, as shown in Figure 122, leads to an increase in electron loss rate with decreasing altitude in this region.

The arguments in favour of the nitric oxide concentrations in the upper mesosphere being considerably lower than indicated by the measurements which have been made using rocket-borne instruments, are that the resultant predictions of the rate of production of NO^+, and hence electrons, must match up with the well-established electron density profile with altitude. Some of the higher 'measured' concentrations of nitric oxide below 80 km altitude correspond to electron densities more than an order of magnitude larger than are measured. This is good support for the low nitric oxide concentrations in the vicinity of 70 km altitude represented in Figure 117. Of the three principal parameters involved in the calculation of the rate of electron production, viz. the nitric oxide concentration, the Lyman α photon flux density, and the photoionization (absorption) coefficient, only the first offers much scope for uncertainty below 80 km altitude.

The hydrated cluster ions have the ability to grow in further reactions represented by the general equation:

$$H^+(H_2O)_n + H_2O + M \longrightarrow H^+(H_2O)_{n+1} + M$$

The rate constants of these three-body reactions decrease steadily with increasing n. However the rate constants for the reverse dissociation process increase with increasing values of n. Since a balance is likely to be established in the upper mesosphere between the forward and reverse processes, these variations of the rate constants explain the increasing rarity of clusters with higher degrees of hydration.

We may conclude our brief survey of the positive-ion chemistry of the upper atmosphere by conceding that major details remain to be established.

10.3 The Major Negative Ionic Species in the Mesosphere

Negative ions have been detected in the upper atmosphere but largely below 90 km altitude. Several rocket-borne mass spectrometric investigations have shown that the composition of these ions is predominantly restricted to the cluster ions of formula $NO_3^-(H_2O)_n$, with n ranging from 0 to 5. In contrast to the positive-ion chemistry, the negative-ion chemistry is well understood on the basis of laboratory studies of the species.

The chemical interpretation can be divided into two stages, firstly the generation mechanism for the nitrate ion and secondly the hydration sequence.

The overwhelming initial formation of negative ions takes place through a three-body electron attachment to molecular oxygen according to the equation:

$$e^- + O_2 + O_2 \longrightarrow O_2^- + O_2$$

A very minor contribution is made by the two-body dissociative attachment of electrons to ozone represented by the equation:

$$e^- + O_3 \longrightarrow O^- + O_2$$

Dissociative attachments of electrons to other neutral species are too slow to contribute significantly to general negative-ion production.

Following the initial generation of O_2^-, there are three competing removal reactions of importance, represented by the equations:

$$O_2^- + O(^3P) \longrightarrow O_3 + e^- \tag{68}$$

$$O_2^- + O_3 \longrightarrow O_3^- + O_2 \tag{69}$$

$$O_2^- + O_2 + O_2 \longrightarrow O_4^- + O_2$$

Now it turns out that the formation of O_3^- and O_4^- provides the gateway to the subsequent production of NO_3^-. It is clear that reaction (68) effectively reverses the initial electron attachment to molecular oxygen. The ratio of concentrations of ground-state oxygen atoms and ozone is therefore critical to the relative importance of reactions (68) and (69). Reference back to Figure 113 shows that above 80 km altitude these concentrations move sharply in opposite directions; $[O(^3P)]$ rises sharply with increasing altitude while $[O_3]$ falls, largely in response to the decrease in total pressure with increasing altitude. It is therefore clear that reaction (68) will become increasingly dominant above 80 km altitude so that the negative ion content of the atmosphere will decrease. A reasonable picture, therefore, is that above 90 km reaction (68) is so dominant that electrons remain substantially unattached.

An interesting sidelight in this connection concerns the transmission of radiowaves over long distances. These move round the Earth by reflection from the underside of the ionosphere. Negative ions are too heavy to oscillate in resonance with the radiowave frequency and so produce little attenuation. However, electrons can oscillate in resonance and so attenuate the transmission severely. The electron density of the ionospheric underside layer falls to near zero at night, not only due to the removal of direct ionization by solar radiation but also due to the disappearance of $O(^3P)$ atoms by combination as discussed in Section 9.6. This results in most of the negative charge being in the form of negative ions rather than electrons, since overnight reaction (69) evidently dominates reaction (68) in the removal of O_3^-. Thus the improvement in long range radio transmission at night can be expained in these fairly simple terms.

Both O_3^- and O_4^- undergo further conversions through a complex series of bimolecular reactions represented in schematic form in Figure 123. We see that NO_3^- is the terminal ion for negatively charged species. Hence it is thsi species and its hydrated forms which persist throughout the night. In the mesosphere, concentrations of carbon dioxide exceed those of nitric oxide by several orders of magnitude, and the reactions in Figure 123 with carbon dioxide as reactant are fast. Hence the eventual formation of nitrite and nitrate ions involves predominantly CO_3^- and CO_4^- precursors. Whether most of these precursors go on to form nitrite and nitrate ions or revert back to O_2^- will depend upon the rates of the reactions represented by the equations:

$$CO_3^- + NO \longrightarrow CO_2 + NO_2^-$$

$$CO_4^- + NO \longrightarrow CO_2 + NO_3^-$$

$$CO_3^- + O(^3P) \longrightarrow CO_2 + O_2^-$$

$$CO_4^- + O(^3P) \longrightarrow CO_3^- + O_2$$

The rate constants of these reactions are all large and of roughly the same order of magnitude. Comparison of the concentration profiles of $O(^3P)$ and nitric oxide in Figures 113 and 117 shows that the former is three to four orders of magnitude more abundant than the latter in the mesosphere. It is therefore clear that passage from the carbon-containing ions to the nitrogen-containing ions must be a relatively inefficient process, dominated by reversion to O_2^-.

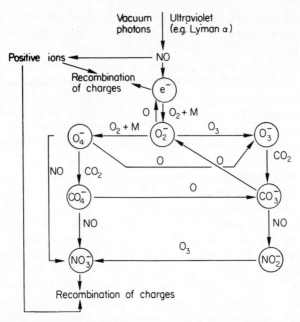

Figure 123. Diagrammatic representation of D-region negative ion conversions.

The subsequent hydration of the terminal nitrate ion is likely to proceed by an indirect mechanism analogous to that invoked for hydration of NO^+ in the preceding subsection. The precise nature of the elementary steps involved in this is uncertain as yet. In any event, the few rocket observation studies which have been made have disagreed on the question of whether most of the nitrate ions are hydrated or not. Further details of the negative ion chemistry in the upper atmosphere must await more definitive direct measurements of the composition.

Bibliography

1. 'Theoretical Models of the D-region', by C. F. Sechrist, *Journal of Atmospheric and Terrestrial Physics,* **34**, 1565–1589 (1972).
2. 'A Generalised Aeronomic Model of the Mesosphere and Lower Thermosphere including Ionospheric Processes', by B. G. Hunt, *Journal of Atmospheric and Terrestrial Physics,* **35**, 1755–1798 (1973).
3. 'D-region Ion Chemistry', by E. E. Ferguson, *Reviews of Geophysics and Space Physics,* **9**, 997–1008 (1971).
4. 'Thermal-Energy Negative Ion—Molecule Reactions', by E. E. Ferguson, *Accounts of Chemical Research,* **3**, 402–408 (1970).

References

J. G. Anderson (1971), *Journal of Geophysical Research,* **76**, 7820.

G. de Beni and C. Marchetti (1972), Symposium on Non-fossil Fuels, *163rd National Meeting of the American Chemical Society,* Boston.

R. A. Cox (1972), *Journal of Physical Chemistry,* **76**, 814.

R. A. Cox (1974), *Journal of Photochemistry,* **3**, 175.

W. A. Daniels (1967), *SAE Transactions,* **76**, 774.

O. T. Denmead, J. R. Simpson, and J R. Freney (1974), *Science,* **185**, 609.

M. M. Eisenstadt and K. E. Cox, *Solar Energy,* **17**, 59.

C. England and W. H. Corcoran (1975), *Industrial and Engineering Chemistry Fundamentals,* **14**, 55.

A. G. Gaydon and H. G. Wolfhard (1950), *Proceedings of the Royal Society,* **A201**, 570.

H. Harrison and T. Larson (1974), *Journal of Geophysical Research,* **79**, 3095.

E. M. Hartley and M. J. Matteson (1975), *Industrial and Engineering Chemistry Fundamentals,* **14**, 67.

J. B. Homer and I. R. Hurle (1972), *Proceedings of the Royal Society,* **A327**, 61.

R. B. Ingersoll (1972), 'The Capacity of Soil as a Natural Sink for Carbon Monoxide', *Stanford Research Institute Report, SR1 LSU-1380 CRC A PRAC CAPA 4 68 6,* 13 pp.

R. E. Inman, R. B. Ingersoll, and E. A. Levy (1971), *Science,* **172**, 1229.

R. H. Kummler and M. H. Bortner (1970) , *Journal of Geophysical Research,* **75**, 3115.

A. L. Lazrus and B. W. Gandrud (1974), *Journal of the Atmospheric Sciences,* **31**, 1102.

A. L. Lazrus, B. W. Gandrud, R. N. Woodward, and W. A. Sedlacek (1976), *Journal of Geophysical Research,* **81**, 1067.

P. S. Liss (1971), *Nature,* **233**, 327–329.

J. E. Lovelock, R. J. Maggs, and R. A. Rasmussen (1972), *Nature,* **237**, 452.

R. C. Millikan and D. R. White (1963), *Journal of Chemical Physics,* **39**, 98.

T. Novakov, S. G. Chang, and A. B. Harker (1974), *Science,* **186**, 259.

T. Novakov, R. K. Mueller, A. E. Alcocer, and J. W. Otvos (1972), *Journal of Colloid and Interface Science,* **39**, 225–234.

C. B. Opal and H. W. Moos (1969), *Journal of Geophysical Research,* **74**, 2398.

J. H. Park and J. London (1974), *Journal of Atmospheric Sciences,* **31**, 1898,

C. T. Pate, B. J. Finlayson, and J. N. Pitts (1974), *Journal of the American Chemical Society,* **96**, 6554.

S. A. Penkett, F. J. Sandalls, and J. E. Lovelock (1975), *Atmospheric Environment,* **9**, 139.

384

D. Pimentel, L. E. Hurd, A. C. Bellotti, M. J. Forster, L. N. Oka, O. D. Sholes, and R. J. Williams (1973), *Science,* **182**, 443.

J. W. Rogers, R. E. Murphy, A. T. Stair, J. C. Ulwick, K. D. Baker, and L. L. Jensen (1973), *Journal of Geophysical Research,* **78**, 7023.

W. D. Scott, D. Lamb, and D. Duffy (1969), *Journal of Atmospheric Sciences,* **26**, 726.

A. M. Squires (1972), *International Journal of Sulphur Chemistry,* **7**, 85–98.

B. L. Tarmy and G. Ciprios (1965), 'The Methanol Fuel Cell Battery', in *Engineering Developments in Energy Conversion,* The American Society of Mechanical Engineers, New York, p. 272–283.

H. A. Wiebe, A. Villa, T. M. Hellman, and J. Heicklen (1973), *Journal of the American Chemical Society,* **95**, 7.

Appendix

Reaction Numbering Scheme and Reaction Index

Reaction equation	Number	Page citations
$H + O_2 \longrightarrow OH + O$	(1)	103, 104, 122
$O + OH \longrightarrow O_2 + H$	(−1)	122, 293, 328, 354, 355, 360
$O + H_2 \longrightarrow OH + H$	(2)	103, 122
$OH + H_2 \longrightarrow H_2O + H$	(3)	103, 122
$H + H + M \longrightarrow H_2 + M$	(4)	104
$O + N_2 \longrightarrow NO + N$	(5)	105, 117
$N + NO \longrightarrow N_2 + O$	(−5)	105, 117, 365, 368, 370, 371
$N + O_2 \longrightarrow NO + O$	(6)	105, 117, 367, 370, 371
$O + NO \longrightarrow N + O_2$	(−6)	105, 117
$CH + N_2 \longrightarrow HCN + N$	(7)	106
$H + HCN \longrightarrow H_2 + CN$	(8)	106
$CN + O_2 \longrightarrow CO + NO$	(9)	106
$CN + OH \longrightarrow CO + NH$	(10)	106
$O + NH \longrightarrow NO + H$	(11)	106
$O + O + M \longrightarrow O_2 + M$	(12)	115–117
$O_2 + M \longrightarrow O + O + M$	(−12)	116
$CO + OH \longrightarrow CO_2 + H$	(13)	122, 255, 263, 328
$H + CO_2 \longrightarrow CO + OH$	(−13)	122
$NO + O_3 \longrightarrow NO_2 + O_2$	(14)	213, 253, 320, 321, 325, 327, 335, 341, 365, 372
$2.NO + O_2 \longrightarrow 2.NO_2$	(15)	252, 253
$O + O_2 + M \longrightarrow O_3 + M$	(16)	254, 259, 291, 294, 324, 354
$H + O_2 + M \longrightarrow HO_2 + M$	(17)	255, 263, 354, 357
$HO_2 + NO \longrightarrow OH + NO_2$	(18)	213, 256, 263, 327
$HCO + O_2 \longrightarrow HO_2 + CO$	(19)	256, 273

Reaction equation	Number	Page citations
$NO + NO_2 + H_2O \longrightarrow 2.HNO_2$	(20)	257
$O(^1D) + H_2O \longrightarrow 2.OH$	(21)	261, 305, 327
$H + NO_2 \longrightarrow OH + NO$	(22)	267
$CH_3 + O_2 \longrightarrow CH_3O_2$	(23)	271
$CH_3O_2 + NO \longrightarrow CH_3O + NO_2$	(24)	272
$CH_3O + O_2 \longrightarrow HCHO + HO_2$	(25)	273
$CH_3O + NO \longrightarrow HCHO + HNO$	(26)	273
$CH_3O + NO \longrightarrow CH_3ONO$	(27)	273
$CH_3O + NO_2 \longrightarrow HCHO + HONO$	(28)	273
$CH_3O + NO_2 \longrightarrow CH_3ONO_2$	(29)	273
$O + SO_2 + M \longrightarrow SO_3 + M$	(30)	278, 334
$OH + SO_2 + M \longrightarrow HSO_3 + M$	(31)	278, 334
$O + O_3 \longrightarrow 2.O_2$	(32)	293, 324, 325
$O(^1D) + N_2O \longrightarrow N_2 + O_2$	(33)	305, 312
$O(^1D) + N_2O \longrightarrow 2.NO$	(34)	305, 312, 330, 340
$HO_2 + HO_2 \longrightarrow H_2O_2 + O_2$	(35)	314
$O + NO_2 \longrightarrow NO + O_2$	(36)	213, 324, 325, 335, 342, 372
$OH + O_3 \longrightarrow HO_2 + O_2$	(37)	325, 328
$O(^1D) + CH_4 \longrightarrow OH + CH_3$	(38)	305, 327, 329
$O + HO_2 \longrightarrow OH + O_2$	(39)	327, 354, 355, 357
$OH + NO_2 + M \longrightarrow HNO_3 + M$	(40)	328
$OH + HO_2 \longrightarrow H_2O + O_2$	(41)	328, 354, 355
$HO_2 + O_3 \longrightarrow OH + 2.O_2$	(42)	328
$OH + NH_3 \longrightarrow H_2O + NH_2$	(43)	213, 331
$HO_2 + SO_2 \longrightarrow OH + SO_3$	(44)	334
$Cl + O_3 \longrightarrow ClO + O_2$	(45)	341
$O + ClO \longrightarrow Cl + O_2$	(46)	341, 342
$NO + ClO \longrightarrow NO_2 + Cl$	(47)	341
$Cl + CH_4 \longrightarrow HCl + CH_3$	(48)	343
$Cl + H_2 \longrightarrow HCl + H$	(49)	343
$OH + HCl \longrightarrow H_2O + Cl$	(50)	344
$H + O_3 \longrightarrow OH + O_2$	(51)	213, 354, 355, 358
$H + HO_2 \longrightarrow H_2O + O$	(52)	354, 355
$H + HO_2 \longrightarrow 2.OH$	(53)	354, 355
$H + HO_2 \longrightarrow H_2 + O_2$	(54)	354, 355
$O(^1S) + O \longrightarrow 2.O$	(55)	361
$O(^1S) + O_2 \longrightarrow O + O_2$	(56)	361
$O + O + O \longrightarrow O(^1S) + O_2$	(57)	362
$e^- + N_2 \longrightarrow 2.N(^4S\ or\ ^2D) + e^-$	(58)	367
$N_2^+ + e^- \longrightarrow 2.N(^4S\ or\ ^2D)$	(59)	367
$O + N_2^+ \longrightarrow NO^+ + N(^4S\ or\ ^2D)$	(60)	367, 376
$O^+ + N_2 \longrightarrow NO^+ + N$	(61)	367

Reaction equation	Number	Page citations
$N + CO_2 \longrightarrow NO + CO$	(62)	367
$N(^2D) + O_2 \longrightarrow NO + O$	(63)	367
$NO^+ + e^- \longrightarrow N + O$	(64)	371
$NO^+ + e^- \longrightarrow N(^2D) + O$	(65)	371
$O + O_4^+ \longrightarrow O_2^+ + O_3$	(66)	378
$NO^+(H_2O)_3 + H_2O \longrightarrow H^+(H_2O)_3 + HNO_2$	(67)	378
$O_2^- + O \longrightarrow O_3 + e^-$	(68)	380
$O_2^- + O_3 \longrightarrow O_3^- + O_2$	(69)	380

All species in these reactions above are ground electronic states unless otherwise indicated.

Index